Theory, Design and Applications of Power Electronics

Theory, Design and Applications of Power Electronics

Editor: Blair Cole

New York

Published by NY Research Press
118-35 Queens Blvd., Suite 400,
Forest Hills, NY 11375, USA
www.nyresearchpress.com

Theory, Design and Applications of Power Electronics
Edited by Blair Cole

International Standard Book Number: 978-1-64725-460-5 (Hardback)

Cataloging-in-Publication Data

Theory, design and applications of power electronics / edited by Blair Cole.
   p. cm.
Includes bibliographical references and index.
ISBN 978-1-64725-460-5
1. Power electronics. 2. Electric power. 3. Electronics. I. Cole, Blair.
TK7881.15 .P75 2023
621.317--dc23

Contents

Preface

Power electronics is an interdisciplinary field within electrical engineering that is concerned with the design, control and conversion of power in its electric form. A power electronic system is a system that transforms electric energy to an electric load by using a control circuit. Applications of power electronics include generation, transmission, distribution, and control of power. In daily life, power electronics is used in induction cookers, fan regulators, air conditioners, battery chargers, vacuum cleaners, and computers. Furthermore, it also has various applications in automation industry, such as trolleys, hybrid electric vehicles, and subways. It is also used for generating renewable energy, such as wind energy and solar energy. This book traces the progress in the field of power electronics and highlights some of its key concepts and applications. A number of latest researches have been included to keep the readers up-to-date with the theory, design, and applications of power electronic. This book aims to serve as a resource guide for students and experts alike, and contribute to the growth of the discipline.

After months of intensive research and writing, this book is the end result of all who devoted their time and efforts in the initiation and progress of this book. It will surely be a source of reference in enhancing the required knowledge of the new developments in the area. During the course of developing this book, certain measures such as accuracy, authenticity and research focused analytical studies were given preference in order to produce a comprehensive book in the area of study.

This book would not have been possible without the efforts of the authors and the publisher. I extend my sincere thanks to them. Secondly, I express my gratitude to my family and well-wishers. And most importantly, I thank my students for constantly expressing their willingness and curiosity in enhancing their knowledge in the field, which encourages me to take up further research projects for the advancement of the area.

Editor

Design, Analysis and Implementation of an Equalizer Circuit for the Elimination of Voltage Imbalance in a Half-Bridge Boost Converter with Power Factor Correction

Jhon Bayona [1],*(iD), Nancy Gélvez [2](iD) and Helbert Espitia [2](iD)

1   Facultad de Ingeniería, Universidad ECCI, Bogotá 111311, Colombia
2   Facultad de Ingeniería, Universidad Distrital Francisco José de Caldas, Bogotá 11021-110231588, Colombia; nygelvezg@udistrital.edu.co (N.G.); heespitiac@udistrital.edu.co (H.E.)
*   Correspondence: jbayonan@ecci.edu.co

Abstract: For the implementation of a boost converter, the half-bridge topology provides a simpler conversion circuit; however, the voltage imbalance between the capacitors is a critical factor since traditional control methodology decreases the power factor when correcting the imbalance. Consequently, this paper proposes a circuit to correct such imbalance keeping the power factor at the same time. Then, this work carries the analysis, simulation, and implementation of a strategy to reduce voltage imbalance in a half-bridge boost converter with correction of the power factor. The first part offers a description of the equalizer circuit; then, an average model is employed to perform the mathematical analysis. Later, a comparison via simulation is undertaken including other conventional converters in different scenarios. Moreover, an experimental laboratory setup is made; the results show that the equalizer circuit reduces voltage imbalance between the capacitors in a half-bridge booster converter.

Keywords: control; half-bridge boost; power electronics; power factor correction; voltage imbalance

1. Introduction

In electrical engineering, it is essential to supply power to the wide variety of devices connected to the network, achieved by energy conversion [1]. The devices connected to the network range from domestic use lamps to more complex technologies such as electrical microgrids [2]. The majority of distribution systems operate with alternating current (AC); thus, it is necessary to use power converters that serve as an interface to supply the electronic loads with direct current (DC). This is done through electronic converters; but means an issue for distribution networks since their operation includes harmonic currents, the injection of reactives to the system, and the increase of the total harmonic distortion (THD) [3].

On the other hand, circuits with power factor correction (PFC) are a way to reduce the inconveniences generated by this type of converters. Circuits with PFC allow limiting harmonic currents while decreasing the THD value seen from the network, improving the efficiency of the conversion circuit [4]. With this approach, the waveform of the current supplied presents a sinusoidal behavior with a small phase angle between the current and the voltage, so that from the point of view of the network the load behaves in an almost resistive way [5,6].

According to [7], the high efficiency and power factor of converters are very important since can improve the electrical system's quality. Boost power factor correction via critical conduction mode (CRM) control, and power on control and valley detection technique are applied to converters with

PFC in electrical appliances, mainly light loads. However, these control schemes have low-efficiency problems due to a sudden increase in the switching frequency with light loads and low power factor associated with the limitation of the power on time.

Different converters with PFC have been implemented evaluating the efficiency under load conditions, such as the cascade buck-boost [8], Ćuk converter [9], flyback circuit [10] and boost converter [11]. In this regard, the boost converter has greater efficiency as it can operate in a wide range of loads in continuous and discontinuous mode [12]. Two of the most widely used variations of this converter are half-bridge and full-bridge topology. The half-bridge provides a simpler conversion circuit using fewer switching devices although it requires the use of an additional capacitor [6].

Regarding the applications using converters with power factor correction, in [10] the power-factor-correction converter of a single-stage forward-flyback with quasi-resonant (QR) control is studied. This uses a flyback and direct converters through a common transformer. Only the flyback sub-converter works when the input voltage is less than the reflected output voltage, while both the flyback sub-converter and the direct sub-converter operate to share the output power in the repose region. Other application for Ćuk-type converters can be seen in [13], where a control system for this converter is used to supply power to a fed brushless DC motor. For the design of the controller, moth-flame optimization (MFO) is used to make the settings of a fuzzy logic controller (FLC). The proposed Ćuk converter works in discontinuous conduction mode (DCM) to achieve better power factor.

An attractive control strategy to implement converters with PFC is predictive control; thus, reference [11] presents a digital predictive control strategy for a single-phase boost PFC converter. Considering the structure of the converter circuit, the values of the output voltage and the inductor current of the next switching cycle are predicted in advance. Duty cycle is calculated only through predicted output voltage and steady-state inductor current values, and the optimized duty cycle is predicted during the dynamic process. Meanwhile, in [14] a predictive control algorithm is proposed including the detection of the conduction mode for the power-factor-correction converter. In converters with PFC, the line current is often distorted due to the characteristics of the proportional-integral (PI) current controller. To improve the quality of the current, the optimum duty cycle is determined by estimating the next current state in the continuous conduction mode (CCM) and the discontinuous conduction mode.

Considering digital control applications, authors of [15] describe a digitally controlled power-factor-correction system based on two interleaved drive converters operating with pulse width modulation (PWM). Both converters are controlled independently by an internal regulator loop based on a discrete-time sliding mode (SM) approach that imposes a loss-free resistor (LFR) behavior. The switching surface implements an average current mode driver so that the power factor (PF) is high. Additional work can be seen in [16], where a mixed-signal control scheme is presented for a boost power-factor-correction rectifier. The digital controller modulates the maximum inductor current to produce low distortion at the AC line current in discontinued conduction mode and continuous conduction mode without detecting the average current.

Concerning other applications for the design of a digital control system for power electronic converters, the reference [17] presents an integrated interleaved dual-mode time-sharing inverter (IIDMI) for the grid-tied transformer-less photovoltaic systems. The IIDMI has low total harmonic distortion of the AC with reduced filtering requirements and reduced current of power devices. The current control, especially transitions between the buck and the boost modes of operation, is made using a fast response dead-beat control (DBC). Likewise, document [18] presents a single-phase transformer-less dual-mode interleaved multilevel inverter (DMIMI) that injects a highly sinusoidal AC to the grid. Dead-beat controllers are developed to calculate the optimal duty cycles directly.

The DMIMI offers high efficiency, and fewer components conducting simultaneously in each operation mode. In addition, the work in [19] proposes the application of the dual-mode time-sharing technique in transformer-less photovoltaic inverters. The indirect current control, especially at transition modes of operation, is improved using a fast dead-beat control scheme. The system obtains a low leakage current, which is a primary concern with the transformer-less PV inverters. On the other hand, the authors of [20] propose an extendable quadratic bidirectional DC-DC converter with an improved voltage transfer ratio (VTR), capable of redundancy and modularity for electric vehicle applications. The converter has a simple structure with the lowest rating of semiconductors in the family of quadratic bidirectional converters.

Regarding LED (Light-Emitting Diode) applications, the work in [21] presents the design and experimental evaluation of a one-stage AC/DC converter with PFC and a hybrid full-bridge rectifier to supply streetlights. The proposed converter consists of an LLC resonant tank, two boost circuits, and a shared inductor. By incorporating a relay switch on the secondary side of the circuit, the output stage can operate as two different types of rectifier: the first is as a full-bridge rectifier and the second type is as a full-bridge voltage doubling rectifier. In addition, paper [22] presents a single-stage controller with smooth switching and PFC functions for LED street lighting applications. The system integrates a PFC buck-boost converter interleaved with coupled inductors and a half-bridge LLC resonant converter in a single-stage power conversion circuit with reduced voltage across the DC connected capacitor and power switches. The inductors coupled in the interleaved PFC buck-boost converter operate in discontinuous conduction mode to achieve PFC. Additional work can be seen in [23] that presents the implementation of a two-stage light-emitting diode driver. The LED driver circuit design drives a 150 W LED module. The controller stages are: AC/DC power-factor-correction stage, and DC/DC power converter stage. The PFC stage implementation uses the NCP1608 integrated circuit, which uses the critical conduction mode to ensure unity power factor with a wide range of input voltages.

Regarding other applications made for power-factor-correction converters, reference [24] presents a built-in photovoltaic (PV) power interface circuit with a buck-boost converter and a full-bridge inverter. The circuit consists of five switches, an input inductor, and LC filters. The buck-boost converter operates at a high switching frequency to make the output current a sine wave, while the full-bridge inverter operates at a low switching frequency of 50–60 Hz. A high power factor is achieved in the output stage without an additional current driver due to the input inductor current operating in a discontinuous conduction mode.

The central aspect to considerer in this paper is that for implementing a boost converter, the half-bridge topology provides a simpler conversion circuit by using fewer switching devices; however, the voltage imbalance between the capacitors is a critical study factor. Previous research [25] aims at finding alternatives to eliminate the cause of imbalance and outlines for designing a power circuit using a fixed band hysteresis current control (HCC) technique. In [26], it is proposed to add a DC bias to each phase's current source to solve the voltage imbalance problem. Elimination strategies are developed by carefully analyzing all possible paths of DC bias currents and their effect on voltage imbalance. Later in [27], these same authors study the adverse effects of the imbalance elimination control circuit on the input power factor. Finally, a related work to consider is presented in [28], which analyzes a four-switch voltage doubler boost rectifier. The authors propose a control scheme to eliminate output voltage imbalance when the load is imbalanced between the two output DC rails. To limit the output voltage imbalance, the controller employs a comparator with hysteresis. Unipolar pulse width modulation switching patterns are used to reduce the discharge rate of the output capacitor for low voltages. Figure 1 shows the traditional control methodology to correct the imbalance; however, injecting a direct component to $i_{L_{REF}}$ or generating a phase difference between i_g and v_g decreases the PF [6]. Therefore, the possibility is open for the development of alternatives to correct the imbalance.

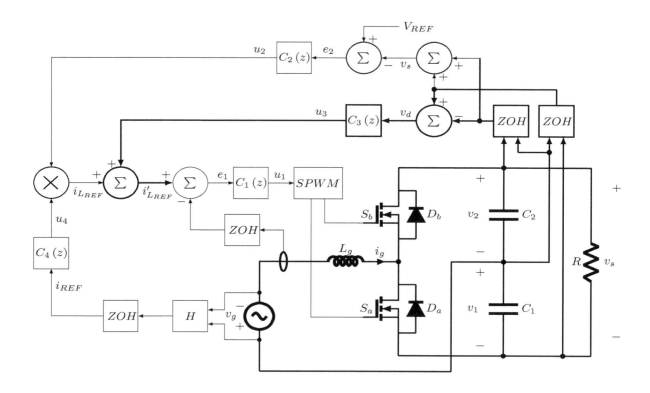

Figure 1. Control diagram with differential voltage loop for a converter with power factor correction in half-bridge configuration.

Table 1 and Figure 2 describe the related works showing the relevance of low power factor converters, emphasizing the half-bridge boost converter. The main issue here is the voltage imbalance; moreover, some applications and control strategies are also displayed.

Table 1. Description of the related works.

Topic	Description	References
Power Factor	General aspects about power quality, power factor and harmonic distortion in electrical systems	[1–7]
Converters PFC	Provision of a general description of PFC converters, cascade buck-boost, Ćuk converter, flyback circuit, and boost converter	[6,8–10,10–12]
Fuzzy logic control	PFC converter control using fuzzy logic in the application to supply power to a fed brushless DC motor	[13]
Predictive control	Predictive control application for PFC converters, the system calculates the future value of the duty cycle	[11,14]
Digital Controller	It shows a brief review of different applications of converters using digital controllers	[15–20]
LED applications	Application of PFC converters for LED lighting using full-bridge and half-bridge rectifiers	[21–23]
Photovoltaic applications	Description of the application of PFC converters in photovoltaic systems	[17,24]
Voltage Imbalance	Describe main works related to the imbalance in half-bridge rectifiers. This issue is addressed in this work	[25–28]

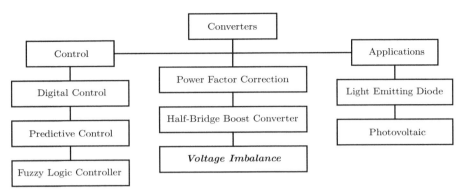

Figure 2. Graphic description of the related works.

This article focuses on showing that the proposed topology allows the reduction of the capacitor imbalance in a half-bridge boost converter. For this, mathematical analysis and experimental tests are carried out. In classical topologies, when correcting the imbalance, the power factor is sacrificed as shown in [25]. Thus, it is presented a switched DC/DC converter circuit (equalizer) operating in discontinuous conduction mode that solves the voltage imbalance problem in the half-bridge boost converter. The proposed equalizer circuit is added to the voltage doubler boost rectifier. It is composed of two transistors, in this case, MOSFETs S_1, S_2 and the respective diodes D_1, D_2, as well as an inductance L. As a result, a voltage doubler boost rectifier with equalizer is originated, which is shown in Figure 3. In addition, h is the duty cycle associated with the switches S_a and S_b of the doubler-rectifier-elevator, and $1 - h$ correspond to the complement.

The document is organized as follows: Section 2 qualitatively describes the equalizer circuit operation, its average circuit model is given in Section 3. Section 4 presents a detailed analysis of the equalizer's voltage elimination imbalance using its average circuit model. Aiming to eliminate the voltage imbalance, Section 5 proposes a hysteretic band control scheme to govern (control) the equalizer. A comparison whit other classic converters via simulation appears in Section 6, later in Section 7 a prototype of the voltage doubler boost rectifier with the proposed equalizer circuit is built to test it and validate the theoretical analysis; finally, the conclusions are given.

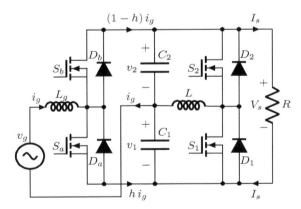

Figure 3. Voltage doubler boost rectifier with equalizer.

2. Description of the Proposed Equalizer Circuit

As previously mentioned, the imbalance of voltages in the capacitors in a half-bridge boost converter is a critical factor, which is why this document analyzes, designs and implements a strategy called equalizer circuit to eliminate the voltage imbalance on the capacitors.

To explain the operation of the proposed equalizer circuit operating in DCM, it is considered that the output capacitors C_1 and C_2 are sufficiently large, so that the voltage ripples produced by switching frequency and line frequency can be neglected; on the other hand, all components of the equalizer circuit are ideal. According to the comparison between the voltages of the capacitors, two modes arise in the operation of the equalizer circuit that affect the state of charge and discharge of both capacitors, as described below:

2.1. Analysis for the Case $V_1 > V_2$

In this mode, the transistor S_1 and the diode D_2 operate, as shown in Figure 4; likewise, the waveforms of the current i' and the voltage v' of the inductance for this mode are shown in Figure 5. The inductance current is equal to zero at the beginning of each switching period T_s, during the first interval $d_1 T_s$, simultaneously the transistor S_1 is turned on and the diode D_2 turns off; the voltage of the inductance is equal to V_1; hence, the current of the inductance increases with a slope equal to $\frac{V_1}{L}$, at the end of this first interval, the inductance current reaches its maximum value given by:

$$I_m = \frac{V_1}{L} d_1 T_s,$$ (1)

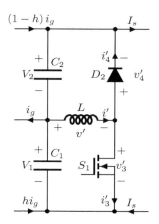

Figure 4. Equalizer circuit for the case $V_1 > V_2$.

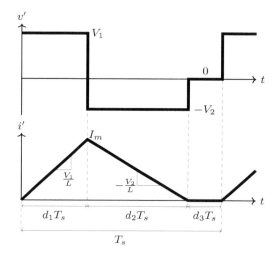

Figure 5. Waveforms of v' and i' associated with the equalizer circuit for the case $V_1 > V_2$.

Therefore, the maximum current value I_m is directly proportional to the voltage of the capacitor C_1, as well as to the duration of the first interval, therefore, the inductance absorbs energy from the capacitor C_1, which is proportional to the square of I_m and is given by:

$$W = \frac{1}{2} L I_m^2$$ (2)

During the second interval $d_2 T_s$ the transistor S_1 turns off. Meanwhile, the diode D_2 turns on, the inductance voltage is equal to $-V_2$, then, the inductance current decreases with a slope equal to $-\frac{V_2}{L}$, at the end of this second interval the diode D_2 is polarized in inverse and the energy that the inductance absorbed during the first interval is transferred to the capacitor C_2. The inductance current and voltage are kept at zero during the third interval $d_3 T_s$. Consequently, the equalizer circuit discharges the capacitor C_1 during the first interval and charges the capacitor C_2 during the second interval in this mode.

2.2. Analysis for the Case $V_2 > V_1$

In Figure 6, waveforms of the current i' and the voltage v' of the inductance are equal to the previous mode. Nevertheless, the direction of the current and the voltage polarity are reversed, as illustrated in Figure 7, where is seen that the transistor S_2 and the diode D_1 now operate. Similar to the previous mode, the inductance current is equal to zero at the beginning of each switching period T_s, during the first interval $d_1 T_s$, the transistor S_2 is turned on, simultaneously, diode D_1 turns off, the inductance voltage equals V_2; therefore, the inductance current increases with a slope equal to $\frac{V_2}{L}$, at the end of this first interval, the inductance current reaches its maximum value given by:

$$I_m = \frac{V_2}{L} d_1 T_s,$$ (3)

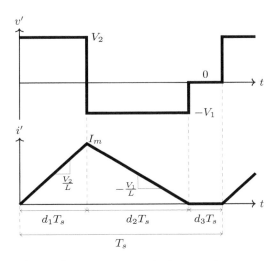

Figure 6. Waveforms of v' and i' associated with the equalizer circuit for the case $V_2 > V_1$.

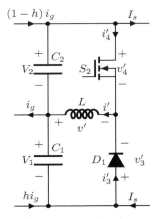

Figure 7. Equalizer circuit for the case $V_2 > V_1$.

Therefore, the maximum value of current I_m is directly proportional to the voltage of the capacitor C_2 and the duration of the first interval; then, the inductance absorbs energy from the capacitor C_2, which is proportional to the square of I_m and is given by (2). During the second interval $d_2 T_s$, the transistor S_2 turns off. Meanwhile, diode D_1 turns on, and inductance voltage is equal to $-V_1$, which is why the inductance current decreases with a slope equal to $-\frac{V_1}{L}$; at the end of this second interval, diode D_1 is reverse biased, and the energy absorbed by the inductance during the first interval is transferred to the capacitor C_1. The inductance current and voltage are kept at zero during the third interval $d_3 T_s$. As a result, the equalizer circuit discharges the capacitor C_2 during the first interval and charges the capacitor C_1 during the second interval in this mode.

From the above, when $V_1 > V_2$, the equalizer absorbs energy from C_1, which is delivered to C_2, then, V_1 decreases, at the same time V_2 increases; likewise, when $V_2 > V_1$, the equalizer absorbs energy from C_2 that is delivered to C_1; hence V_2 decreases while V_1 increases; to conclude, the equalizer circuit eliminates the voltage imbalance that occurs in the capacitors. In Section 4, a more detailed analysis is made using the average circuit model of the equalizer circuit.

3. Average Circuit Model

The average circuit model of the equalizer in each operating mode is obtained by following the average switch-modeling technique [29], which generates an equivalent circuit that models the average of the waveforms at the terminals of the switch network. In relation to the equalizer when $V_1 > V_2$, Figure 4 shows the equivalent circuit of the transistor S_1 and diode D_2, which is obtained by averaging the respective waveforms of the inductance current, capacitor voltage, and duty cycle.

The average voltage associated with S_1 is v_3, this is found by averaging the waveform v_3' shown in Figure 8.

$$v_3 = d_2 \left(V_1 + V_2 \right) + d_3 V_1 , \tag{4}$$

from any of the waveforms, it follows that d_3 can be expressed as:

$$d_3 = 1 - d_1 - d_2 , \tag{5}$$

replacing (5) in (4) is obtained:

$$v_3 = d_2 V_2 + \left(1 - d_1 \right) V_1 , \tag{6}$$

likewise, i_3 is the average current of S_1 and it is found by averaging the waveform i_3' shown in Figure 8.

$$i_3 = \frac{1}{2} d_1 I_m , \tag{7}$$

substituting (1) in (7), yield:

$$i_3 = \frac{1}{2} \frac{d_1^2 V_1 T_s}{L} \tag{8}$$

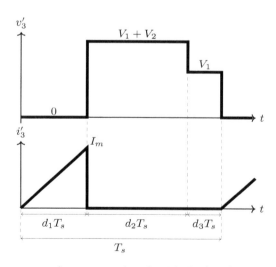

Figure 8. Voltage and current waveforms associated with S_1 for the equalizer circuit when $V_1 > V_2$.

The average voltage of D_2 is v_4, this voltage is found by averaging the waveform v_4' shown in Figure 9.

$$v_4 = d_1 (V_1 + V_2) + d_3 V_2 , \tag{9}$$

substituting (5) in (9) produces:

$$v_4 = d_1 V_1 + (1 - d_2) V_2 \tag{10}$$

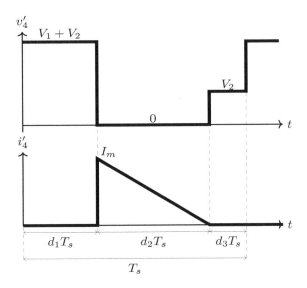

Figure 9. Voltage and current waveforms associated with D_2 for the equalizer circuit when $V_1 > V_2$.

Similarly, i_4 is the average current of D_2 and it is found by averaging the waveform i_4' shown in Figure 9.

$$i_4 = \frac{1}{2} d_2 I_m , \tag{11}$$

substituting Equation (1) in (11), yield:

$$i_4 = \frac{1}{2} \frac{d_1 d_2 V_1 T_s}{L} \tag{12}$$

On the other hand, Figure 5 shows no change in current i' over a commutation period, having as a result:

$$\frac{L}{T_s} \left[i' (t + T_s) - i' (t) \right] = \frac{1}{T_s} \int_t^{t+T_s} v' d\tau = 0 \tag{13}$$

and since:

$$v = \frac{1}{T_s} \int_t^{t+T_s} v' d\tau , \tag{14}$$

the average voltage of the inductance v is equal to zero and the average of the waveform v' shown in Figure 5, is given by:

$$v = d_1 V_1 - d_2 V_2 = 0 , \tag{15}$$

solving d_2 then:

$$d_2 = d_1 \frac{V_1}{V_2} , \tag{16}$$

therefore, the average of each of the waveforms is obtained, at the terminals of S_1 and D_2, substituting (16) in (6), (8), (10), and (12), is obtained:

$$v_3 = V_1 \tag{17}$$

$$v_4 = V_2 \tag{18}$$

$$i_3 = \frac{1}{2}\frac{d^2 v_3 T_s}{L} \tag{19}$$

$$i_4 = \frac{1}{2}\frac{d^2 v_3^2 T_s}{L v_4} \tag{20}$$

where d is the duty cycle equal to d_1; in conclusion, Equations (17)–(20) are simple expressions that represent the average of the waveforms of the transistor S_1 and diode D_2 of the equalizer circuit in DCM, when $V_1 > V_2$.

To find a circuit that models the S_1 and D_2 waveforms, it is considered that S_1 absorbs and D_2 delivers apparent average power. According to Figure 4, S_2 absorbs apparent power, since its current enters through the positive value of its voltage, at the same time, D_2 delivers apparent power, since its current leaves through the positive value of its voltage; in addition, the apparent average power of S_1 and D_2 are equal and expressed as:

$$v_3 i_3 = v_4 i_4 = \frac{1}{2}\frac{d^2 V_1^2 T_s}{L} \tag{21}$$

Also, the apparent average power of S_1 and D_2 is deduced considering the energy stored by the inductance and expressing its maximum current reached I_m, during the first interval of time dT_s in terms of the voltage V_1; therefore, replacing (1) in (2) is obtained:

$$W = \frac{1}{2}\frac{(dT_s V_1)^2}{L}, \tag{22}$$

dividing both sides of (22) by the commutation period T_s, it is obtained the power absorbed by the inductance during the first time interval, equal to the expression given in (21). As a result, it can be stated that during the first time interval, the power is transferred from the capacitor C_1 to the inductance L through the terminals of the transistor S_1. During the second time interval the inductance L releases all the energy stored to the capacitor C_2, through the terminals of diode D_2; then, the average power consumed by the transistor and the diode equals zero, i.e., the transistor S_1 behaves like a power source that absorbs P_1. Simultaneously, the diode D_2 behaves like a power source that supplies P_2. The power P_2 is dependent on the power P_1, since the voltage and current of the S_1 terminals are independent of the voltage and current from the D_2 terminals.

Therefore, the equalizer average model when $V_1 > V_2$ is represented by two power sources that substitute transistor S_1 and diode D_2, as shown in Figure 10. Source P_1 absorbs power that is later transferred to source P_2, whereby source P_1 absorbs power from capacitor C_1 that flows to the source P_2 and then to capacitor C_2. Also, in Figure 10 is seen the replacement of the inductance by a short circuit considering average voltage equal to zero; likewise, the current i that flows through this short circuit is the average current of inductance i'.

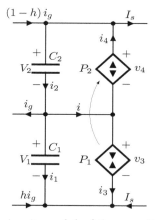

Figure 10. Average circuit model of the equalizer when $V_1 > V_2$.

The equalizer circuit when $V_2 > V_1$ is illustrated in Figure 7, this represents the equivalent circuit composed by the transistor S_2 and diode D_1. By averaging the waveforms in terms of the equalizer input and state variables is obtained:

$$i_3 = \frac{1}{2}\frac{d^2 v_4^2 T_s}{L v_3} \tag{23}$$

$$i_4 = \frac{1}{2}\frac{d^2 v_4 T_s}{L} \tag{24}$$

The average circuit model of the equalizer when $V_2 > V_1$ consists of two power sources that replace the transistor S_2 and diode D_1 as shown in Figure 11. The source P_2 absorbs power that is transferred to the source P_1; hence, the source P_2 absorbs power from the capacitor C_2 that flows to P_1, and then to C_1. The inductance is replaced by a short circuit in the same way than in the average equalizer circuit when $V_1 > V_2$.

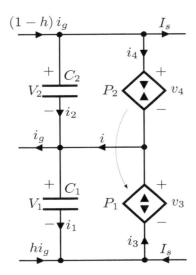

Figure 11. Average circuit model of the equalizer when $V_2 > V_1$.

4. Voltage Imbalance Analysis

To start the analysis in the voltage doubler boost rectifier, it is considered that the voltage variations in both capacitors are produced by the line frequency. The line voltage varies in a sinusoidal way and to obtain a unitary PF; also, the line current is sinusoidal, having:

$$v_g = V_p \sin(\theta) \tag{25}$$

$$i_g = I_p \sin(\theta) \tag{26}$$

On the other hand, the average circuit models of the proposed equalizer obtained in Section 3 are used to calculate the steady-state voltage imbalance in the voltage doubler boost rectifier capacitors, in each mode of operation of the equalizer proposed.

4.1. Analysis for the Case $V_1 > V_2$

In this mode, the average equalizer circuit is connected to the voltage doubling step-up rectifier capacitors, as shown in Figure 10, from which the average currents i_1 and i_2 are expressed as:

$$i_1 = -h i_g - I_s - i_3 \tag{27}$$

$$i_2 = (1 - h) i_g - I_s + i_4 \tag{28}$$

12
Theory, Design and Applications of Power Electronics

By substituting (17) and (18) in (19) and (20), the average currents i_3 and i_4 are obtained in terms of the average voltages of the capacitors V_1 and V_2 having:

$$i_3 = \frac{d^2 V_1 T_s}{2L} \tag{29}$$

$$i_4 = \frac{d^2 V_1^2 T_s}{2L V_2} \tag{30}$$

The average voltage of the inductance L_g, in the voltage doubler boost rectifier [5,30] is given by:

$$L_g \frac{di_g}{dt} = v_g + V_1 - (1-h) V_s , \tag{31}$$

where $V_s = V_1 + V_2$, and replacing (25) and (26) in (31) yield:

$$\omega L_g I_p \cos(\theta) = V_p \sin(\theta) + V_1 - (1-h) V_s , \tag{32}$$

solving h from (32) is obtained:

$$h = \frac{\omega L_g I_p}{V_s} \cos(\theta) - \frac{V_p}{V_s} \sin(\theta) - \frac{V_1}{V_s} + 1 , \tag{33}$$

comparing magnitude for sine and cosine signals $\frac{V_p}{V_s} \gg \frac{\omega L_g I_p}{V_s}$, it is established that $V_p \gg \omega L_g I_p$, then Equation (33) can be approximated to:

$$h = -\frac{V_p}{V_s} \sin(\theta) - \frac{V_1}{V_s} + 1 , \tag{34}$$

replacing (26), (29), (30), and (34) in (27), and (28), is obtained:

$$i_1 = \frac{V_p I_p}{V_s} \sin^2(\theta) - \frac{V_2 I_p}{V_s} \sin(\theta) - I_s - \frac{d^2 T_s V_1}{2L} \tag{35}$$

$$i_2 = \frac{V_p I_p}{V_s} \sin^2(\theta) + \frac{V_1 I_p}{V_s} \sin(\theta) - I_s + \frac{d^2 T_s V_1^2}{2L V_2} \tag{36}$$

Considering that the duty cycle d is constant, then, the variations of the voltages in the capacitors V_1 and V_2 over a switching period are given by:

$$\delta V_1 = \frac{1}{\omega C_1} \int_0^{2\pi} i_1 d\theta = \frac{\pi}{\omega C_1} \left(\frac{V_p I_p}{V_s} - 2I_s - \frac{d^2 T_s V_1}{L} \right) \tag{37}$$

$$\delta V_2 = \frac{1}{\omega C_2} \int_0^{2\pi} i_2 d\theta = \frac{\pi}{\omega C_2} \left(\frac{V_p I_p}{V_s} - 2I_s + \frac{d^2 T_s V_1^2}{L V_2} \right) \tag{38}$$

It is assumed that C_1 and C_2 are equal to C; hence, the variation of the voltage difference over a switching period, with the proposed equalizer when $V_1 > V_2$ is given by:

$$\delta V_2 - \delta V_1 = \frac{\pi d^2 T_s}{\omega L C} \left(\frac{V_1^2}{V_2} + V_1 \right) \tag{39}$$

4.2. Analysis for the Case $V_2 > V_1$

According to Figure 11, the average currents i_1 and i_2 of the average equalizer circuit are given by:

$$i_1 = -h i_g - I_s + i_3 \tag{40}$$

$$i_2 = (1 - h) i_g - I_s - i_4 \tag{41}$$

On the other hand, by replacing (17) and (18) in (23) and (24), respectively, the average currents i_3 and i_4 are obtained in terms of the average voltages of the capacitors V_1 and V_2 being:

$$i_3 = \frac{d^2 V_2^2 T_s}{2LV_1} \tag{42}$$

$$i_4 = \frac{d^2 V_2 T_s}{2L} \tag{43}$$

Substituting (26), (34), (42), and (43) in (40) and (41) yield:

$$i_1 = \frac{V_p I_p}{V_s} \sin^2(\theta) - \frac{V_2 I_p}{V_s} \sin(\theta) - I_s + \frac{d^2 T_s V_2^2}{2LV_1} \tag{44}$$

$$i_2 = \frac{V_p I_p}{V_s} \sin^2(\theta) + \frac{V_1 I_p}{V_s} \sin(\theta) - I_s - \frac{d^2 T_s V_2}{2L} \tag{45}$$

Since the duty cycle d is constant, the variations of the voltages in the capacitors V_1 and V_2 over a switching period are expressed as:

$$\delta V_1 = \frac{1}{\omega C_1} \int_0^{2\pi} i_1 d\theta = \frac{\pi}{\omega C_1} \left(\frac{V_p I_p}{V_s} - 2I_s + \frac{d^2 T_s V_2^2}{LV_1} \right) \tag{46}$$

$$\delta V_2 = \frac{1}{\omega C_2} \int_0^{2\pi} i_2 d\theta = \frac{\pi}{\omega C_2} \left(\frac{V_p I_p}{V_s} - 2I_s - \frac{d^2 T_s V_2}{L} \right) \tag{47}$$

Since C_1 and C_2 are equal to C, the variation of the voltage differs over a switching period, with the proposed equalizer when $V_2 > V_1$ is expressed as:

$$\delta V_1 - \delta V_2 = \frac{\pi d^2 T_s}{\omega LC} \left(\frac{V_2^2}{V_1} + V_2 \right) \tag{48}$$

From this analysis, it is stated that the variations in the difference voltage expressed in (39) and (48) are always positive; as a result, the voltage imbalance is eliminated in both modes of operation, using the proposed equalizer circuit. This elimination is achieved regardless of the load conditions in the voltage doubler boost rectifier.

5. Control Scheme

The proposed control scheme that governs the equalizer circuit in DCM, is shown in Figure 12, this scheme is a circuit that allows the passage of a carrier signal v_x to the transistor S_1 or S_2.

When $V_1 > V_2$, the carrier pass to S_1 is enabled; simultaneously, S_2 is set to zero and then the equalizer circuit operates in mode $V_1 > V_2$, absorbing energy from C_1 which flows towards C_2, so that V_1 decreases and V_2 increases. On the other hand, if $V_2 > V_1$, the carrier pass to S_2 is enabled, at the same time, S_1 is set to zero; therefore, the equalizer circuit operates in the mode $V_2 > V_1$, in which it absorbs energy from C_2 which then flows to C_1; as a result, V_2 decreases and V_1 increases. The carrier signal v_x establishes the duty cycle d and the switching period T_s of the equalizer, in addition, the duty cycle d is constant and its value is chosen to ensure the operation on DCM of the equalizer; therefore, the equalizer in mode $V_1 > V_2$ operates on DCM when:

$$d < \frac{V_2}{V_1 + V_2}, \tag{49}$$

on the other hand, the equalizer in the mode $V_2 > V_1$ operates in DCM if:

$$d < \frac{V_1}{V_1 + V_2},$$ (50)

The conclusion from (49) and (50) is that the duty cycle d must be less than 0.5 to ensure the equalizer operation in DCM for both modes; also, the duty cycle d determines the energy absorbed by the inductance. On the other hand, the hysteresis band of the comparator must be selected carefully, a wide band produces a high voltage imbalance; likewise, as there is ripple in the voltages V_1 and V_2, a narrow band can cause malfunctions in the proposed control scheme.

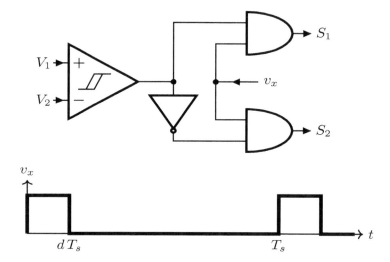

Figure 12. Control scheme of the equalizer circuit.

The hysteresis band was selected considering the variation that can be held in the voltage capacitors. In an experimental way using simulation, different values, displayed in Table 2, are obtained to determine this band, where BH is the hysteresis band and $V_d = V_1 - V_2$ is the difference voltage value between the capacitors. In Table 2, as BH increases, V_d also increases; therefore, it is appropriate to have a small value of BH; however, the operation of the elements is not ideal implying practical limitations. Therefore, using the values of Table 2 in the real circuit, the smallest useful value of BH is 3 V, because for low values of BH undesired oscillations that do not appear in simulation are present in the real circuit.

Table 2. Hysteresis band values obtained in simulation.

BH (V)	V_d (V)
0.25	± 1.58
0.5	± 1.98
1	± 1.71
2	± 2.47
3	± 3
4	± 3.5

6. Simulation Results

This section presents the comparison of the proposed circuit versus other classic configurations via simulation. The first part of the results show that the proposed circuit does not affect drastically the power factor, the harmonic distortion and the power efficiency. Later it is observed the effect that the variation of the capacitors has on the PF for the simulated circuits. The converters considered are:

- CV1: Converter proposed with equalizer circuit. The simulation scheme is show in Figure 13.
- CV2: Classic PFC converter using half-bridge boost topology developed in [25]. The diagram of the simulation is show in Figure 14.
- CV3: Half-bridge PFC Boost converter with digital PID controller and pre-compensation loop developed in [6]. The scheme of the simulation is presented in Figure 15.

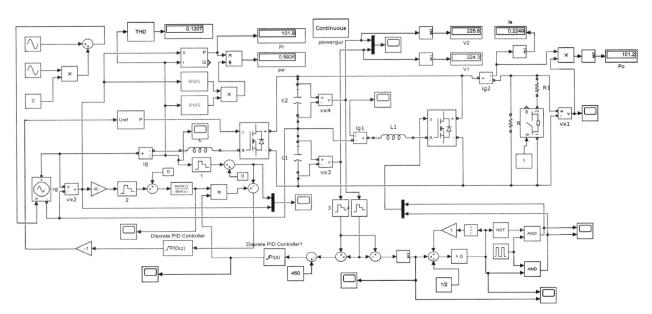

Figure 13. Simulation scheme for CV1.

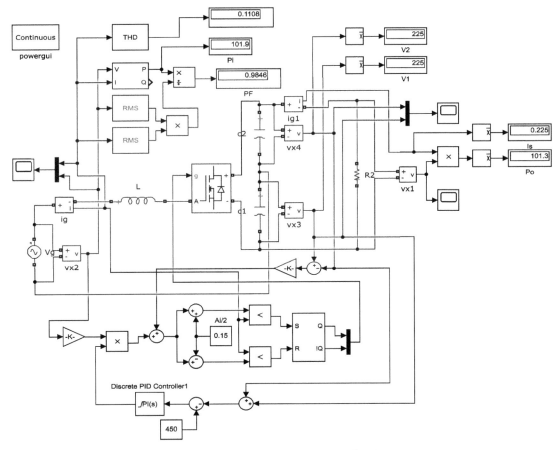

Figure 14. Simulation diagram for CV2.

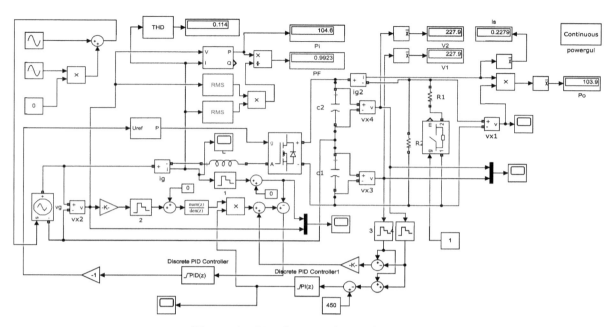

Figure 15. Simulation scheme for CV3.

For comparison, different load current I_s values are considered $0.225A$, $0.168A$ and $0.112A$; thus Table 3 shows the root mean square (RMS) voltage of V_1 and V_2 which corresponds to voltages in capacitors C_1 and C_2. It also includes power factor, total harmonic distortion, the input power P_i, output power P_o, and the power efficiency η.

Table 3. Comparison results without capacitors variation.

Circuit	I_s (A)	V_1 (V)	V_2 (V)	PF	THD	P_o (W)	P_i (W)	η (%)
	0.225	225	225	0.9926	0.1203	101.2	101.9	99.31
CV1	0.168	225	225	0.9819	0.1623	75.9	76.4	99.34
	0.112	225	225	0.9506	0.2315	50.6	50.9	99.41
	0.225	225	225	0.9846	0.1108	101.3	101.9	99.41
CV2	0.168	225	225	0.9778	0.1466	75.95	76.3	99.54
	0.112	225	225	0.9669	0.2111	50.6	50.8	99.60
	0.2249	224.9	224.9	0.9909	0.1175	101.2	101.8	99.41
CV3	0.168	225	225	0.9758	0.1617	75.9	76.3	99.47
	0.112	225	225	0.9412	0.2298	50.6	50.8	99.60

Table 3 shows that for a load current of $0.225A$ and $0.168A$, the best PF is obtained with CV1. In other cases, the best values for PF, THD and efficiency are obtained with the circuit CV2; nevertheless, the difference with circuit CV1 for the worst case was 0.0163 to PF when $I_s = 0.168$, THD of 0.0204 with a load current of $I_s = 0.112$. Finally, difference in power efficiency of 0.19% when $I_s = 0.168$. Thus, circuit CV1 displays results similar to those of CV2 with a slight variation in PF, THD and power efficiency. It should be noted that the results show no variation in the condensers; therefore, the voltage is equally distributed, which is an ideal case. Thus, Table 3 shows that the values for V_1 and V_2 are the same.

Figure 16 displays the results for CV1 to observe the wave's shape at full load $I_s = 0.225A$. Likewise, Figure 17 shows the results for CV2 and Figure 18 for CV3, where i_g and v_g are the source current and voltage signals. From the results, when having converters with power factor correction, the current signal i_g tends to be in phase with the voltage signal v_g.

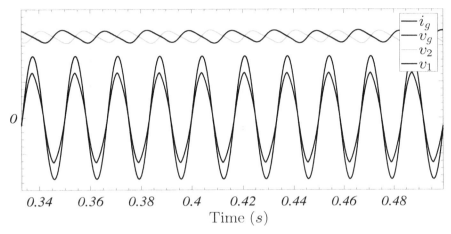

Figure 16. Waveforms obtained for CV1; i_g: 1 A/div, v_g, v_1 and v_2: 100 V/div.

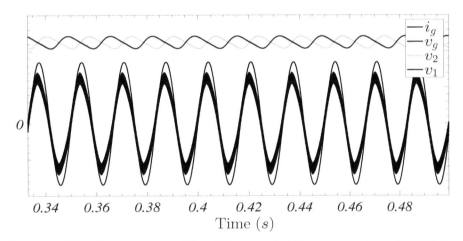

Figure 17. Waveforms obtained for CV2; i_g: 1 A/div, v_g, v_1 and v_2: 100 V/div.

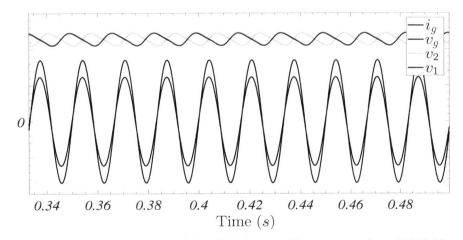

Figure 18. Waveforms obtained for CV3; i_g: 1 A/div, v_g, v_1 and v_2: 100 V/div.

In the previous results, the condenser's value variations are disregarded; therefore, the following experimental test focuses on observing the effect when the condensers have value variations toward $\pm 5\%$ of the nominal value. Table 4 displays such results when considering the circuit operation at full load.

Table 4. Comparison results with capacitors variation.

Circuit	C_1 (µF)	C_2 (µF)	V_1 (V)	V_2 (V)	PF
CV1	95	105	225	225	0.9925
	105	95	225	225	0.9925
CV2	95	105	223.9	226.2	0.9845
	105	95	226.2	223.9	0.9845
CV3	95	105	227.9	227.9	0.9923
	105	95	227.9	227.9	0.9923

Table 4 shows that the CV1 configuration reduces the imbalance by presenting the best PF while CV2 displays the largest imbalance, also affecting PF. Moreover, circuit CV3 fulfills equality in the values V_1 and V_2; nevertheless, this is done employing a value greater than 225, which should be the operation value. It is noticeable that CV3 incorporates a feedback loop to correct the imbalance.

7. Experimental Results

A prototype of the voltage doubler boost rectifier with the proposed equalizer circuit was built in the laboratory (Figure 19), with the circuit parameters shown in Table 5. The selection is made considering the recommendations of [5], L is first calculated using the equation:

$$L = \frac{V_s}{4 f_s \delta i_{L,p-p}} \tag{51}$$

where $\delta i_{L,p-p} = 0.4$ A maximum ripple current (value desired), $f_s = 50$ kHz switching frequency, $V_s = 450$ V output voltage; using these values the inductance value obtained is $L \approx 5$ mH. Afterwards, the capacitors $C_1 = C_2 = C$ are calculated using equation:

$$C = \frac{1}{\delta v_{s,p-p} V_s} \sqrt{\frac{(V_p + r_\rho I_p)^2 I_p^2}{\omega^2} + L^2 I_p^4} \tag{52}$$

where $\delta v_{s,p-p} = 10$ V is the peak-to peak capacitors voltage variation (value desired), $V_p = 120\sqrt{2}$ V the source peak voltage, $I_p = 0.9927$ A peak current, $\omega = 2\pi 60$ rad/s line frequency, $r_\rho \approx 0$ loss resistance associated with components. Using these values, the capacitors C_1 and C_2 are 100 µF.

Figure 19. Prototype of the voltage doubler boost rectifier.

Likewise, the semiconductors employed are diodes MUR460, transistors STP13NK60ZFP (ST Microelectronics), and the integrated circuited HCPL-3120 as the trigger driver. The controller was implemented with the digital signal processor (DSP) Texas Instruments TMDX28069USB (Piccolo F28069). In addition, the measurement equipment used are:

- Oscilloscope: Agilent MSO-X 3014A.
- Current probe: Keysight 1146B 100 kHz/100 A.
- Current probe: N2783B 100 MHz/30 Arms AC/DC.

Table 5. Parameters of the voltage doubling step-up rectifier with the proposed equalizer circuit.

Circuit Parameter	Symbol	Value
Peak line voltage	V_p	170 V
Output voltage	V_s	450 V
Equalizer inductance	L	500 µH
Rectifier inductance	L_g	5 mH
Output capacitors	$C = C_1 = C_2$	100 µF
Load resistance	R	2 kΩ
Line frequency	f_L	60 Hz
Switching frequency rectifier and equalizer	f_s	50 kHz
Duty cycle of S_1 and S_2	d	0.125
Hysteresis band of the comparator	-	±3 V

The experimental results show how the proposed equalizer circuit eliminates the voltage imbalance. Figures 20–23 illustrate the waveforms of v_1, v_2, S_1 and S_2 for values I_s of 222 mA, 198 mA, 153 mA and 110 mA, respectively. The signals of S_1 and S_2 are measured at the output of the DSP. The operation of the proposed equalizer circuit can be verified by the absence of voltage imbalance in the output capacitors for the load current values. Moreover, by the way S_1 and S_2 work, it can be seen that when $V_1 > V_2$, S_1 and D_2 commute so that L absorbs energy from C_1 which is delivered to C_2; then, V_1 decreases at the same time V_2 increases.

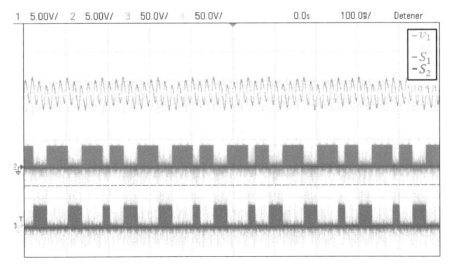

Figure 20. Waveforms of v_1, v_2, S_1 and S_2 for $I_s = 222$ mA.

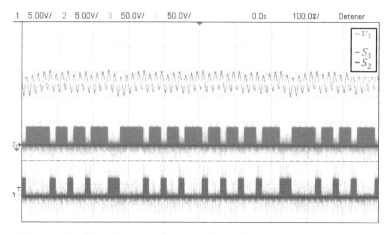

Figure 21. Waveforms of v_1, v_2, S_1 and S_2 for $I_s = 198$ mA.

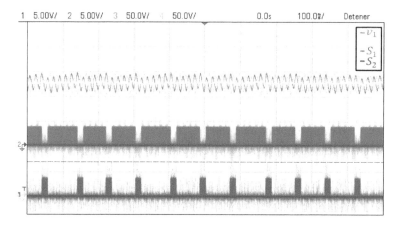

Figure 22. Waveforms of v_1, v_2, S_1 and S_2 for $I_s = 153$ mA.

Additionally, when $V_2 > V_1$, S_2 and D_1 commute so that L absorbs energy from C_2 which is delivered to C_1; hence, V_2 decreases and simultaneously V_1 increases; consequently, that mentioned in Sections 2 and 4 is confirmed regarding the circuit operation and elimination of voltage imbalance.

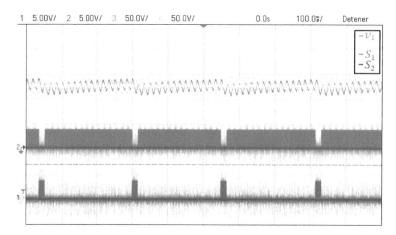

Figure 23. Waveforms of v_1, v_2, S_1 and S_2 for $I_s = 110$ mA.

On the other hand, Figures 24 and 25 shows how the input current i_g follows input voltage v_g, for values I_s of 222 mA and 198 mA, respectively, so that the proposed equalizing circuit does not affect the operation of the voltage doubling step-up rectifier; in addition, high power factor values were obtained under steady-state conditions, these values were 0.995 and 0.992 for values I_s of 222 mA and 198 mA.

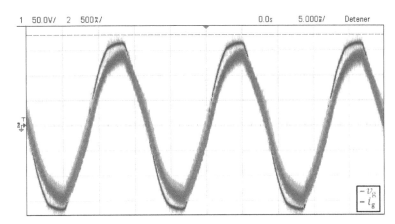

Figure 24. Waveforms of v_g and i_g for $I_s = 222\,\text{mA}$.

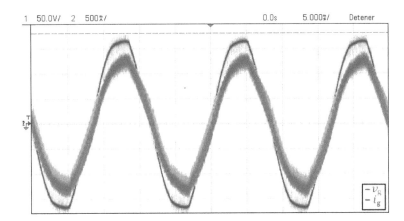

Figure 25. Waveforms of v_g and i_g for $I_s = 198\,\text{mA}$.

As discussion, it is observed that the proposed circuit allows reducing the imbalance in the voltage of the condensers without affecting the power factor. However, it is mandatory to employ additional components with a control system, enhancing the construction and cost complexities.

8. Conclusions

An equalizing converter circuit operating on DCM was proposed to solve the voltage imbalance problem. The equalizer circuit operation was described qualitatively; moreover, an average circuit model was obtained consisting of two power sources. The model was used to analyze in detail how the equalizer eliminates the voltage imbalance in the capacitors; for this, the expressions that define the average currents that flow through both capacitors were calculated, obtaining the voltage variations, as well as the variation of the voltage difference over a switching period, in each operating mode of the equalizer. In addition, a simple hysteretic band control scheme was proposed that manages the "on" and "off" of the equalizer switches.

The analysis results were verified by experiments carried out on a prototype built in the laboratory. The results show how the equalizer circuit eliminates the voltage imbalance in steady state for different load conditions; thus, a high power factor is demonstrated at full load. Therefore, the equalizer circuit does not affect the operation of the voltage doubler boost rectifier.

The comparison made concerning to other classical topologies allows appreciation that the proposed circuit does not present a decrease in the power factor when there is variation in the capacitor's value; however, it is necessary to use additional components, which increases its complexity.

In further work, it is expected another controller be developed to eliminate voltage imbalance across the converter capacitors using the model of the two power sources proposed in this paper.

Author Contributions: Conceptualization, J.B., N.G. and H.E.; Methodology, J.B., N.G. and H.E.; Project administration, J.B.; Supervision, J.B.; Validation, J.B.; Writing—original draft, J.B. and H.E.; Writing—review & editing, J.B., N.G. and H.E. All authors have read and agreed to the published version of the manuscript.

Acknowledgments: The authors express gratitude to the Universidad Distrital Francisco José de Caldas, and also to the Universidad ECCI.

References

1. Shukla, A.; Pradhan, M.K. A neural network controller for high power factor in fly back converters. In Proceedings of the 2016 International Conference on Research Advances in Integrated Navigation Systems (RAINS), Bangalore, India, 6–7 May 2016.
2. Rocabert, J.; Luna, A.; Blaabjerg, F.; Rodríguez, P. Control of power converters in ac microgrids. *IEEE Trans. Power Electron.* **2012**, *27*, 4734–4749. [CrossRef]
3. Hung, C.-Y.; Wu, J.-C.; Chen, Y.-L.; Jou, H.-L. A gridconnected battery charger with power factor correction. In Proceedings of the 2016 IEEE 11th Conference on Industrial Electronics and Applications (ICIEA), Hefei, China, 5–7 June 2016.
4. Bodetto, M.; Aroudi, A.E.; Cid-Pastor, A.; Calvente, J.; Salamero, L.M. Design of ac dc pfc high-order converters with regulated output current for low-power application. *IEEE Trans. Power Electron.* **2016**, *31*, 2012–2025. [CrossRef]
5. Bayona, J.; Chamorro, H.; Sánchez, A.; Aguillón J.; Rubio, D. Linear control of a power factor correction rectifier in half-bridge configuration. In Proceedings of the IEEE CACIDI 2016—IEEE Conference on Computer Sciences, Buenos Aires, Argentina, 30 November–2 December 2016; pp. 1–6.
6. Bolívar, F.; Díaz, N.; Bayona, J. Design and implementation of a digital PID controller with pre-compensation loop for a half-bridge PFC Boost converter. *Rev. UIS Ing.* **2020**, *19*, 179–192.
7. Cho, K.S.; Lee, B.K.; Kim, J.S. CRM PFC Converter with New Valley Detection Method for Improving Power System Quality. *Electronics* **2020**, *9*, 38. [CrossRef]
8. Kang, B.-G.; Kim, C.-E.; Black, J.; Kim, D.-K.; Moon, G.-W. A high power density and power factor cascade buck-boost pfc under expanded high line voltage. In Proceedings of the IEEE Transportation Electrification Conference and Expo, Asia-Pacific (ITEC AsiaPacific), Busan, Korea, 1–4 June 2016.
9. Sheela, V.; Gopinath, M. Effiency analysis of bridgeless cuk converter for pfc applications. In Proceedings of the International Conference on Information Communication and Embedded Systems (ICICES), Chennai, India, 21–22 February 2013.
10. Xie, X.; Li, J.; Peng, K.; Zhao, C.; Lu, Q. Study on the singlestage forward-flyback pfc converter with qr control. *IEEE Trans. Power Electron.* **2016**, *31*, 430–442. [CrossRef]
11. Sun, D.; Xu, S.; Sun, W.; Lu, S. A new digital predictive control strategy for boost pfc converter. *IEICE Electron. Express* **2015**, *12*, 20150726. [CrossRef]
12. Dung, N.A.; Hieu, P.P.; Chiu, H.-J.; Hsieh, Y.-C.; Lin, J.-Y. A new digital control strategy of boost pfc at high-line light-load condition. In Proceedings of the IEEE International Conference on Industrial Technology (ICIT), Taipei, Taiwan, 14–17 March 2016.
13. Kamalapathi, K.; Priyadarshi, N.; Padmanaban, S.; Holm-Nielsen, J.B.; Azam, F.; Umayal, C.; Ramachandaramurthy, V.K. A Hybrid Moth-Flame Fuzzy Logic Controller Based Integrated Cuk Converter Fed Brushless DC Motor for Power Factor Correction. *Electronics* **2018**, *7*, 288. [CrossRef]
14. Park, J.H.; Kim, D.J.; Lee, K.B. Predictive Control Algorithm Including Conduction-Mode Detection for PFC Converter. *IEEE Trans. Ind. Electron.* **2016**, *63*, 5900–5911. [CrossRef]
15. Marcos-Pastor, A.; Vidal-Idiarte, E.; Cid-Pastor, A.; Martinez-Salamero, L. Interleaved Digital Power Factor Correction Based on the Sliding-Mode Approach. *IEEE Trans. Power Electron.* **2016**, *31*, 4641–4653. [CrossRef]
16. Fernandes, R.; Trescases, O. A Multimode 1-MHz PFC Front End With Digital Peak Current Modulation. *IEEE Trans. Power Electron.* **2016**, *31*, 5694–5708. [CrossRef]
17. Heydari, H.; Monfared, M. An Integrated Interleaved Dual-Mode Time-Sharing Inverter for Single Phase Grid Tied Applications. *IEEE Trans. Ind. Electron.* **2019**, *66*, 286–296. [CrossRef]
18. Pourfaraj, A.; Monfared, M.; Heydari, H. Single-Phase Dual-Mode Interleaved Multilevel Inverter for PV Applications. *IEEE Trans. Ind. Electron.* **2020**, *67*, 2905–2915. [CrossRef]
19. Abbaszadeh, M.A.; Monfared, M.; Heydari, H. High-buck in Buck and High-boost in Boost Dual-Mode Inverter (Hb^{2}DMI). *IEEE Trans. Ind. Electron.* **2020**, doi:10.1109/TIE.2020.2988240. [CrossRef]

20. Hosseini, S.H.; Ghazi, R.; Heydari, H. An Extendable Quadratic Bidirectional DC-DC Converter for V2G and G2V Applications. *IEEE Trans. Ind. Electron.* **2020**, doi:10.1109/TIE.2020.2992967. [CrossRef]

21. Samsudin, N.A.; Ishak, D.; Ahmad, A.B. Design and experimental evaluation of a single-stage AC/DC converter with PFC and hybrid full-bridge rectifier. *Eng. Sci. Technol. Int. J.* **2018**, *21*, 189–200. [CrossRef]

22. Cheng, C.A.; Chang, C.H.; Cheng, H.L.; Chang, E.C.; Chung, T.Y.; Chang, M.T. A Single-Stage LED Streetlight Driver with Soft-Switching and Interleaved PFC Features. *Electronics* **2019**, *8*, 911. [CrossRef]

23. Tung, N.T.; Tuyen, N.D.; Huy, N.M.; Phong, N.H.; Cuong, N.C.; Phuong, L.M. Design and Implementation of 150 W AC/DC LED Driver with Unity Power Factor, Low THD, and Dimming Capability. *Electronics* **2020**, *9*, 52. [CrossRef]

24. Kang, F.; Park, S.J.; Cho, S.E.; Kim, J.M. Photovoltaic power interface circuit incorporated with a buck-boost converter and a full-bridge inverter. *Appl. Energy* **2005**, *82*, 266–283. [CrossRef]

25. Srinivasan, R.; Oruganti, R. A unity power factor converter using half-bridge boost topology. *IEEE Trans. Power Electron.* **1998**, *13*, 487–500. [CrossRef]

26. Lo, Y.K.; Chiu, H.J.; Song, T.H. Elimination of voltage imbalance between the split capacitors in three-phase half-bridge switch-mode rectifiers. In Proceedings of the 4th IEEE International Conference on Power Electronics and Drive Systems, Denpasar, Indonesia, 25 October 2001.

27. Lo, Y.K.; Song, T.H.; Chiu, H.J. Analysis and elimination of voltage imbalance between the split capacitors in half-bridge boost rectifiers. *IEEE Trans. Ind. Electron.* **2002**, *49*, 1175–1177.

28. Lo, Y.K.; Ho, C.T.; Wang, J.M. Elimination of the Output Voltage Imbalance in a Half-Bridge Boost Rectifier. *IEEE Trans. Power Electron.* **2007**, *22*, 1352–1360. [CrossRef]

29. Erickson, R.W.; Maksimovic, D. *Fundamentals of Power Electronics*; Springer: Berlin/Heidelberg, Germany, 2001.

30. Bayona, J.; Guarnizo, J.; Gelvez, N. Pulse Width Prediction Control Technique Applied to a Half-Bridge Boost. *Tecciencia* **2018**, *13*, 48–54.

2

Analysis on Displacement Angle of Phase-Shifted Carrier PWM for Modular Multilevel Converter

Qian Cheng [ID], **Chenchen Wang and Jian Wang** *[ID]

School of Electrical Engineering, Beijing Jiaotong University, Beijing 100044, China; 17117415@bjtu.edu.cn (Q.C.); chchwang@bjtu.edu.cn (C.W.)
* Correspondence: jwang4@bjtu.edu.cn
† This paper is an extended version of our paper published in 2019 IEEE Energy Conversion Congress and Exposition (ECCE), Baltimore, MD, USA, 2019; pp. 4801–4807.

Abstract: This paper provides theoretical and experimental discussions on the characteristics of the modular multilevel converter (MMC) when phase-shifted carrier sinusoidal pulse-width modulation (PSC-SPWM) is applied. Harmonic-cancellation characteristics of output voltage and circulating current are analyzed on the basis of a general implementation of PSC-SPWM with two freedom displacement angles. Five available PSC-SPWM schemes with different carrier displacement angles were obtained, and a detailed performance comparison about output voltage and circulating current harmonic characteristics is presented. On the basis of the equivalent circuit with ideal transformer representation of the SMs, capacitor voltages affected by PSC-SPWM schemes are also briefly analyzed. The proposed PSC-SPWM schemes can unify two different cases of odd and even SM situations for output voltage and circulating current harmonic minimization, respectively. Lastly, the optimal schemes for practical MMC application were verified by simulation and experiments on an MMC prototype.

Keywords: modular multilevel converter (MMC); phase-shifted carrier SPWM (PSC-SPWM); displacement angle; harmonic characteristics; capacitor voltage

1. Introduction

The modular multilevel converter (MMC) has been widely studied because it presents great advantages such as a transformer-less and modular structure, common DC bus, and good harmonic characteristics [1–3]. These advantages make MMC the most attractive topology for various medium- and high-voltage, high-power applications, such as high-voltage direct-current (HVDC) transmission systems [4,5], static synchronous compensators (STATCOM) [6,7], high-voltage isolated DC/DC [8], and medium-voltage motor drive [9–12].

The MMC allows for the flexible selection of inserted submodules (SMs), so there are various modulation methods that directly affect the harmonic characteristics of output voltage and circulating current. The staircase wave-modulation methods in [13] are preferable in applications that require an extremely high number of SMs, such as HVDC applications. Some optimization schemes for reducing power losses are mentioned in [14,15]. The selective harmonic eliminating modulation technology [16], space vector pulse-width modulation (SVPWM) [17,18], and phase-shifted carrier sinusoidal pulse-width modulation (PSC-SPWM) technology [19–23] are also considered as modulation methods for MMC applications that need fewer submodules. Among these methods, PSC-SPWM technology is more used for fewer SMs. Each power unit has the same switching frequency with PSC-SPWM, which is a benefit for heat-dissipation structure design. When PSC-SPWM is applied, the output voltage of MMC, as shown in Figure 1, has $N + 1$ or $2N + 1$ levels with different displacement-angle assignments between the carriers of each SM [21]. The displacement angle affects the harmonic characteristics of output voltage and circulating current. An available implementation scheme of PSC-SPWM technology for MMC was presented in [22]. The displacement angle is set to π/N when N is even, and 0 when N is odd for maximal harmonic cancellation of output voltage. On the other hand, to achieve the greatest harmonic cancellation of circulating current, the displacement angle should be π/N when N is odd, and 0 when N is even. However, the angle irregularly changes according to the parity of SM number. Furthermore, how the displacement angle between the adjacent SMs in the same arm influences the performance of harmonics and the voltage-balancing issue of capacitors is not considered. Energy distribution is only relatively balanced between submodules when the right displacement angle is selected. Four available PSC-SPWM schemes were presented with performance comparison about output voltage and circulating current harmonic characteristics in our previous paper [23], whereas the selection of displacement angle does not give theoretical derivation.

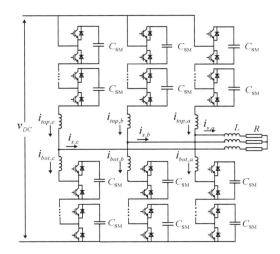

Figure 1. Structure of three-phase modular multilevel converter (MMC).

This paper presents a theoretical derivation of displacement-angle selection based on harmonic-cancellation characteristics. All possible solutions for harmonic cancellation were obtained, and the optimal PSC-SPWM schemes were selected on the basis of harmonic-cancellation characteristics and voltage stability. The proposed PSC-SPWM schemes unify two different cases of odd and even SM situations. The rest of the paper is organized as follows: In Section 2, the MMC topology is introduced, and the mathematical model is established. Then, a general PSC-SPWM scheme with two freedom displacement angles (displacement angle between adjacent SMs in the same arm, and the displacement angle between top and bottom SMs) is presented in Section 3, and optimal PSC-SPWM schemes for MMC were obtained on the basis of mathematical analysis of the harmonic-cancellation characteristics of output voltage and circulating current affected by the displacement angles. Moreover, the capacitor voltage of SMs affected by displacement angles was analyzed on the basis of the equivalent circuit with the ideal transformer representation of the SMs in this section. Section 4 gives the simulation and experiment results with a detailed performance comparison about the characteristics of output voltage and circulating current, which show that the selected displacement-angle assignment was the most suitable PSC-SPWM scheme for practical MMC application. Lastly, conclusions are summarized in Section 5.

2. Basic MMC Operational Principle

2.1. MMC Structure

A schematic diagram of a three-phase MMC is shown in Figure 1. The main circuit consists of six arms. Each arm is composed of N half-bridge SMs, which are connected with a separated arm inductor in series. Each SM is formed by a DC capacitor C_{SM} and two power switches. Three-phase RL load is respectively connected to the midpoint of three phases.

2.2. Model of MMC

The mathematical model of MMC can be obtained by Kirchhoff's voltage law and Kirchhoff's current law as

$$
\begin{cases}
v_{DC} = v_{top,j}^{sub} + v_{bot,j}^{sub} + 2L_{arm}\dfrac{di_{z,j}}{dt} \\
v_{s,j} = \dfrac{1}{2}(v_{bot,j}^{sub} - v_{top,j}^{sub}) - \dfrac{1}{2}L_{arm}\dfrac{di_{s,j}}{dt},
\end{cases}
\tag{1}
$$

where $v_{top,j}^{sub}$, $v_{bot,j}^{sub}$ are voltages across N top SMs and bottom SMs in phase j ($j \in \{a, b, c\}$), respectively. $v_{s,j}$, $i_{s,j}$ are the output voltage and current of phase j, respectively. L_{arm} is arm inductance, and v_{DC} is the DC-link voltage. The circulating current circulates through both the top and bottom arms of phase j, which is expressed as

$$
i_{z,j} = \frac{1}{2}\left(i_{top,j} + i_{bot,j}\right)
\tag{2}
$$

where $i_{top,j}$, $i_{bot,j}$ are the top and bottom arm current of phase j, respectively.

The voltage across arm inductors $v_{L,j}$ is :

$$
v_{L,j} = 2L_{arm}\frac{di_{z,j}}{dt} = v_{DC} - \left(v_{top,j}^{sub} + v_{bot,j}^{sub}\right)
\tag{3}
$$

For the SMs of MMC, the output voltage of the kth SM in the top and bottom arms can be expressed by $u_{top,j}(k)$, $u_{bot,j}(k)$ respectively:

$$
\begin{cases}
u_{top,j}(k) = S_{top,j}(k)v_{top,j}(k) \\
u_{bot,j}(k) = S_{bot,j}(k)v_{top,j}(k),
\end{cases}
\tag{4}
$$

where $v_{top,j}(k)$, $S_{top,j}(k)$ and $v_{bot,j}(k)$, $S_{bot,j}(k)$ are the capacitor voltage and switching function of kth SM of the top and bottom arms, respectively.

The dynamic equation of SM capacitors can be obtained as

$$
\begin{cases}
C\dfrac{dv_{top,j}(k)}{dt} = S_{top,j}(k)i_{z,j} + 0.5S_{top,j}(k)i_{s,j} \\
C\dfrac{dv_{bot,j}(k)}{dt} = S_{bot,j}(k)i_{z,j} - 0.5S_{bot,j}(k)i_{s,j}
\end{cases}
\tag{5}
$$

According to (1), (4), and (5), one phase circuit of MMC can be redrawn by the equivalent circuit on the basis of the ideal transformer representation of SMs [24], as shown in Figure 2. The notation of turns-ratio term is abbreviated as the switching function of SMs, which fully represents the switching action. On the basis of the equivalent circuit, energy was transferred from the DC side to the AC side through SM capacitors, which acted as intermediate storage elements. In this equivalent circuit, external and internal characteristics of the MMC could be reflected well.

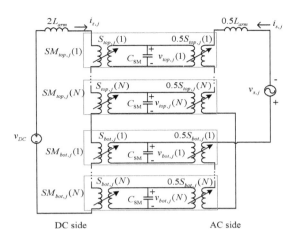

Figure 2. Equivalent circuit of single MMC phase.

3. Phase-Shifted PWM Schemes for MMC

In general, the PSC-SPWM scheme of MMC (N SMs each arm) with two freedom displacement angles is shown in Figure 3. Each SM is assigned with a specific reference signal and a triangular carrier with a displacement angle. The state of the SM is determined by the relationship between reference signal and carrier. When the reference is larger than the carrier value, the corresponding switching function $S_{top,j}(k)$ or $S_{bot,j}(k)$ equals to 1; when the reference is smaller than the carrier value, the corresponding switching function $S_{top,j}(k)$ or $S_{bot,j}(k)$ equals to 0.

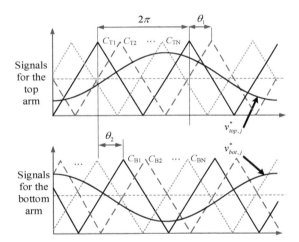

Figure 3. General MMC phase-shifted carrier sinusoidal pulse-width modulation (PSC-SPWM) schemes.

θ_1, θ_2 are the displacement angles between the adjacent carriers of the same arm, and the displacement angle between the carriers of the top and bottom arms, respectively. Reference voltages are given as

$$\begin{cases} v^*_{top,j} = \dfrac{v_{DC}}{2N} - \dfrac{Mv_{DC}}{2N}\cos\left(\omega_0 t + \varphi_j\right) \\ v^*_{bot,j} = \dfrac{v_{DC}}{2N} + \dfrac{Mv_{DC}}{2N}\cos\left(\omega_0 t + \varphi_j\right) \end{cases} \tag{6}$$

where M is the modulation index, ω_0 is the fundamental angle frequency, φ_j is the phase angle, and v_{DC} is the voltage of DC bus.

Each SM corresponding to a specific triangular carrier with different displacement angle. The general carrier function of the top and bottom SMs can be calculated as:

$$f_{CT}(k) = \dfrac{v_{DC}}{N}\left(\dfrac{1}{2} + \dfrac{1}{2}\cdot\dfrac{2}{\pi}\arcsin\left[\sin\left(\omega_c t + (k-1)\,\theta_1\right)\right]\right) \tag{7}$$

$$f_{CB}(k) = \frac{v_{DC}}{N} \left(\frac{1}{2} + \frac{1}{2} \cdot \frac{2}{\pi} \arcsin\left[\sin\left(\omega_c t + (k-1)\,\theta_1 + \theta_2 \right) \right] \right) \tag{8}$$

3.1. Fourier Series Representation of Switching Functions

The double Fourier series based on analysis in [25] is presented to obtain the harmonic feature of output voltage and circulating current of MMC. In (7) and (8), there are two freedom displacement degrees, θ_1 and θ_2, which constitute different PSC-SPWM schemes. The principle of how θ_1 and θ_2 affect the harmonics of output voltage and circulating current is presented in this section.

From (6–8), the switching function can be expressed by Fourier representation as

$$
\begin{aligned}
S_{top,j}(k) = {} & \frac{1}{2} - \frac{M}{2} \cos\left(\omega_o t + \varphi_j \right) + \sum_{m=1}^{\infty} \sum_{n=-\infty}^{\infty} \frac{2}{m\pi} \\
& \times \sin\left[\frac{(m+n)\pi}{2} \right] \times J_n\left(\frac{Mm\pi}{2} \right) \\
& \times \cos\left[m\left(\omega_c t + (k-1)\theta_1 \right) + n\left(\omega_o t + \varphi_j \right) \right]
\end{aligned}
\tag{9}
$$

$$
\begin{aligned}
S_{bot,j}(k) = {} & \frac{1}{2} + \frac{M}{2} \cos\left(\omega_o t + \varphi_j \right) + \sum_{m=1}^{\infty} \sum_{n=-\infty}^{\infty} \frac{2}{m\pi} \\
& \times \sin\left[\frac{(m+n)\pi}{2} \right] \times J_n\left(\frac{Mm\pi}{2} \right) \\
& \times \cos\left[m\left(\omega_c t + (k-1)\theta_1 + \theta_2 \right) + n\left(\omega_o t + \varphi_j + \pi \right) \right]
\end{aligned}
\tag{10}
$$

where m is carrier index variable ($m = 1, 2, \ldots, \infty$), n is the baseband index variable ($n = -\infty, \ldots, -1, 0, 1, \ldots, \infty$), and $J_n(x)$ is the Bessel coefficient.

For simplicity, it was assumed that all the capacitor voltage of SMs are naturally balanced and the voltage fluctuation is ignored. From (1), (9), and (10), the equivalent output voltage of phase j can be obtained as

$$
\begin{aligned}
u_{s,j} = {} & \frac{M v_{DC}}{2} \cos\left(\omega_o t + \varphi_j \right) - \frac{1}{2} \sum_{m=1}^{\infty} \sum_{n=-\infty}^{\infty} \frac{2 v_{DC}}{m\pi N} \\
& \times \sin\left[\frac{(m+n)\pi}{2} \right] \times J_n\left(\frac{Mm\pi}{2} \right) \\
& \times 2\sin\left[\frac{m\theta_2 + n\pi}{2} \right] \times \sum_{k=1}^{N} \cos\left(\frac{m(N-1-2(k-1))\theta_1}{2} \right) \\
& \times \sin\left[m\omega_c t + n\left(\omega_o t + \varphi_j \right) + \frac{m\theta_2 + n\pi}{2} + \frac{m(N-1)\pi}{2} \right]
\end{aligned}
\tag{11}
$$

From (11), the harmonic characteristic of output voltage is affected by the following items, which are the function of θ_1, θ_2 and N:

$$
\begin{cases}
k_1 = \displaystyle\sum_{k=1}^{N} \cos\left(\dfrac{m(N-1-2(k-1))\theta_1}{2} \right) \\[4mm]
k_2 = \sin\left[\dfrac{(m+n)\pi}{2} \right] \\[4mm]
k_3 = \sin\left[\dfrac{m\theta_2 + n\pi}{2} \right]
\end{cases}
\tag{12}
$$

Voltage across the arm inductors of phase j can be obtained as

$$
\begin{aligned}
u_{L,j}(k) = & \sum_{m=1}^{\infty} \sum_{n=-\infty}^{\infty} \frac{2v_{DC}}{m\pi N} \times J_n\left(\frac{Mm\pi}{2}\right) \times \sin\left[\frac{(m+n)\pi}{2}\right] \\
& \times 2\cos\left[\frac{m\theta_2 + n\pi}{2}\right] \times \sum_{k=1}^{N} \cos\left(\frac{m(N-1-2(k-1))\theta_1}{2}\right) \\
& \times \cos\left[m\omega_c t + n\left(\omega_o t + \varphi_j\right) + \frac{m\theta_2 + n\pi}{2} + \frac{m(N-1)\pi}{2}\right]
\end{aligned}
\tag{13}
$$

Similarly, the harmonic characteristic of voltage across the arm inductors is affected by

$$
\begin{cases}
k_1 = \sum_{k=1}^{N} \cos\left(\dfrac{m(N-1-2(k-1))\theta_1}{2}\right) \\
k_2 = \sin\left[\dfrac{(m+n)\pi}{2}\right] \\
k_4 = \cos\left[\dfrac{m\theta_2 + n\pi}{2}\right]
\end{cases}
\tag{14}
$$

3.2. Harmonic-Cancellation Characteristics of MMC with PSC-SPWM

According to (12) and (14), if one of k_1, k_2 and k_3 is equal to 0, the corresponding output-voltage harmonic components ($m\omega_c + n\omega_0$) can be completely eliminated. The corresponding circulating current harmonic components can be completely eliminated if one of k_1, k_2 and k_4 is equal to 0 in the same way. Specifically, if index m is odd/even, and index n is odd/even, k_2 equals to 0 in the expression of both Equations (11) and (13). As a result, odd side-band harmonics are eliminated in the output waveform and voltage of arm inductors around the odd carrier frequency multiples and the even side-band harmonics are eliminated in the output waveform and voltage of arm inductors around the even carrier-frequency multiples.

Harmonic-cancellation characteristics of k_1, k_3 and k_4 are affected by θ_1 and θ_2. Item k_1 can eliminate the harmonics of the whole group $m\omega_c$ with proper θ_1, and the relationship between k_1 and θ_1 is shown in Figure 4. If θ_1 equals to π/N, all the even multiples carrier-frequency harmonic groups except those at $2N$-multiples and its integral multiple are eliminated. For example, the harmonics of group $2\omega_c$ and $4\omega_c$ are eliminated when $N=3$. The harmonics of group $2\omega_c$, $4\omega_c$ and $6\omega_c$ are eliminated when $N=4$. If θ_1 equals to $2\pi/N$, all multiples carrier-frequency harmonic groups except those at N-multiples and its integral multiples are eliminated. For example, the harmonics of group ω_c, $2\omega_c$, $4\omega_c$ and $5\omega_c$ are eliminated when $N=3$. The harmonics of group ω_c, $2\omega_c$, $3\omega_c$, $5\omega_c$, $6\omega_c$ and $7\omega_c$ are eliminated when $N=4$.

After eliminating the harmonics of some whole group $m\omega_c$ with proper θ_1, proper displacement angle θ_2 can be selected to eliminate more harmonic components. The harmonic components of circulating current are the same as the voltage across the arm inductor. However, according to the relationship of trigonometric functions, k_3 in (12) and k_4 in (14) are completely different. For a particular harmonic group $m\omega_c$, output-voltage harmonics and the circulating current harmonic vary with θ_2 at the completely opposite tendency. Displacement angle θ_2 should be selected to obtain a zero k_3 or k_4 item coordinate with the harmonic-cancellation capability of k_2 to achieve maximal harmonic elimination of output voltage or circulating current, respectively.

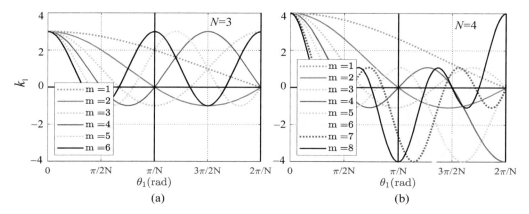

Figure 4. Harmonics of group cancellation characteristics with PSC-SPWM schemes. **(a)** $N = 3$. **(b)** $N = 4$.

For MMC, output voltage with lower harmonic distortion means smaller and lower-cost filters, and better output characteristics, especially when N is small (e.g., in motor drive applications). On the other hand, the harmonics of circulating current causes higher loss and current stress. For a great number of SM applications (e.g., HVDC application), the harmonics of circulating current become the main problem. In this paper, five potential PSC-SPWM schemes (PSC1 to PSC5) for MMC that achieve maximal voltage harmonic elimination or maximal circulating current harmonic elimination are presented, as shown in Figure 5. They are described with a modulation wave frequency of 50 Hz, a carrier frequency of 200 Hz, and four SMs in each arm. Displacement angle θ_1, θ_2 of each scheme is given as follows.

PSC1:

$$\begin{cases} \theta_1 = \dfrac{2\pi}{N} \\ \theta_2 = \dfrac{\pi}{N} + \pi \end{cases} \tag{15}$$

PSC2:

$$\begin{cases} \theta_1 = \dfrac{2\pi}{N} \\ \theta_2 = \dfrac{\pi}{N} \end{cases} (N \text{ is even}), \begin{cases} \theta_1 = \dfrac{2\pi}{N} \\ \theta_2 = 0 \end{cases} (N \text{ is odd}) \tag{16}$$

PSC3:

$$\begin{cases} \theta_1 = \dfrac{\pi}{N} \\ \theta_2 = 0 \end{cases} \tag{17}$$

PSC4:

$$\begin{cases} \theta_1 = \forall\theta \in (0, 2\pi/N] \\ \theta_2 = \pi \end{cases} \tag{18}$$

PSC5:

$$\begin{cases} \theta_1 = \dfrac{2\pi}{N} \\ \theta_2 = 0 \end{cases} (N \text{ is even}), \begin{cases} \theta_1 = \dfrac{2\pi}{N} \\ \theta_2 = \dfrac{\pi}{N} \end{cases} (N \text{ is odd}) \tag{19}$$

where PSC2 and PSC5 were mentioned in [22]. PSC1 and PSC3 were first proposed in [23], and PSC4 was added into this paper for discussion. Theoretically, PSC1, PSC2, and PSC3 can achieve maximal voltage harmonic elimination, and PSC4 and PSC5 can achieve maximal circulating current harmonic elimination. Compared with PSC2 and PSC5, the proposed PSC1, PSC3, and PSC4 unify two different cases of odd and even SM situations.

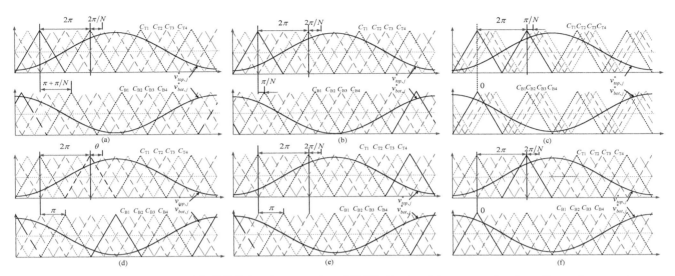

Figure 5. Proposed PSC-SPWM schemes for MMC that achieve maximal voltage harmonic elimination or maximal circulating current harmonic elimination. (**a**) PSC-SPWM1. (**b**) PSC-SPWM2. (**c**) PSC-SPWM3. (**d**) PSC-SPWM4. (**e**) PSC-SPWM4 when $\theta_1 = 2\pi/N$. (**f**) PSC-SPWM5.

For a specific index m, if both even n and odd n can result in a zero k_2 or k_3 item, the corresponding harmonic groups are eliminated in output voltage. Similarly, if both even n and odd n can result in a zero k_2 or k_4 item, the corresponding harmonic groups are eliminated in circulating current. For PSC1 and PSC2, all harmonics except those even side-band harmonics around the $(2l+1)N$-multiples carrier frequency and the odd side-band harmonics around the $(2l)N$-multiples carrier frequency remain with $\theta_1 = 2\pi/N$. When the m summation index is equal to $(2l+1)N$, k_2 and k_3 of PSC1 and PSC2, it can be represented as follows.

PSC1:

$$\begin{cases} k_2 = \sin \dfrac{((2l+1)N+n)\pi}{2} \\ k_3 = \sin \dfrac{((2l+1)(N+1)+n)\pi}{2} \end{cases} \tag{20}$$

In (20), both even n and odd n can result in an even $((2l+1)N+n)$ or even $((2l+1)(N+1)+n)$, which obtains a zero k_2 or k_3 item. In other words, the harmonics of $(2l+1)N$-multiples carrier frequency are totally eliminated in output voltage with PSC1, resulting in maximal voltage harmonic elimination.

PSC2:

$$\begin{cases} k_2 = \sin \dfrac{((2l+1)N+n)\pi}{2} \\ k_3 = \sin \dfrac{(2l+1+n)\pi}{2} \end{cases} (N \text{ is even}) \\ \begin{cases} k_2 = \sin \dfrac{((2l+1)N+n)\pi}{2} \\ k_3 = \sin \dfrac{n\pi}{2} \end{cases} (N \text{ is odd}) \tag{21}$$

In (21), both even n and odd n can result in an even $((2l+1)N+n)$ or $((2l+1+n)$ when N is even, and an even $((2l+1)N+n)$ or even n when N is odd, which obtains a zero k_2 or k_3 item. As a result, the harmonics of $(2l+1)N$-multiples carrier frequency are totally eliminated in output voltage with PSC2 the same with PSC1.

On the other hand, when the m summation index is equal to $(2l)N$, the k_2 and k_4 of PSC1 and PSC2 can be represented as

PSC1:

$$\begin{cases} k_2 = \sin \dfrac{(2lN + n)\,\pi}{2} \\ k_4 = \cos \dfrac{(2l\,(N+1)+n)\,\pi}{2} \end{cases} \tag{22}$$

PSC2:

$$\begin{cases} k_2 = \sin \dfrac{(2lN+n)\,\pi}{2} \\ k_4 = \cos \dfrac{(2l+n)\,\pi}{2} \end{cases} (N \ \text{is even}) \\ \begin{cases} k_2 = \sin \dfrac{(2lN+n)\,\pi}{2} \\ k_4 = \cos \dfrac{n\pi}{2} \end{cases} (N \ \text{is odd}) \tag{23}$$

In (22), both even n and odd n can result in an even $(2lN + n)$ or odd $(2l(N+1)+n)$, which obtains a zero k_2 or k_4 item. Similarly, in (23), both even n and odd n can result in an even $(2lN + n)$ or odd $(2l + n)$ when N is even, and an even $(2lN + n)$ or odd n when N is odd, which obtains a zero k_2 or k_4 item too. As a result, the extra harmonics of group $(2l)N$-multiples carrier frequency are eliminated in circulating current with PSC1 or PSC2.

For PSC3, all harmonics of odd multiples and $(2l)N$-multiples carrier frequency are remained when $\theta_1 = \pi/N$. When the m summation index is equal to $(2l + 1)$, k_2 and k_3 can be represented as
PSC3:

$$\begin{cases} k_2 = \sin \dfrac{(2l+1+n)\,\pi}{2} \\ k_3 = \sin \dfrac{n\pi}{2} \end{cases} \tag{24}$$

In (24), both even n and odd n can result in a zero k_2 or k_3 item. As a result, all odd multiples carrier-frequency harmonic groups are eliminated in output voltage with PSC3, resulting in maximal voltage harmonic elimination, the same as in PSC1 and PSC2.

On the other hand, when the m summation index is equal to $(2l)N$, the k_2 and k_4 of PSC3 can be represented as
PSC3:

$$\begin{cases} k_2 = \sin \dfrac{(2lN+n)\,\pi}{2} \\ k_4 = \cos \dfrac{n\pi}{2} \end{cases} \tag{25}$$

Both odd n and even n can result in a zero k_2 or k_4. The extra harmonics of group $(2l)N$-multiples carrier frequency are eliminated in circulating current with PSC3.

For PSC4, k_2 and k_4 can be represented as
PSC4:

$$\begin{cases} k_2 = \sin \dfrac{(m+n)\,\pi}{2} \\ k_4 = \cos \dfrac{(m+n)\,\pi}{2} \end{cases} \tag{26}$$

All multiples carrier-frequency harmonic groups are eliminated in the circulating current with PSC4, while k_2 and k_3 can be represented as
PSC4:

$$\begin{cases} k_2 = \sin \dfrac{(m+n)\,\pi}{2} \\ k_3 = \sin \dfrac{(m+n)\,\pi}{2} \end{cases} \tag{27}$$

Only even side-band harmonics around the even multiples and the odd side-band harmonics are around the odd multiples are eliminated in the output voltage with PSC4 when $\theta_1 \neq \pi/N$ and $\theta_1 \neq 2\pi/N$.

For PSC5, when the m summation index is equal to lN, k_2 and k_4 can be represented as PSC5:

$$\begin{cases} k_2 = \sin\dfrac{(lN+n)\,\pi}{2} \\ k_4 = \cos\dfrac{(lN+n)\,\pi}{2} \end{cases} (N \text{ is odd}), \\ \begin{cases} k_2 = \sin\dfrac{(lN+n)\,\pi}{2} \\ k_4 = \cos\dfrac{n\pi}{2} \end{cases} (N \text{ is even}) \tag{28}$$

All the harmonics of multiples carrier frequency are eliminated in the circulating current with a zero item k_2 or k_4 when PSC5 is applied.

On the other hand, when the m summation index is equal to lN, k_2 and k_3 can be represented as PSC5:

$$\begin{cases} k_2 = \sin\dfrac{(lN+n)\,\pi}{2} \\ k_3 = \sin\dfrac{(lN+n)\,\pi}{2} \end{cases} (N \text{ is odd}), \\ \begin{cases} k_2 = \sin\dfrac{(lN+n)\,\pi}{2} \\ k_3 = \sin\dfrac{n\pi}{2} \end{cases} (N \text{ is even}) \tag{29}$$

There are no extra harmonics that can be eliminated in output voltage with PSC5.

From the above analysis, the conclusion is that only odd side-band harmonics around $(2l)N$-multiples carrier frequency are retained in output voltage with PSC1, PSC2, and PSC3 applied. The extra harmonic group of $(2l)N$-multiples carrier frequency is eliminated in circulating current with PSC1 and PSC2, while all even side-band harmonics around odd multiples carrier frequency remain with PSC3. All carrier-frequency harmonics are eliminated in the circulating current with both PSC4 and PSC5 applied. Simultaneously, maximal extra harmonic cancellation in output voltage can be obtained with PSC4 when $\theta_1 = 2\pi/N$, which is equivalent to PSC5.

Carrier angles of each SM for five PSC-SPWM schemes when $N = 3$ (odd number SMs example) and $N = 4$ (even number SMs example) are listed in Figures 6 and 7. PSC1, PSC2, and PSC3 are the minimal voltage harmonic schemes. PSC4 (when $\theta_1 = 2\pi/N$) and PSC5 are the minimal circulating current harmonic schemes. Considering that the status of each SM in the same arm is identical, PSC1 and PSC2, PSC4, and PSC5 are equivalent except that SMs in the bottom arm corresponding to the top arms are different. SMs with the same status (carrier angle) are filled with the same color and connected by red arrows. Obviously, the proposed PSC1 and PSC4 could automatically distribute the carrier angle.

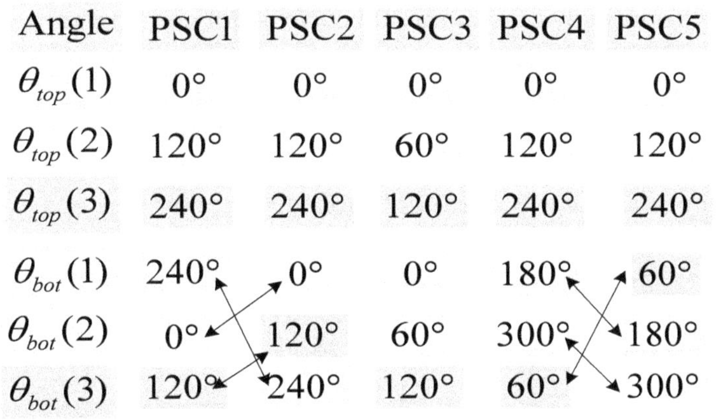

Angle	PSC1	PSC2	PSC3	PSC4	PSC5
$\theta_{top}(1)$	0°	0°	0°	0°	0°
$\theta_{top}(2)$	120°	120°	60°	120°	120°
$\theta_{top}(3)$	240°	240°	120°	240°	240°
$\theta_{bot}(1)$	240°	0°	0°	180°	60°
$\theta_{bot}(2)$	0°	120°	60°	300°	180°
$\theta_{bot}(3)$	120°	240°	120°	60°	300°

Figure 6. Carrier angles of each submodule (SM) for five PSC-SPWM schemes when $N = 3$.

Angle	PSC1	PSC2	PSC3	PSC4	PSC5
$\theta_{top}(1)$	0°	0°	0°	0°	0°
$\theta_{top}(2)$	90°	90°	45°	90°	90°
$\theta_{top}(3)$	180°	180°	90°	180°	180°
$\theta_{top}(4)$	270°	270°	135°	270°	270°
$\theta_{bot}(1)$	225°	45°	0°	180°	0°
$\theta_{bot}(2)$	315°	135°	45°	270°	90°
$\theta_{bot}(3)$	45°	225°	90°	0°	180°
$\theta_{bot}(4)$	135°	315°	135°	90°	270°

Figure 7. Carrier angles of each SM for five PSC-SPWM schemes when $N = 4$.

The specific PSC-SPWM scheme is determined according to the number of SMs and aforementioned harmonic-cancellation characteristics analysis. Considering the harmonic performance of both output voltage and circulating current, the optimal maximal output voltage and circulating current harmonic elimination schemes are PSC1 and PSC4 when $\theta_1 = 2\pi/N$, respectively, for which implementation does not need irregular change with parity of submodule number.

3.3. Capacitor Voltage of SMs Affected by PSC-SPWM Schemes

The above analysis is based on the assumption that all SM capacitor voltages are naturally balanced and voltage fluctuations are ignored. However, SM voltage is also affected by modulation methods. From (5), capacitor voltage varies with switching function, load current, and circulating current. The voltage fluctuation of SM capacitors in one carrier cycle is given as

$$\begin{cases} \Delta v_{top,j}(k) = \dfrac{1}{C_{SM}} \displaystyle\int_0^{T_s} \left(S_{top,j}(k)i_{z,j} + 0.5S_{top,j}(k)i_{s,j} \right) dt \\ \Delta v_{bot,j}(k) = \dfrac{1}{C_{SM}} \displaystyle\int_0^{T_s} \left(S_{bot,j}(k)i_{z,j} - 0.5S_{bot,j}(k)i_{s,j} \right) dt \end{cases} \tag{30}$$

where T_s is a fundamental period. SM capacitor voltage is affected by switching function, circulating current and the load current. Natural balancing of the 2-Cell modular multilevel converter with traditional PSC-SPWM schemes was discussed in [24], which is closely related to harmonic distribution of output current and circulating current. If the switching function and circulating current contain the same harmonic components, or the switching function and phase current contain the same harmonic components in the orthogonal trigonometric system, the right side of (21) could contain a non-negligible DC component. Voltage variation in a fundamental cycle is not equal to zero, and the capacitor voltage of each SM deviates from the rated value.

According to the above analysis, even side-band harmonics around the carrier frequency are remained in circulating current with PSC3. However, these harmonic components are also contained in switching functions $S_{top,j}$ and $S_{bot,j}$. In an orthogonal trigonometric function system, these large harmonic components could cause a non-negligible DC current in SM capacitors, which results in voltage deviation. The capacitor-voltage deviation phenomenon is verified by simulation and experiments in next section. Similarly, even side-band harmonics around the carrier frequency remain in output current with PSC4 when $\theta_1 = \pi/N$, which also results in voltage deviation. Therefore, from the perspective of the effect on capacitor voltage, $\theta_1 = 2\pi/N$ should be met for PSC-SPWM schemes.

4. Simulation and Experiment Results

Simulation and experiment results were obtained to verify the analysis of the PSC-SPWM strategy in a three-phase MMC with four half-bridge SMs. Simulation results were obtained from

MATLAB/Simulink, and the experiment results were obtained from a low-power laboratory prototype as shown in Figure 8. A digital-signal processor (DSP; TI TMS32028335) was used to generate the reference voltage, while two field-programmable gate arrays (FPGAs; Spartan-6 XC6SLX9)were adopted to generate the triangular carriers and data processing, respectively. PWM signals were transmitted to SMs via optical fibers. The main parameters of the converter are given in Table 1.

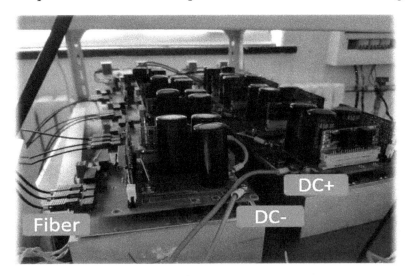

Figure 8. Laboratory prototype.

Table 1. MMC parameters.

Quantity	Symbol	Values
DC bus voltage	v_{DC}	200 V
Submodule capacitance	C_{sm}	3.6 mF
Number of submodules	N	4
Bridge inductance	L_{arm}	2 mH
Carrier frequency	f_c	1000 Hz
Load inductance	L	5 mH
Load resistance	R	24 Ω
Modulation index	M	0.8

4.1. Simulation Results

Figures 9–13 shows the simulation results of output-phase voltage with five PSC-SPWM strategies. The equivalent output-phase voltage reached $2N + 1$ levels with PSC1, PSC2, and PSC3, which had maximal output-voltage harmonic-cancellation ability, and voltage total harmonic distortion (THD) was 14.7% with the main odd side-band harmonics around $2N$-multiples carrier frequency and its integral multiple harmonic groups. The simulation results of PSC4, when $\theta_1 = 2\pi/N$, and PSC5 are shown in Figures 12 and 13, respectively. Both of these PSC-SPWM schemes had the same output characteristics. Equivalent output voltage reached $N + 1$ levels, and voltage THD was much larger than that of PSC1, PSC2, and PSC3, with 36.223%. The main harmonic components were around N times the carrier frequency and its integral multiples.

Simulation results of circulating current under five PSC-SPWM strategies are shown in Figures 14–18. The circulating current included the extra N-multiples carrier frequency and its odd multiple harmonic groups with PSC1 and PSC2 applied, all odd carrier multiple harmonic groups remained with PSC3 applied. The circulating current was mainly composed of the DC component and the secondary component with PSC4 when $\theta_1 = 2\pi/N$ and PSC5. All high-frequency harmonics were completely eliminated, with maximal harmonic-cancellation ability. On the other hand, the harmonic characteristic of output-phase voltage was decreased. Crculating current harmonic and output-voltage harmonics varied with θ_2 at a completely opposite tendency.

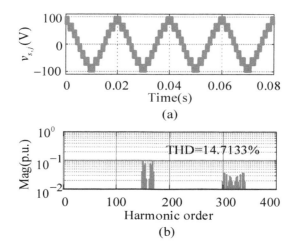

Figure 9. Theoretical output voltage of PSC1. (**a**) Waveform of equivalent output-phase voltage; (**b**) fast Fourier transform (FFT) analysis.

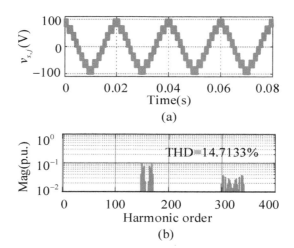

Figure 10. Theoretical output voltage of PSC2. (**a**) Waveform of equivalent output-phase voltage; (**b**) FFT analysis.

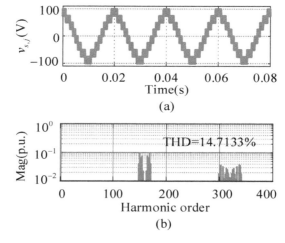

Figure 11. Theoretical output voltage of PSC3. (**a**) Waveform of equivalent output-phase voltage; (**b**) FFT analysis.

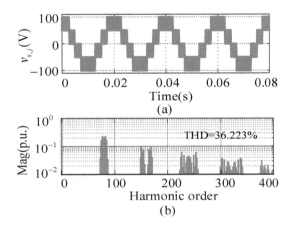

Figure 12. Theoretical output voltage of PSC4. (**a**) Waveform of equivalent output-phase voltage; (**b**) FFT analysis.

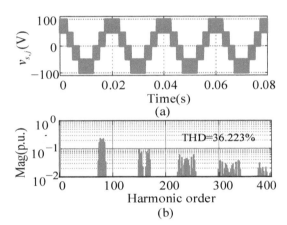

Figure 13. Theoretical output voltage of PSC5. (**a**) Waveform of equivalent output-phase voltage; (**b**) FFT analysis.

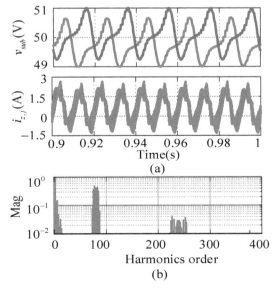

Figure 14. Simulation results of PSC1. (**a**) Capacitor voltage of SMs and circulating current. (**b**) FFT analysis of circulating current.

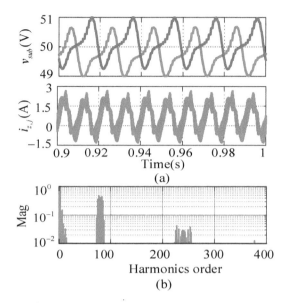

Figure 15. Simulation results of PSC2. (**a**) Capacitor voltage of SMs and circulating current. (**b**) FFT analysis of circulating current.

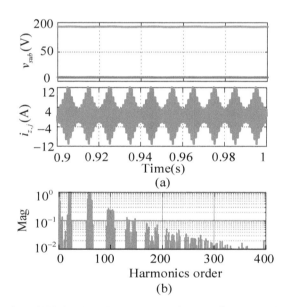

Figure 16. Simulation results of PSC3. (**a**) Capacitor voltage of SMs and circulating current. (**b**) FFT analysis of circulating current.

As shown in Figure 16, PSC3 caused the deviation and unbalance of capacitor voltage because large even side-band harmonics around the carrier frequency (especially h18, h20, and h22) remained in circulating current with PSC3, which were also contained in switching function of SMs. These harmonic components caused a non-negligible DC current in the capacitor of the SMs. This phenomenon can also be intuitively understood from the number of submodules connected to each phase. Unlike other solutions, 0 to $2N$ SMs could be put into a phase with PSC3. From (1), when $2N$ SMs were inserted, voltage across the arm inductors was v_{DC}, causing high circulating current and unstable SM capacitor voltages. Therefore, the selection of θ_1 and θ_2 needs to consider the harmonic-cancellation characteristic of both output voltage and circulating current. PSC1 and PSC4 when $\theta_1 = 2\pi/N$ are the optimal output-voltage harmonic minimization PSC-SPWM scheme and the optimal circulating current harmonic minimization PSC-SPWM scheme, respectively, with a stable and balanced capacitor voltage.

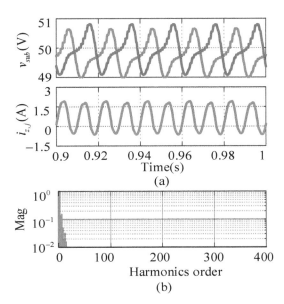

Figure 17. Simulation results of PSC4. (**a**) Capacitor voltage of SMs and circulating current. (**b**) FFT analysis of circulating current.

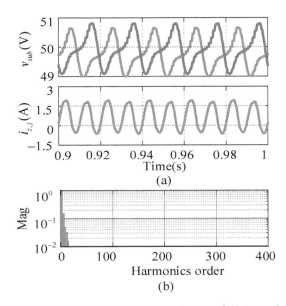

Figure 18. Simulation results of PSC5. (**a**) Capacitor voltage of SMs and circulating current. (**b**) FFT analysis of circulating current.

4.2. Experiment Results

Figure 19 shows the experiment results with PSC1, where displacement angles were selected as $\theta_1 = 90°$ and $\theta_2 = 225°$. Experiment results were almost identical to the simulation results, including output-voltage levels, the harmonic distribution of output voltage and circulating current, and the voltage stability of capacitors; $2N + 1$ voltage levels were generated, and output THD was 13.75%, which was close to the theoretical 14.71%. Simultaneously, high-frequency harmonic groups (mainly side-band harmonics around $(2l + 1)N$-multiples carrier frequency) remained in the circulating current, which was calculated by a mathematical function of an oscilloscope according to (2). The peak value of circulating current was approximately equal to 1.5 A capacitor voltage of SMs in the top and bottom arms, which were naturally balanced with small fluctuation.

Figure 20 shows the experiment results with PSC2, where displacement angles were selected as $\theta_1 = 90°$ and $\theta_2 = 45°$. PSC2 achieved the same characteristics as those of PSC1. The equivalence of PSC1 and PSC2 was verified as analyzed before, and only SMs corresponding to the top and bottom arms were different; the proposed PSC1 unified two different cases of odd and even SM situations for maximal harmonic elimination of output voltage.

Figure 21 shows the experiment results with PSC3, where displacement angles were selected as $\theta_1 = 45°$ and $\theta_2 = 0°$. Only three levels of output voltage were generated. Theoretical $2N + 1$ output-voltage levels could not be achieved because the capacitor voltage of SMs deviated to v_{DC} or 0, which was consistent with the simulation results. This is a validation that PSC3 is not acceptable for industrial applications, and $\theta_1 = 2\pi/N$ should be guaranteed for PSC-SPWM schemes.

Figure 22 shows the experiment results with PSC4, where displacement angles were selected as $\theta_1 = 90°$ and $\theta_2 = 180°$. Experiment results were almost identical to the simulation results; $N + 1$ voltage levels were generated, and output THD was 32.42%, which was close to the theoretical 36.23%. Simultaneously, all high-frequency harmonic groups were completely eliminated in the circulating current, and the peak value of the circulating current was approximately equal to 0.5 A, which was smaller than that with PSC1 and PSC2. Capacitor voltages of SMs in the top and bottom arms were naturally balanced with small fluctuation. PSC4 when $\theta_1 \neq 2\pi/N$ caused the deviation and unbalance of capacitor voltage because of the same reason as that for PSC3.

Figure 23 shows the experiment results with PSC5, where displacement angles were selected as $\theta_1 = 90°$ and $\theta_2 = 0°$. PSC5 achieved the same characteristics as those of PSC4. PSC4 and PSC5 were equivalent as analyzed before, and only SMs corresponding to the top and bottom arms were different. The proposed PSC4 when $\theta_1 = 2\pi/N$ could unify two different cases of odd and even SM situations for the maximal harmonic elimination of the circulating current.

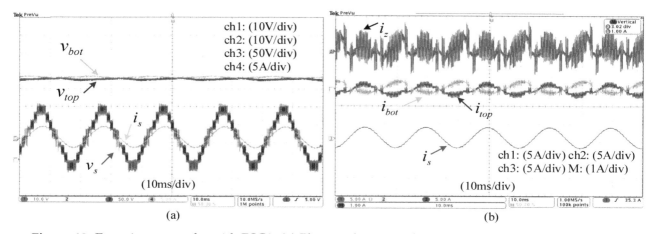

Figure 19. Experiment results with PSC1. (**a**) Phase voltage v_s, phase current i_s, capacitor voltage of SM v_{top} and v_{bot}; (**b**) circulating current i_z, phase current i_s, arm current i_{top}, and i_{bot}.

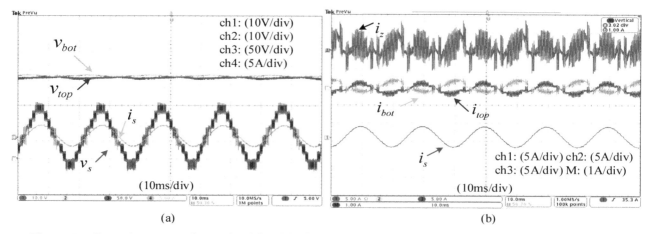

Figure 20. Experiment results with PSC2. (**a**) Phase voltage v_s, phase current i_s, capacitor voltage of SM v_{top} and v_{bot}; (**b**) circulating current i_z, phase current i_s, arm current i_{top} and i_{bot}.

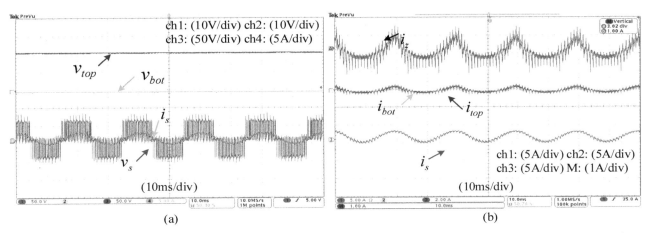

Figure 21. Experiment results with PSC3. (**a**) Phase voltage v_s, phase current i_s, capacitor voltage of SM v_{top} and v_{bot}; (**b**) circulating current i_z, phase current i_s, arm current i_{top} and i_{bot}.

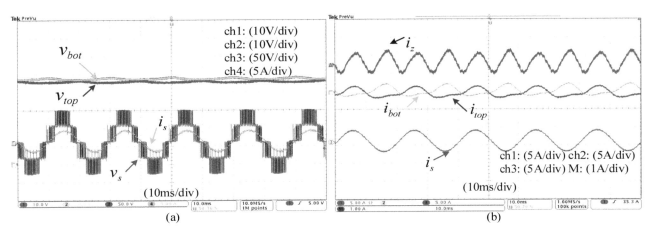

Figure 22. Experiment results with PSC4 when $\theta_1 = 2\pi / N$. (**a**) Phase voltage v_s, phase current i_s, capacitor voltage of SM v_{top} and v_{bot}; (**b**) circulating current i_z, phase current i_s, arm current i_{top} and i_{bot}.

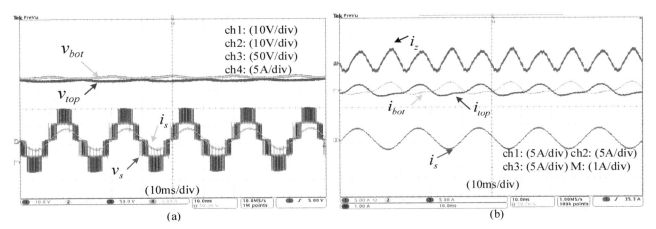

Figure 23. Experiment results with PSC5. (**a**) Phase voltage v_s, phase current i_s, capacitor voltage of SMs v_{top} and v_{bot}; (**b**) circulating current i_z, phase current i_s, arm current i_{top} and i_{bot}.

The PSC-SPWM of MMC is different from that of CHB, and only those SMs in the identical arm are equivalent. In order to avoid deviation of capacitor voltage, the selection of θ_1 and θ_2 needs to consider harmonic cancellation characteristics of both output voltage and circulating current. A comparison of five PSC-SPWM schemes is listed in Table 2. Considering these facts and implementations, PSC1

($\theta_1 = 2\pi/N$, $\theta_2 = \pi + \pi/N$) is the most suitable PSC-SPWM scheme in practical applications when the number of SMs is small. Only odd side-band harmonics around $(2l)N$-multiples carrier frequency remained in output voltage, and even side-band harmonics around $(2l-1)N$-multiples carrier frequency remained in circulating current; at the same time, the capacitor voltage of SMs was stable and balanced. On the other hand, PSC4 ($\theta_1 = 2\pi/N$, $\theta_2 = \pi$) is the most suitable PSC-SPWM scheme in practical applications when the number of SMs is large. Only side-band harmonics around $(l)N$-multiples carrier frequency remained in output voltage, and all carrier-frequency harmonics were eliminated in circulating current; at the same time, the capacitor voltage of SMs was stable and balanced. Moreover, the implementations of both PSC1 and PSC4 when $\theta_1 = 2\pi/N$ do not need irregular change with the parity of submodule number.

Table 2. Comparison of five PSC-SPWM schemes.

	PSC1	PSC2	PSC3	PSC4	PSC5
Output level	$2N+1$	$2N+1$	$2N+1$	$N+1$	$N+1$
Output-voltage harmonics	$2lNf_c$	$2lNf_c$	$2lNf_c$	lNf_c	lNf_c
Circulating-current harmonics	$(2l-1)Nf_c$	$(2l-1)Nf_c$	$(2l-1)f_c$	0	0
Expression of displacement angles	fixed	Different with parity of N	fixed	fixed	Different with parity of N
Voltage stability of capacitors	stable	stable	unstable	stable	stable

5. Conclusions

This paper provided theoretical and experimental discussions on the characteristics of the modular multilevel converter (MMC) when phase-shifted carrier sinusoidal pulse width modulation (PSC-SPWM) is applied. The harmonic components of output phase voltage and circulating current were derived in detail. We also analyzed how the displacement angle between SM carriers affects the harmonic-cancellation characteristics of output voltage, circulating current, and capacitor voltage of SMs. Displacement angles were found to affect the harmonic magnitudes of both output voltage and circulating current. On the basis of analysis, five potential PSC-SPWM schemes with maximal harmonic-cancellation ability for output voltage or circulating current were proposed. Furthermore, capacitor voltages affected by PSC-SPWM schemes were analyzed. Optimal displacement angles were selected for minimizing the harmonics of output voltage and circulating current that are suitable for application with a low and high number of SMs. Lastly, the mathematical analysis and proposed schemes were verified by simulation and experiments on a low-power MMC prototype.

Author Contributions: Conceptualization, C.W. and Q.C.; methodology, Q.C.; software, Q.C. and J.W.; validation, Q.C.; formal analysis, Q.C.; investigation, Q.C.; resources, J.W.; data curation, Q.C.; writing—original-draft preparation, Q.C.; writing—review and editing, Q.C., C.W. and J.W.; visualization, Q.C.; supervision, C.W.; project administration, C.W.; funding acquisition, C.W. All authors have read and agreed to the published version of the manuscript.

Acknowledgments: We thank all the journal editors and the reviewers for their valuable time and constructive comments that have contributed to improving this manuscript.

References

1. Lesnicar, A.; Marquardt, R. An innovative modular multilevel converter suitable for a wide power range. In Proceedings of the 2003 IEEE Bologna Power Tech Conference Proceedings, Bologna, Italy, 23–26 June 2003; Volume 3, p. 6.
2. Dekka, A.; Wu, B.; Fuentes, R.L.; Perez, M.; Zargari, N.R. Evolution of Topologies, Modeling, Control Schemes, and Applications of Modular Multilevel Converters. *IEEE Trans. Emerg. Sel. Topics Power Electron.* **2017**, *5*, 1631–1656. [CrossRef]

3. Zhang, J.; Xu, S.; Din, Z.; Hu, X. Hybrid Multilevel Converters: Topologies, Evolutions and Verifications. *Energies* **2019**, *12*, 615. [CrossRef]

4. Solas, E.; Abad, G.; Barrena, J.A.; Aurtenetxea, S.; Cárcar, A.; Zając, L. Modular Multilevel Converter With Different Submodule Concepts—Part II: Experimental Validation and Comparison for HVDC Application. *IEEE Trans. Ind. Electron.* **2013**, *60*, 4536–4545. [CrossRef]

5. Yang, X.; Xue, Y.; Chen, B.; Lin, Z.; Mu, Y.; Zheng, T.Q.; Igarashi, S.; Li, Y. An enhanced reverse blocking MMC with DC fault handling capability for HVDC applications. *Electr. Power Syst. Res.* **2018**, *163*, 706–714. [CrossRef]

6. Bina, M.T. A Transformerless Medium-Voltage STATCOM Topology Based on Extended Modular Multilevel Converters. *IEEE Trans. Power Electron.* **2011**, *26*, 1534–1545.

7. Cupertino, A.F.; Farias, J.V.M.; Pereira, H.A.; Seleme, S.I.; Teodorescu, R. Comparison of DSCC and SDBC Modular Multilevel Converters for STATCOM Application During Negative Sequence Compensation. *IEEE Trans. Ind. Electron.* **2019**, *66*, 2302–2312. [CrossRef]

8. Chen, Y.; Zhao, S.; Li, Z.; Wei, X.; Kang, Y. Modeling and Control of the Isolated DC–DC Modular Multilevel Converter for Electric Ship Medium Voltage Direct Current Power System. *IEEE Trans. Emerg. Sel. Topics Power Electron.* **2017**, *5*, 124–139. [CrossRef]

9. Hagiwara, M.; Nishimura, K.; Akagi, H. A Medium-Voltage Motor Drive With a Modular Multilevel PWM Inverter. *IEEE Trans. Power Electron.* **2010**, *25*, 1786–1799. [CrossRef]

10. Sau, S.; Fernandes, B.G. Modular Multilevel Converter Based Variable Speed Drive With Reduced Capacitor Ripple Voltage. *IEEE Trans. Ind. Electron.* **2019**, *66*, 3412–3421. [CrossRef]

11. Ma, F.; Xu, Q.; He, Z.; Tu, C.; Shuai, Z.; Luo, A.; Li, Y. A Railway Traction Power Conditioner Using Modular Multilevel Converter and Its Control Strategy for High-Speed Railway System. *IEEE Trans. Transp. Electrif.* **2016**, *2*, 96–109. [CrossRef]

12. Wang, Z.; Zheng, Z.; Li, Y.; Li, G. Modulation and Control Strategy for Electric Traction Drive System of Rail Transit Vehicles. *Trans. China Electrotech. Soc.* **2016**, *31*, 223–232.

13. Konstantinou, G.; Pou, J.; Ceballos, S.; Darus, R.; Agelidis, V.G. Switching Frequency Analysis of Staircase-Modulated Modular Multilevel Converters and Equivalent PWM Techniques. *IEEE Trans. Power Del.* **2016**, *31*, 28–36. [CrossRef]

14. Samajdar, D.; Bhattacharya, T.; Dey, S. A Reduced Switching Frequency Sorting Algorithm for Modular Multilevel Converter With Circulating Current Suppression Feature. *IEEE Trans. Power Electron.* **2019**, *34*, 10480–10491. [CrossRef]

15. Tu, Q.; Xu, Z.; Xu, L. Reduced Switching-Frequency Modulation and Circulating Current Suppression for Modular Multilevel Converters. *IEEE Trans. Power Del.* **2011**, *26*, 2009–2017.

16. Pérez-Basante, A.; Ceballos, S.; Konstantinou, G.; Pou, J.; Andreu, J.; de Alegría, I.M. (2N+1) Selective Harmonic Elimination-PWM for Modular Multilevel Converters: A Generalized Formulation and A Circulating Current Control Method. *IEEE Trans. Power Electron.* **2018**, *33*, 802–818. [CrossRef]

17. Dekka, A.; Wu, B.; Zargari, N.R.; Fuentes, R.L. A Space-Vector PWM-Based Voltage-Balancing Approach With Reduced Current Sensors for Modular Multilevel Converter. *IEEE Trans. Ind. Electron.* **2016**, *63*, 2734–2745. [CrossRef]

18. Ronanki, D.; Williamson, S.S. A Simplified Space Vector Pulse Width Modulation Implementation in Modular Multilevel Converters for Electric Ship Propulsion Systems. *IEEE Trans. Transp. Electrif.* **2019**, *5*, 335–342. [CrossRef]

19. Ilves, K.; Harnefors, L.; Norrga, S.; Nee, H. Analysis and Operation of Modular Multilevel Converters with Phase-Shifted Carrier PWM. *IEEE Trans. Power Electron.* **2015**, *30*, 268–283. [CrossRef]

20. Zhou, D.; Yang, S.; Tang, Y. Model-Predictive Current Control of Modular Multilevel Converters With Phase-Shifted Pulsewidth Modulation. *IEEE Trans. Ind. Electron.* **2019**, *66*, 4368–4378. [CrossRef]

21. Sasongko, F.; Sekiguchi, K.; Oguma, K.; Hagiwara, M.; Akagi, H. Theory and Experiment on an Optimal Carrier Frequency of a Modular Multilevel Cascade Converter With Phase-Shifted PWM. *IEEE Trans. Power Electron.* **2016**, *31*, 3456–3471. [CrossRef]

22. Li, B.; Yang, R.; Xu, D.; Wang, G.; Wang, W.; Xu, D. Analysis of the Phase-Shifted Carrier Modulation for Modular Multilevel Converters. *IEEE Trans. Power Electron.* **2015**, *30*, 297–310. [CrossRef]

23. Cheng, Q.; Wang, C. Comparison of Phase-Shifted Carrier PWM Schemes for Modular Multilevel Converter. In Proceedings of the 2019 IEEE Energy Conversion Congress and Exposition (ECCE), Baltimore, MD, USA, 29 September–3 October 2019; pp. 4801–4807.

24. Van der Merwe, W. Natural Balancing of the 2-Cell Modular Multilevel Converter. *IEEE Trans. Ind Appl.* **2014**, *50*, 4028–4035. [CrossRef]

25. Lipo, T.; Holmes, D. *Pulse Width Modulation for Power Converters: Principles and Practice*; Wiley: Hoboken, NJ, USA, 2003.

3

Improved Step Load Response of a Dual-Active-Bridge DC–DC Converter

Yifan Zhang, Xiaodong Li *, Chuan Sun and Zhanhong He

Faculty of Information Technology, Macau University of Science and Technology, Macau, China;
zhangyifan0329@outlook.com (Y.Z.); sunchuanmust@163.com (C.S.); calvinhzh@163.com (Z.H.)
* Correspondence: xdli@must.edu.mo

Abstract: This paper proposes a fast load transient control for a bidirectional dual-active-bridge (DAB) DC/DC converter. It is capable of maintaining voltage–time balance during a step load change process so that no overshoot current and DC offset current exist. The transient control has been applied for all possible transition cases and the calculation of intermediate switching angles referring to the fixed reference points is independent from the converter parameters and the instantaneous current. The results have been validated by extended experimental tests.

Keywords: transient control; DC–DC conversion; bidirectional converter

1. Introduction

In recent decades, with the increasing concern in environment issue and energy crisis, the power conversion systems (PCSs) have been using widely in renewable generation facilities. The dual-active-bridge (DAB) converter seems to be a preferred choice of PCSs in various bidirectional DC/DC applications, such as energy storage systems, electric vehicles (EV) and solid state transformers, because of its high power density, low cost and zero-voltage-switching (ZVS) features [1–14].

Interfacing two different DC sources, the DAB converter is a kind of bidirectional DC–DC converter, which consists of two full bridges linked by a high-frequency (HF) transformer with the turn ratio of $n_t : 1$, whose circuit layout is shown in Figure 1. Including the leakage inductance of the transformer, the inductor L_s is connected on the primary side as the main energy transfer device. The resistor r_s, which normally is small enough to be neglected in steady-state analysis, is an equivalent resistance of L_s and the total winding resistance of HF transformer. The eight active switches can be controlled by their gating signals with 50% duty cycle and fixed switching frequency. The voltage gain of DAB converter is defined as $M = n_t V_2 : V_1$, where V_1 is the input voltage and V_2 is the output voltage. v_p and v_s are two HF voltages generated on the primary side and the secondary side, respectively. The power is manipulated by controlling the phase-shift angles among each switch arms.

Depending on the number of varying phase-shifts, there are several control schemes for a DAB converter, which are single-phase-shift (SPS) control, extended-phase-shift control (EPS), dual-phase-shift (DPS) control and triple-phase-shift (TPS) control. The SPS control is the simplest control strategy that is easy to implement. However, it has many disadvantages such as high circulating power and loss of ZVS if the converter gain is away from unity [1,2]. In addition, the TPS control has three independent phase-shifts to be controlled, which makes it costly and more complex in real implementation [3,4]. In contrast, EPS control and DPS control are compromised ones with both enough flexibility and easy implementation [5–14]. To make a step load change in a DAB converter, one or more phase-shift angles should be adjusted accordingly. The detailed procedure to adjust those phase-shift angles have direct influence on the transient responses. Improper transient control may cause temporal overcurrent and DC offset in inductor current which can arise extra losses and saturation in magnetic components. Therefore, it is necessary to propose some methods to eliminate DC offset and minimize the load transient period.

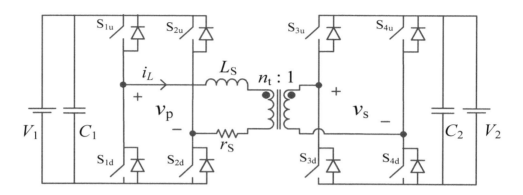

Figure 1. A dual-active-bridge DC/DC converter.

Inserting a capacitor before the transformer is an easy way to block DC bias in the HF transformer current with the increase in the cost and the size of the circuit. Current oscillation might be induced due to change of stored energy in the capacitor at the step-change of load. An effective solution by means of a special magnetic sensor and active compensator is presented in [15], which is quite costly and complicated regardless of the excellent performance. Based on SPS control, a transient control named asymmetric-double-side modulation proposed in [16] distributes the required phase-shift adjustment to both two bridges according to an optimized ratio to depress the DC bias current in the transition process. The same technique is then extended to EPS control in [17,18]. However, the obtained results in [17] can not deal with the operation when the converter gain is close to unity. Another solution under SPS control in [19] manipulates the gating signals of two switch arms in the same bridge with different phase-shifts during the transient process. The method was applied in a three-phase DAB converter too [20]. Although the calculation of this method is easy and is independent on the converter gain, it can be proved that a single current pulse can be induced for specified load transient conditions. In [21], a novel approach to keep transient voltage–time balance is to introduce a small zero-voltage duration in one of the two HF voltages. It is capable of eliminating DC bias in both inductor current and magnetizing current of the transformer. Generally, it is seen that most of the reported solutions are limited in the scenario of SPS control in a steady state. However, two or more phase-shifts are needed for power manipulation in applications with wide variation in converter gain. Thus, to explore new transient control with multiple phase-shifts will be meaningful. In this work, a load transient modulation for EPS control will be proposed for depressing DC bias current in the load-changing process, which stems from the approach in [21]. The proposed transient control will be applied to different transition cases between the two steady-state EPS modes. It will be shown that it is able to not only reduce transient period, but also depress the DC bias effectively.

The paper is organized as follows. In Section 2, the proposed transient control method will be analyzed in detail for each transition case. The values of all gating signal angles would be given before, during and after the transition process. The theoretical analysis is then verified by experimental tests on a lab prototype converter in Section 3. The final conclusion is presented in Section 4.

2. The Proposed Transient Control Method

Under the EPS control scheme, there are two phase shift angles to be used: φ_1—the phase delay between the turn-on moment of S_{1d} and that of S_{2u}; φ_2—the phase delay between the turn-on moment of S_{1d} and that of S_{3d}, S_{4u}. While working under EPS, the converter may have two different steady-state modes according to the different relationship of φ_1, φ_2 shown in Figure 2. Mode A is defined with $0 \leq \varphi_1 \leq \varphi_2 \leq \pi$ and mode B is defined with $0 \leq \varphi_2 \leq \varphi_1 \leq \pi$.

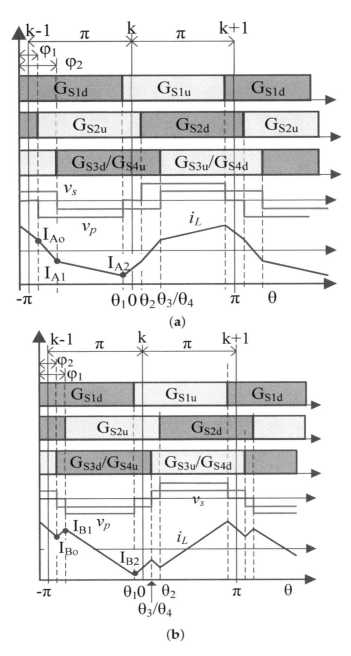

Figure 2. Defined control timing in (**a**) mode A and (**b**) mode B.

2.1. Reference Points and Definition of Switching Angles

By referring to the fixed reference point, $\theta_1, \theta_2, \theta_3, \theta_4$ are defined as the switching angles where $S_{1u}/S_{1d}, S_{2u}/S_{2d}, S_{3u}/S_{3d}, S_{4u}/S_{4d}$ are turned on, respectively.

If the original mode is mode A, the reference points are defined in the mid of φ_1 and the interval between each of them is π. Thus, the switching angles in mode A are expressed as follows:

$$\theta_1 = -\frac{\varphi_1}{2}, \theta_2 = \frac{\varphi_1}{2}; \; \theta_3 = \theta_4 = \varphi_2 - \frac{\varphi_1}{2}. \tag{1}$$

If the original mode is mode B, the reference points are defined in the mid of φ_2. Thus, the switching angles in mode B can be expressed as follows:

$$\theta_1 = -\frac{\varphi_2}{2}, \theta_2 = \varphi_1 - \frac{\varphi_2}{2}; \; \theta_3 = \theta_4 = \frac{\varphi_2}{2}. \tag{2}$$

It is known that in each time interval the change of inductor current is proportional to the voltage difference across it: $\Delta i_L = \frac{v_p - v_s}{\omega_s L_s} \Delta \theta$. Therefore, the instantaneous inductor currents at switching moments in mode A and B can be found as:

$$I_{Ao}(\varphi_1, \varphi_2) = \pi - \varphi_1 + M(2\varphi_2 - 2\varphi_1 - \pi), \tag{3}$$

$$I_{A1}(\varphi_1, \varphi_2) = \pi + \varphi_1 - 2\varphi_2 - M\pi, \tag{4}$$

$$I_{A2}(\varphi_1, \varphi_2) = \varphi_1 - \pi + M(\pi - 2\varphi_2), \tag{5}$$

$$I_{Bo}(\varphi_1, \varphi_2) = \pi - \varphi_1 - M\pi, \tag{6}$$

$$I_{B1}(\varphi_1, \varphi_2) = \pi - \varphi_1 + M(2\varphi_1 - 2\varphi_2 - \pi), \tag{7}$$

$$I_{B2}(\varphi_1, \varphi_2) = \varphi_1 - \pi + M(\pi - 2\varphi_2). \tag{8}$$

When the load level is changed abruptly, there are four condition of DAB converter in EPS control to be dealt with as shown in Table 1.

Table 1. Different load transition conditions of EPS control.

	Initial Mode	Final Mode
Condition 1	Mode A	Mode A
Condition 2	Mode A	Mode B
Condition 3	Mode B	Mode B
Condition 4	Mode B	Mode A

2.2. Load Transient Control within Mode A

As shown in Figure 3, a step load transition happened around the kth reference point. In other words, the switching behaviour referring to $(k-1)^{th}$ reference point is the original steady state with phase angles φ_1, φ_2, while the switching behaviour referring to $(k+1)^{th}$ reference point is the destination steady state with phase angles φ_1', φ_2'. The phase angles referring to the kth reference point should be selected properly to complete the transition process as fast as possible.

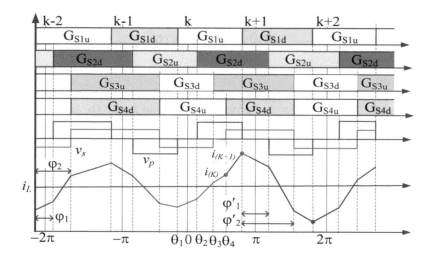

Figure 3. Gate signals, voltage and current waveforms of transient control within mode A.

With the unknown $\theta_1 \sim \theta_4$, two important instant currents are calculated then:

$$i_{(k)} = (-\pi - 2\theta_1 - 2\theta_2 + 2\theta_4) + M(\pi + \varphi_1 - 2\varphi_2 + 2\theta_3), \tag{9}$$

$$i_{(k+1)} = (\pi - 2\theta_1 - 2\theta_2 - \varphi_1') + M(-\pi + \varphi_1 - 2\varphi_2 + 2\theta_3 + 2\theta_4 + \varphi_1'), \tag{10}$$

where $i_{(k)}$ is the instant current at the last switching point of kth reference point, and $i_{(k+1)}$ is the first current at the first switching point of $(k+1)^{th}$ reference point.

To keep voltage–time balance, the average inductor current should be zero while the inductor current enter into the new steady state at once, which indicates:

$$i_{(k)} = -I_{A1}(\varphi_1', \varphi_2'), \; i_{(k+1)} = -I_{A2}(\varphi_1', \varphi_2'). \tag{11}$$

Therefore, the following conditions can be obtained by substituting Equations (4), (5), (9) and (10) into (11):

$$\theta_1 = -\theta_2, \; \theta_3 = \varphi_2 - \frac{\varphi_1}{2}, \; \theta_4 = \varphi_2' - \frac{\varphi_1'}{2}. \tag{12}$$

It is seen that v_s is set to zero during the interval $[\theta_3 \; \theta_4]$. As shown in Equation (12), the phase shift angles θ_1 and θ_2 during transitions are free to be chosen. However, they should also satisfy the requirement that $|\theta_1| = |\theta_2| < \min\{\pi - \varphi_2 + \varphi_1/2, \varphi_2' - \varphi_1'/2\}$ lets the converter work in mode A.

2.3. Load Transient Control within Mode B

In this condition, the DAB converter is working in mode B from beginning to end, and the reference points are defined in the mid of φ_2. As is shown in Figure 4, the transition is done at the kth reference point and the converter is expected in the destination steady state at the $(k+1)^{th}$ reference point. The instant currents after the intermediate adjustment are calculated as:

$$i_{(k)} = (-\pi - 2\theta_1 - \varphi_2 + \varphi_1) + M(\pi - \varphi_2 + 2\theta_3 + 2\theta_4 - 2\theta_2), \tag{13}$$

$$i_{(k+1)} = (\pi - 2\theta_1 - 2\theta_2 - \varphi_2' + \varphi_1 - \varphi_2) + M(-\pi + \varphi_2' - \varphi_2 + 2\theta_3 + 2\theta_4). \tag{14}$$

To meet such an expectation, the followed equation should be satisfied:

$$i_{(k)} = -I_{B1}(\varphi_1', \varphi_2'), \; i_{(k+1)} = -I_{B2}(\varphi_1', \varphi_2'). \tag{15}$$

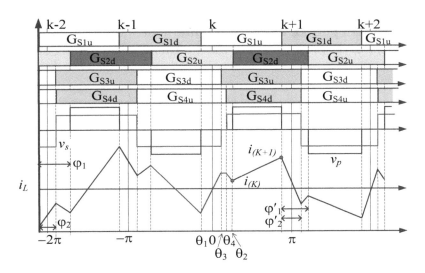

Figure 4. Gate signals, voltage and current waveforms of the proposed control method in within mode B.

Combining Equations (7), (8), (13), (14) and (15), the following switching angles referring to the kth reference point can be calculated to achieve a fast transition within mode B:

$$\theta_1 = \frac{\varphi_1 - \varphi_2 - \varphi_1'}{2}, \theta_2 = \varphi_1' - \frac{\varphi_2'}{2}, \; \theta_3 + \theta_4 = \frac{\varphi_2 + \varphi_2'}{2}. \tag{16}$$

It is seen that v_s is set to zero during the interval $[\theta_3 \ \theta_4]$. Equation (16) reveals that θ_3 and θ_4 can be selected flexibly. However, they should also satisfy $\max\{\theta_3, \theta_4\} < \pi - \varphi'_2/2$ to let the converter work in mode B.

2.4. Load Transient Control from Mode A to Mode B

In this condition, as shown in Figure 5, the phase shift angles are changed from mode A to mode B by means of the intermediate adjustment around the kth reference point. Different from the previous cases, the definition of switching angles are changed since that the original mode is not the same as the destination one. The instant currents after the intermediate adjustment are calculated as:

$$i_{(k)} = (-\pi - 2\theta_1) + M(\pi + \varphi_1 - 2\varphi_2 - 2\theta_2 + 2\theta_3 + 2\theta_4), \tag{17}$$

$$i_{(k+1)} = (\pi - 2\theta_1 - 2\theta_2 - \varphi'_2) + M(-\pi + \varphi_1 - 2\varphi_2 + 2\theta_3 + 2\theta_4 + \varphi'_2). \tag{18}$$

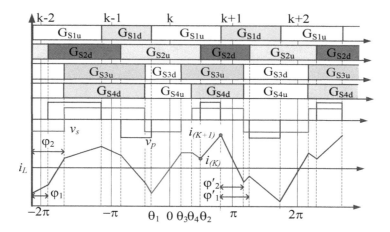

Figure 5. Gate signal, voltage and current waveforms of the proposed control method from mode A to mode B.

To minimize the transient period, the same requirement shown in Equation (15) should be met. Combining Equations (7), (8), (17), (18) and (15), the intermediate switching angles for transition from mode A to mode B are given as:

$$\theta_1 = -\frac{\varphi'_1}{2}, \ \theta_2 = \varphi'_1 - \frac{\varphi'_2}{2}, \ \theta_3 + \theta_4 = \frac{\varphi'_2}{2} + \varphi_2 - \frac{\varphi_1}{2}. \tag{19}$$

It is seen that v_s is set to zero during the interval $[\theta_3 \ \theta_4]$. From the results, it can be found that switching angels θ_3 and θ_4 can be determined flexibly. Meanwhile, the condition of mode boundary should be met too: $\max\{\theta_3, \theta_4\} < \pi - \varphi'_1/2$.

2.5. Load Transient Control from Mode B to Mode A

In this condition shown in Figure 6, the original state is mode B while the final state is mode A. Therefore, the instant currents after the kth reference point can be calculated starting from mode B:

$$i_{(k)} = (-\pi - 2\theta_1 + \varphi_1 - \varphi_2) + M(\pi - \varphi_2 - 2\theta_2 + 2\theta_3 + 2\theta_4), \tag{20}$$

$$i_{(k+1)} = (\pi - 2\theta_1 - 2\theta_2 - \varphi'_1 + \varphi_1 - \varphi_2) + M(-\pi + \varphi'_1 - \varphi_2 + 2\theta_3 + 2\theta_4). \tag{21}$$

As the switching angles are redefined in mode A at the $(k+1)^{th}$ reference point, the requirement for the expected fast transition is the same as (11). Substituting Equations (4), (5), (20), (21) into Equation (11), switching angles during transient process referring to the kth reference point for the transition from mode B to mode A are:

$$\theta_1 = \frac{\varphi_1 + \varphi_1' - \varphi_2 - 2\varphi_2'}{2}, \; \theta_2 = \varphi_2' - \frac{\varphi_1'}{2}, \; \theta_3 + \theta_4 = \frac{\varphi_2}{2} + \varphi_2' - \frac{\varphi_1'}{2}. \tag{22}$$

In this case, θ_1 and θ_2 can be chosen flexibly, but is subject to $|\theta_1| < \pi - (\varphi_1 - \varphi_2/2)$ and $\theta_2 < \pi - \varphi_2'/2$.

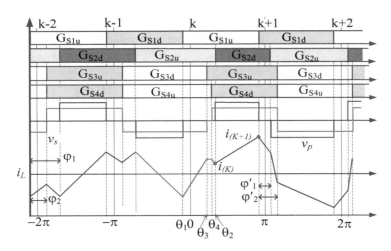

Figure 6. Gate signal, voltage and current waveforms of the proposed control method from mode B to mode A.

3. Validation by Experimental Results

In order to validate the theoretical results, load transition cases using the proposed control method were tested on a lab prototype DAB converter. Table 2 shows the specifications of the converter used in the experiments. The circuit adopts four power MOSFETs (STP40NF20, 200 V, 40 A, 0.038 Ω) on the primary side and the other four power MOSFETs (IPP200N15N3G, 150 V, 50 A, 0.020 Ω) on the secondary side as the switches. The input terminals were connected to a DC power supply, while the output terminals were connected to a DC electronic load. The inductor is made of a toroidal CM400125 MPP core with litz wire winding. The proposed transient control is implemented in a TI-F2812 DSP development board (Texas Instruments, Dallas, USA) and the flowchart is shown in Figure 7. The converter power level is monitored continuously. If no change is to be made to the power, the current φ_1, φ_2 are used to generate $\theta_1 \sim \theta_4$. If a new power command is received and confirmed, the destination will be calculated based on some preset algorithm optimized for better efficiency, which is out of scope of the current work. Then, $\theta_1 \sim \theta_4$ for the next reference point will be updated by $\varphi_1, \varphi_2, \varphi_1', \varphi_2'$ by using one of Equations (12), (16), (19), (22).

Table 2. Specifications of the prototype converter.

Parameters	Value
DC input voltage V_1	120 V
DC output voltage V_2	72 V
Transformer turns ratio $n_t : 1$	1:1
Transformer ferrite core	PC40ETD49
Series inductor L_s	121.875 μH
HF filter capacitance C_1, C_2	330 μF
Primary-side MOSFETs	STP40NF20
Secondary-side MOSFETs	IPP200N15N3G
Switching frequency f_{sw}	100 kHz

MOSFET: metal-oxide-semiconductor field-effect transistor

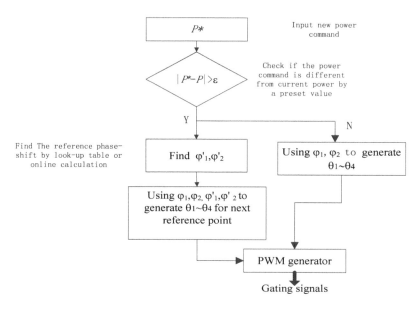

Figure 7. Flowchart for implementation of the proposed transient control method.

The detailed parameters of phase-shift angles of each mode in experiment are shown in Tables 3–6. For each tested load-change condition, a condensed view of the transition is given firstly with a time scale of 300 µs/div while the details are shown later with a time scale of 5 µs/div. In each plot, captured waveforms of v_p, v_s (100 V/div) and i_L (2 A/div) are shown from top to bottom.

Table 3. Transition within mode A.

Phase-Shift-Angles	Initial	Transient	Final
ϕ_1	30°		47.28°
ϕ_2	60°		112.8°
θ_1	−15°	−15°	−23.64°
θ_2	15°	15°	23.64°
θ_3	45°	45°	89.16°
θ_4	45°	89.16°	89.16°

Table 4. Transition within mode B.

Phase-Shift-Angles	Initial	Transient	Final
ϕ_1	60°		88.8°
ϕ_2	42°		82.32°
θ_1	−21°	−35.4°	−41.16°
θ_2	39°	47.64°	47.64°
θ_3	21°	21°	41.16°
θ_4	21°	41.16°	41.16°

Table 5. Transition from mode A to mode B.

Phase-Shift-Angles	Initial	Transient	Final
ϕ_1	30°		90.48°
ϕ_2	60°		81.6°
θ_1	−15°	−45.24°	−40.8°
θ_2	15°	49.68°	49.68°
θ_3	45°	49.68°	40.8°
θ_4	45°	45°	40.8°

Table 6. Transition from mode B to mode A.

Phase-Shift-Angles	Initial	Transient	Final
ϕ_1	114°		30°
ϕ_2	79.2°		60°
θ_1	−39.6°	−27.6°	−15°
θ_2	74.4°	45°	15°
θ_3	39.6°	39.6°	45°
θ_4	39.6°	45°	45°

As an example for comparison, the first case tested (Figure 8) is a transition within mode A by directly changing the phase-shifts. The phase-shift angles φ_1, φ_2 are changed from 30°, 60° to 47.28°, 112.8°. The inductor peak current is expected to rise from 1.56 A to 2.13 A. However, an abnormal peak current 2.95 A results in a transient process and it takes about 20 HF cycles to be absorbed. In the duration of transient process, a temporal DC bias current decays from about 0.69 A until zero.

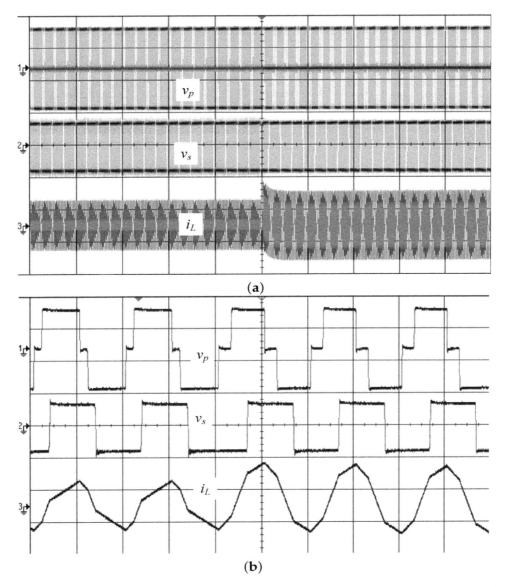

(a)

(b)

Figure 8. Experimental plots of transition within mode A without proposed control method. The signals shown from top to bottom are: v_p (100 V/div), v_s (100 V/div) and i_L (2 A/div). (a) condensed view (300 μs/div); (b) expanded view (5 μs/div).

The same transition is then repeated with proposed transient control in Figure 9. According to (12), θ_3, θ_4 are calculated as $45°$ and $89.16°$ during the transient process. While satisfying (12), the switch angles θ_1 and θ_2 are selected as $-\phi_1/2$ and $\phi_1/2$ during transient for the purpose of convenience. In addition, the final values of $\theta_1 \sim \theta_4$ are $-23.64°$, $23.64°$, $89.16°$ and $89.16°$. The transient process now can be completed almost instantly as shown in Figure 9. It is seen that there is no noticeable overshoot current.

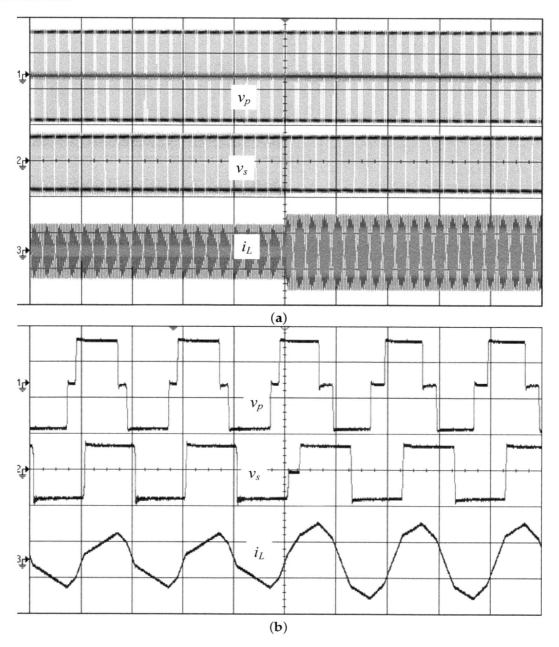

Figure 9. Experimental plots of transition within mode A with proposed control method. The signals shown from top to bottom are: v_p (100 V/div), v_s (100 V/div) and i_L (2A/div). (**a**) condensed view (300 μs/div); (**b**) expanded view (5 μs/div).

For the load transition from mode B to mode B in Figure 10, the phase-shift ϕ_1, ϕ_2 are changed from $60°$, $42°$ to $88.8°$, $82.32°$. After the transient modulation, the switching angles θ_1, θ_2, θ_3 and θ_4 are changed from $-21°$, $39°$, $21°$, $21°$ to $-41.16°$, $47.64°$, $41.161°$ and $41.16°$, respectively. During the transient process, θ_1, θ_2 are calculated $-35.4°$, $47.64°$ directly according to (16). In addition, the transient θ_3 and θ_4 are selected as $\phi_2/2 = 21°$ and $\phi_2'/2 = 41.16°$, respectively.

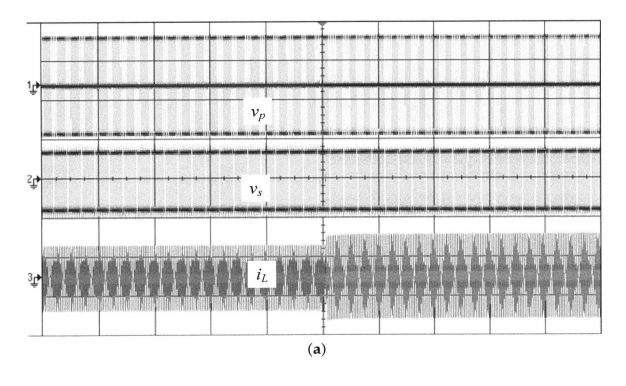

(a)

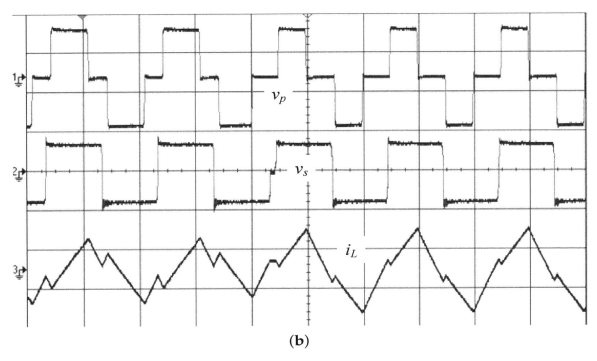

(b)

Figure 10. Experimental plots of transition within mode B with proposed control method. The signals shown from top to bottom are: v_p (100 V/div), v_s (100 V/div) and i_L (2 A/div). (**a**) condensed view (300 μs/div); (**b**) expanded view (5 μs/div).

In the condition of transition from mode A to mode B in Figure 11, the phase-shift ϕ_1, ϕ_2 are changed from 30°, 60° to 90.48°, 81.6°. The initial angles θ_1, θ_2, θ_3 and θ_4 are −15°, 15°, 45°, 45° and the final angles should be −40.8°, 49.68°, 40.8° and 40.8°, respectively. At the kth reference point, θ_1, θ_2 are determined to be −45.24° and 49.68° based on (19). Under the constraint given in (19), the switching angles θ_3 and θ_4 are selected as $\phi_2'/2 = 49.68°$ and $\phi_2 - \phi_1/2 = 45°$, respectively, for the purpose of convenience.

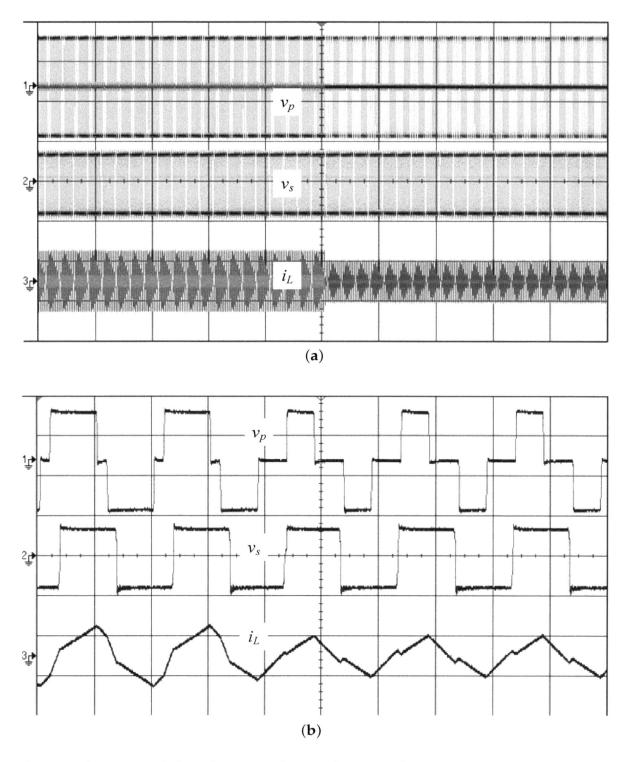

Figure 11. Experimental plots of transition from mode A to mode B with proposed control method. The signals shown from top to bottom are: v_p (100 V/div), v_s (100 V/div) and i_L (2 A/div). (**a**) condensed view (300 μs/div); (**b**) expanded view (5 μs/div).

In Figure 12, the condition of transition from mode B to mode A is presented, in which the phase-shift ϕ_1, ϕ_2 are changed from $114°$, $79.2°$ to $30°$, $60°$. The initial angles θ_1, θ_2, θ_3 and θ_4 are $-39.6°, 74.4°, 39.6°, 39.6°$ and the final angles should be $-15°, 15°, 45°$ and $45°$, respectively. At the kth reference point, θ_1, θ_2 are determined to be $-27.6°$ and $45°$ based on (22). Under the constrain given in (22), the switch angles θ_3 and θ_4 are selected as $\phi_2'/2 = 39.6°$ and $\phi_2 - \phi_1/2 = 45°$ for the purpose of convenience.

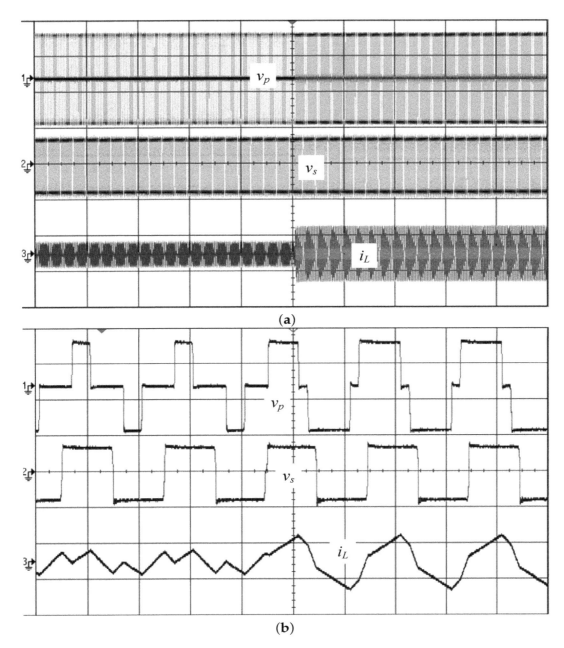

Figure 12. Experimental plots of transition from mode B to mode A with proposed control method. The signals shown from top to bottom are: v_p (100 V/div), v_s (100 V/div) and i_L (2 A/div). (**a**) condensed view (300 μs/div); (**b**) expanded view (5 μs/div).

4. Conclusions

In this work, a fast transient control is proposed for a DAB converter that is able to improve the step-load transient response in terms of response time and overshoot current. This transient control is implemented based on the definition of switching angles for each switch arm, which makes it easy for implementation in pulse-width-modulation (PWM) units of common micro-controller platforms. A small duration of zero-voltage is introduced in the transformer voltage during the transition process to keep the voltage-second balance of the inductor. All the transient switching angles can be calculated from the original and final phase-shift angles directly and are not affected by the converter parameters. Though the proposed control method aims to modulate the transient inductor current, no information about the instantaneous current is needed. With this proposed transient control method, the DAB converter can transfer from one steady state to another quickly and smoothly and causes no DC offset in inductor current, which has been validated successfully by a series of experimental tests.

Author Contributions: Y.Z. did most of the theoretical analysis, derivation and paper writing. C.S. contributed to circuit implementation and experimental test. X.L. is responsible for planning, coordination and proofreading. Z.H. contributed to drawing figures and formatting paper.

References

1. Kheraluwala, M.H.; Gascoigne, R.; Divan, D.M.; Baumann, E. Performance characterization of a high power dual active bridge DC-to-DC converter. *IEEE Trans. Ind. Appl.* **1992**, *28*, 1294–1301. [CrossRef]
2. De Doncker, R.W.; Divan, D.M.; Kheraluwala, M.H. A three-phase soft-switched high-power-density DC/DC converter for high-power applications. *IEEE Trans. Ind. Appl.* **1991**, *27*, 63–73. [CrossRef]
3. Huang, J.; Wang, Y.; Li, Z.; Lei, W. Unified triple-phase-shift control to minimize current stress and achieve full soft-switching of isolated bidirectional DC–DC converter. *IEEE Trans. Ind. Electron.* **2016**, *63*, 4169–4179. [CrossRef]
4. Wu, K.; de Silva, C.W.; Dunford, W.G. Stability analysis of isolated bidirectional dual active full-bridge DC–DC converter with triple phase-shift control. *IEEE Trans. Power Electron.* **2012**, *27*, 2007–2017. [CrossRef]
5. Moonem, M.A.; Pechacek, C.L.; Hernandez, R.; Krishnaswami, H. Analysis of a multilevel dual active bridge (ML-DAB) DC–DC converter using symmetric modulation. *Electronics* **2015**, *4*, 239–260. [CrossRef]
6. Khan, M.A.; Zeb, K.; Sathishkumar, P.; Ali, M.U.; Uddin, W.; Hussian, S.; Ishfaq, M.; Khan, I.; Cho, H.-G.; Kim, H.-J. A Novel Supercapacitor/Lithium-Ion Hybrid Energy System with a Fuzzy Logic-Controlled Fast Charging and Intelligent Energy Management System. *Electronics* **2018**, *7*, 63. [CrossRef]
7. Wang, Y.-C.; Ni, F.-M.; Lee, T.-L. Hybrid Modulation of Bidirectional Three-Phase Dual-Active-Bridge DC Converters for Electric Vehicles. *Energies* **2016**, *9*, 492. [CrossRef]
8. Zhao, B.; Song, Q.; Liu, W.; Sun, Y. Overview of dual-active-bridge isolated bidirectional DC–DC converter for high-frequency-link power-conversion system. *IEEE Trans. Power Electron.* **2014**, *29*, 4091–4106. [CrossRef]
9. Krismer, F.; Kolar, J. Accurate power loss model derivation of a highcurrent dual active bridge converter for an automotive application. *IEEE Trans. Ind. Electron.* **2010**, *57*, 881–891. [CrossRef]
10. Oggier, G.G.; Garcia, G.O.; Oliva, A.R. Modulation strategy to operate the dual active bridge DC–DC donverter under soft switching in the whole operating range. *IEEE Trans. Power Electron.* **2011**, *26*, 1228–1236. [CrossRef]
11. Zhao, B.;Yu, Q.; Sun, W. Extended-phase-shift control of isolated bidi-rectional DC–DC converter for power distribution in microgrid. *IEEE Trans. Power Electron.* **2012**, *27*, 4667–4680. [CrossRef]
12. Jain, A.K.; Ayyanar, R. PWM control of dual active bridge: Comprehensive analysis and experimental verification. *IEEE Trans. Power Electron.* **2011**, *26*, 1215–1227. [CrossRef]
13. Bai, H.; Mi, C. Eliminate reactive power and increase system efficency of isolated bidirectional dual-active-bridge DC–DC converters using novel dual- phase-shift control. *IEEE Trans. Power Electron.* **2008**, *23*, 2905–2914. [CrossRef]
14. Zhao, B.; Song, Q.; Liu, W. Efficiency characterization and optimization of isolated bidirectional DC–DC converter based on dual-phase-shift control for DC distribution application. *IEEE Trans. Power Electron.* **2013**, *28*, 1711–1727. [CrossRef]
15. Ortiz, G.; Fassler, L.; Kolar, J.W.; Apeldoorn, O. Flux Balancing of Isolation Transformers and Application of "The Magnetic Ear" for Closed-Loop Volt-Second Compensation. *IEEE Trans. Power Electron.* **2014**, *29*, 4078–4090. [CrossRef]
16. Li, X.; Li, Y.-F. An optimized phase-shift modulation for fast transient response in a dual-active-bridge converter. *IEEE Trans. Power Electron.* **2014**, *29*, 2661–2665. [CrossRef]
17. Lin, S.-T.; Li, X.; Sun, C.; Tang, Y. Fast transient control for power adjustment in a dual-active-bridge converter. *Electron. Lett.* **2017**, *53*, 1130–1132. [CrossRef]
18. Sun, C.; Li, X. Fast Transient Modulation for a Step Load Change in a Dual-Active-Bridge Converter with Extended-Phase-Shift Control. *Energies* **2018**, *11*, 1569. [CrossRef]
19. Zhao, B.; Song, Q.; Liu, W.; Zhao, Y. Transient DC Bias and Current Impact Effects of High-Frequency-Isolated Bidirectional DC–DC Converter in Practice. *IEEE Trans. Power Electron.* **2016**, *31*, 3203–3216. [CrossRef]
20. Engel, S.P.; Soltau, N.; Stagge, H.; De Doncker, R.W. Dynamic and Balanced Control of Three-Phase High-Power Dual-Active Bridge DC–DC Converters in DC-Grid Applications. *IEEE Trans. Power Electron.* **2013**, *28*, 1880–1889. [CrossRef]
21. Takagi, K.; Fujita, H. Dynamic Control and Performance of a Dual-Active-Bridge DC–DC Converter. *IEEE Trans. Power Electr.* **2017**, *33*, 7858–7866. [CrossRef]

4

Exploring the Limits of Floating-Point Resolution for Hardware-in-the-Loop Implemented with FPGAs

Alberto Sanchez [1,*], **Elías Todorovich** [2,3] and **Angel de Castro** [1]

[1] HCTLab Research Group, Universidad Autonoma de Madrid, 28049 Madrid, Spain; angel.decastro@uam.es

[2] Facultad de Ciencias Exactas, Universidad Nacional del Centro de la Provincia de Buenos Aires, Tandil B7001BBO, Argentina; etodorov@exa.unicen.edu.ar

[3] Faculty of Engineering, FASTA University, Mar del Plata B7600, Argentina

[*] Correspondence: alberto.sanchezgonzalez@uam.es

Abstract: As the performance of digital devices is improving, Hardware-In-the-Loop (HIL) techniques are being increasingly used. HIL systems are frequently implemented using FPGAs (Field Programmable Gate Array) as they allow faster calculations and therefore smaller simulation steps. As the simulation step is reduced, the incremental values for the state variables are reduced proportionally, increasing the difference between the current value of the state variable and its increments. This difference can lead to numerical resolution issues when both magnitudes cannot be stored simultaneously in the state variable. FPGA-based HIL systems generally use 32-bit floating-point due to hardware and timing restrictions but they may suffer from these resolution problems. This paper explores the limits of 32-bit floating-point arithmetics in the context of hardware-in-the-loop systems, and how a larger format can be used to avoid resolution problems. The consequences in terms of hardware resources and running frequency are also explored. Although the conclusions reached in this work can be applied to any digital device, they can be directly used in the field of FPGAs, where the designer can easily use custom floating-point arithmetics.

Keywords: hardware-in-the-loop; floating-point; fixed-point; real-time emulation; field programmable gate array

1. Introduction

Digital control for power converters has been growing during the past two decades [1–5]. Despite all the advantages of digital control, the debugging process of this type of control is more complex because the power converter is an analog system while the control is digital. Hardware-in-the-loop (HIL) is a technique that consists in the hardware implementation of mathematical models that represent a real system. HIL simulation presents numerous advantages such as having a safe environment to test controllers, allowing the use of the controller in its final implementation, even before building the real plant to be controlled. HIL techniques are being increasingly implemented using computers [6–11] and also digital devices like FPGAs (Field Programmable Gate Array) [12–17]. The latter make it possible to perform complex calculations faster. Thus, it is not surprising that several companies have released commercial HIL products [18–20].

Arithmetics used in HIL systems have a noteworthy impact in speed, hardware resources needed for the model, the complexity to design the model, and the accuracy of the system. Fixed-point arithmetics provide optimized operations in terms of area and speed. In [21], a comparison between fixed-point and floating-point arithmetics, in the context of FPGA-based HIL systems, was presented. Results showed that floating-point required ten times as many logic resources as well as it ran 10 times slower than fixed-point. For that reason, many HIL systems are based on fixed-point arithmetics when there are hard temporal restrictions [22–26].

The main drawback of fixed-point is that the implementation is more complex because the designer has to define the number of bits of the integer and fractional parts. Thus, the maximum representable value and the required resolution need to be calculated for every signal. However, in floating-point arithmetics, the designer does not take this definition into consideration, as an IEEE-754 single-precision floating-point number can store values up to $\pm 2^{127}$, and the resolution is optimized in every calculation. This is accomplished by the floating-point libraries which automatically adapt the point location through the exponent field. Because of this remarkable advantage of floating-point, most HIL models actually use floating-point arithmetics [9,27,28], including commercial implementations [18–20].

Floating-point arithmetics for FPGAs were not viable in the past as there were no support libraries, and all the logic had to be implemented by the designer. However, with the release of floating-point support libraries, such as float_pkg of the VHDL-2008 Standard, it is easy to include floating-point arithmetics in a VHDL design. Lucia et al. [27] presented one of the first examples of a HIL model using floating-point in an FPGA.

In the literature not many cases of floating-point numerical issues for HIL systems have been reported, as the earliest purposes of HIL was to simulate complex systems with relatively low natural frequencies and integration steps of tens or hundreds of microseconds. With the advances in FPGAs, HIL technique started to be used for new applications, such as power electronics. Firstly, it was applied to converter models with low switching frequencies (kHz or tens of kHz). However, to simulate converters with medium to high switching frequency (hundreds of kHz or MHz), the integration step should be reduced accordingly and the system may present numerical problems, and as a result, obtain wrong simulations. The numerical issues are not related to overflows, as the exponent is automatically adapted. The problem is that, as the integration step is reduced, the increments of every step are smaller, and resolution issues may arise.

This paper explores the limits of floating-point arithmetics for HIL systems, and how to predict the floating-point format needed for accurate simulations. It explores not only the standard formats but also custom formats that can be used thanks to the VHDL-2008 standard libraries or any other libraries.

The rest of the paper is organized as follows. Section 2 shows the application example to illustrate the resolution problems for floating-point HIL simulations. Section 3 explains where the limits for single precision floating-point arithmetics are and how many bits would be necessary to increase the accuracy if needed. Sections 4 and 5 show the experimental and synthesis results respectively. Finally, Section 6 gives the conclusions.

2. Application Example

In this paper a PFC (Power Factor Correction) boost converter is used as an application example. The PFC technique allows regulating the output voltage while reducing the input current harmonics, so the converter behaves as a resistor emulator to the mains. The schematic of a boost converter is shown in Figure 1, excluding the previous diode bridge for ac/dc operation. The parameters of this plant are shown in Table 1. This boost configuration is proposed in an Infineon Design Note [29].

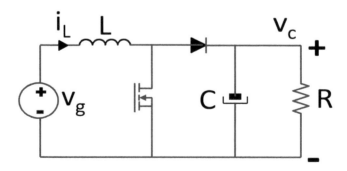

Figure 1. Boost converter topology.

Table 1. Boost converter parameters used in Section 4.

Parameter	C	L	V_{in}	V_{out}	Power
Value	540.5 μF	416.5 μH	230 V	400 V	400 W

The converter can be modeled using the state variables of the system: the inductor current and the capacitor voltage. Therefore, both variables should be updated every simulation step.

The behavior of the inductor and the capacitor can be described using the following equations:

$$v_L = L \cdot \frac{di_L}{dt} \tag{1}$$

$$i_C = C \cdot \frac{dv_{out}}{dt} \tag{2}$$

These equations can be discretized using different numerical methods but the simplest method to be used is explicit Euler [30]. While this method presents several disadvantages such as greater local and global error and risk of instability, these problems are negligible whenever very small integration steps (below microseconds) are used, so it is frequently used for HIL systems in power electronics [13,31–33]. Therefore, the previous Equations (1) and (2) can be discretized and the state variables can be defined as:

$$i_L(n) = i_L(n-1) + \frac{\Delta t}{L} \cdot v_L(n-1)$$
$$v_c(n) = v_c(n-1) + \frac{\Delta t}{C} \cdot i_C(n-1) \tag{3}$$

where dt has been converted into Δt, which is the simulation step, i.e. the time between two calculations of the model.

As can be seen in the previous equations, the state variables depend on the inductor voltage and the capacitor current. These values depend on the conduction state of the switch and the diode, so several states should be considered.

If the switch is closed, the inductor voltage is $v_g - 0$, while the capacitor current is $-i_R$. In this case, the state variables are defined as:

$$i_L(n) = i_L(n-1) + \frac{\Delta t}{L} \cdot v_g(n-1)$$
$$v_c(n) = v_c(n-1) - \frac{\Delta t}{C} \cdot i_R(n-1) \tag{4}$$

If the switch is open, the conduction state of the diode depends on the inductor current. If the current is positive, the diode is conducting (called CCM or Continuous Current Mode) and the inductor voltage is $v_g - v_c$, while the capacitor current is $i_L - i_R$, so the state variables are:

$$i_L(n) = i_L(n-1) + \frac{\Delta t}{L} \cdot (v_g(n-1) - v_c(n-1))$$
$$v_c(n) = v_c(n-1) + \frac{\Delta t}{C} \cdot (i_L(n-1) - i_R(n-1)) \tag{5}$$

Finally, if the inductor is fully discharged, the diode stops conducting (called DCM or Discontinuous Current Mode), so the capacitor current is $-i_R$:

$$i_L(n) = 0$$
$$v_c(n) = v_c(n-1) - \frac{\Delta t}{C} \cdot i_R(n-1) \tag{6}$$

The explicit Euler approach has been chosen in order to simplify the equations and get minimum simulation step. As the proposed system is based on an FPGA, both equations (inductor current and capacitor voltage) can be evaluated and updated in parallel, simplifying the resolution of the system. This is an important advantage compared with software-based HIL systems, in which equations need to be solved sequentially, one after another. Therefore, parallelization is one of the main reasons for the acceleration obtained using FPGAs. Taking advantage of this, every time step (Δt), all the state variables are updated. The accuracy of the system relies on the small value of the simulation step. However, a small time step implies tiny increments for the state variables, as both are proportional. This can lead to resolution issues as it is explained in the next section.

3. Numerical Resolution

As it was explained in the previous section, the model updates its state variables every simulation step. There is a relation between the simulation step and the switching frequency, because the model should be updated with enough intermediate steps inside a switching period. This update is necessary to detect accurately the state of the switch and the diode of the converter so, the smaller the switching period is, the smaller the simulation step should be. In other words, the relation between the switching period and the simulation step is the resolution of the duty cycle. Therefore, the simulation step, and then the increments for the state variables, should be quite small. As HIL systems were firstly applied to low switching frequency converters, numerical issues have not been thoroughly studied in the literature. However, as HIL systems are being used for higher frequency converters, resolution problems arise because the variables width is limited to be able to run the model in real-time [34].

Obtaining low numerical resolution leads to poor accuracy simulations or even an unpredictable simulation behavior. For example, if the system is in steady state and it suffers any small change in the input voltage or load, it is possible that the output voltage changes with the opposite sign rather than the expected one. Besides, the problem is difficult to detect because the system may be able to detect bigger changes in the input conditions, but not the smaller ones.

The optimal solution is to increase the variables width. However, if the width is increased, the calculations are more complex, more hardware resources must be used, and the combinational paths between the flip-flops inside the FPGA get longer. In conclusion, increasing the width leads to longer simulation steps, which have a negative impact on the accuracy of the simulation.

Taking all the previous considerations into account, a trade-off between the simulation step and resolution should be reached. The minimum number of bits can be estimated considering the relation between the maximum expected value of a variable x, $max(x)$, and the increment that should be added, Δx [34]. Besides, some extra bits, n, should be included to store the increment value with more than 1 bit of resolution:

$$width_x = \lceil \lceil \log_2 max(x) \rceil - \log_2 \Delta x \rceil + n \tag{7}$$

The first log_2 operation calculates the number of integer bits needed to store $max(x)$, that is, the exponent. The second log_2 gives the bits needed to store the increment. That second term may be negative, as the increment is usually below 0, indicating that fractional digits should be included. Therefore the subtraction calculates the number of bits needed for the significand field, as it has to store both the value of x and its increments simultaneously.

The maximum values of the state variables are easy to calculate because they are defined by the limits of the converter design. Regarding the increments, it is also easy if they are stable, e.g., in the case of a dc-dc converter in steady state, but otherwise further analysis must be done.

For a boost-based PFC configuration, which is the example of this paper, the minimum incremental values for the output capacitor voltage are reached when i_R is similar to i_L, as shown in Equations (4)–(6). Likewise, the minimum incremental values for the inductor current are reached when v_g is near 0 (in the case of closed switch).

As the number of bits is limited, infinitesimal incremental values will not be properly computed, but the designer can estimate the number of bits that will be necessary to obtain accurate resolutions. In [34], a fixed-point based HIL model for a PFC converter was presented. In that article, the model presented high accuracy when it had enough bits to store the incremental values during 95% of the ac period. We have to take into account that the remaining 5% has a minimum impact on the overall simulation because during that time the increments are smaller than the rest of the time and, therefore, almost negligible.

Following the aforementioned rule of 95%, and given the characteristics of the PFC converter proposed in Table 1 and the current and voltage waveforms from Figure 2, it is possible to calculate the minimum incremental values considered for the state variables. Figure 2 shows points A1 and A2 which correspond to the minimum considered input voltage. Likewise, points B1-4 correspond to the minimum considered difference between currents. Using these points, the minimum considered increments are as follows:

$$\frac{\Delta t}{L} \cdot v_g(n-1) = \frac{50 \text{ ns}}{416.5 \text{ } \mu F} \times 25.52 \text{ V} = 3.064 \text{ mA}$$

$$\frac{\Delta t}{C} \cdot (i_L(n-1) - i_R(n-1)) = \frac{50 \text{ ns}}{540.5 \text{ } \mu F} \times 0.0848 \text{ A} = 7.844 \text{ } \mu V \tag{8}$$

In the previous equations, a value of 50 ns was used as the simulation step (Δt). With the results of the previous equations, it is possible to calculate the number of bits needed for both variables, which are included in Section 4.

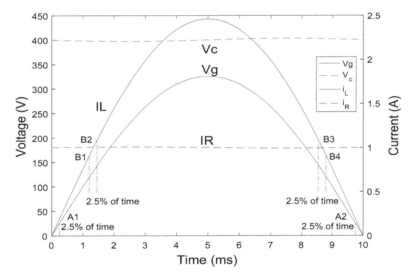

Figure 2. Current and voltages waveforms of the PFC boost converter.

Once the required significand width is obtained, it is possible to estimate the needed floating-point format. IEEE-754 floating-point format [35] defines three binary-based floating-point formats with 32, 64, and 128 bits, also known as Single, Double, and Quadruple precision, respectively. In the field of FPGAs, most cases use single-precision floating-point, as the hardware cost of Double and Quadruple precision formats, in terms of area and speed, makes it inviable to use them.

Custom floating-point formats can be used in order to reach a trade-off between speed, area and numerical resolution. The floating-point formats include separated fields for the sign, exponent, and significand, as can be seen in Figure 3. The problem in power converter HIL models is not the number of exponent bits, as the magnitudes are not extremely big or small, but the number of bits of the significand. Therefore, only the significand field should be enlarged reaching the number of bits calculated using Equations (7) and (8).

In the case of VHDL, this custom floating-point format can be defined using the floating-point package included in the VHDL-2008 standard [36]. Using this package, the codification of Equations (4)–(6) is trivial, so the format of the model variables only has to be taken into account while declaring the signals.

Section 4 compares the accuracy of the PFC boost model using different floating-point formats, and Section 5 shows the hardware results once the model is synthesized in an FPGA.

	sign	exponent	significand
Single	1 bit	8 bits	24 bits
Double	1 bit	11 bits	53 bits
Quadruple	1 bit	15 bits	113 bits

Figure 3. IEEE-754 floating-point formats.

4. Simulation Results

The previous section showed that the designer must reach a trade-off between speed, hardware resources, and accuracy. Therefore, the number of bits cannot be increased as desired. This section compares the accuracy of different floating-point formats regardless of speed and hardware resources.

It is important to note that the designed model has to be tested in open loop, without using any feedback from the control loop. If closed loop were used, the regulator would compensate the numerical errors of the model, so the whole system would probably get the desired current and voltage values at steady state. Simulations in open loop for power factor correction can be done using pre-calculated duty cycles for the PWM signal. This technique has been used previously in the literature because of its low cost, since it gets rid of the current sensor in the case of PFC converters [37–39]. Although using pre-calculated duty cycles also presents disadvantages, such as sensitivity to non-nominal conditions, it can be perfectly used to quantitatively measure the accuracy of the model, as any drift of the model will not be compensated, because it allows open-loop operation for PFC.

Table 2 presents the different experimental scenarios that have been tested, including different output loads, and cases starting at nominal steady state (400 V) and also with small capacitor voltage transients. In the case of the transients, the system will move slowly towards the nominal state following the dynamics of the chosen PFC/Boost converter, as the duty cycles are not modified in these simulations. All the scenarios have been simulated during 100 ms (10 ac semi-cycles) in order to allow the evolution of the output voltage, especially in the case of the small transients (cases 2, 4 and 6). The models have been compared with a double-precision floating-point model (53 bits for the significand field), which implements the same equations. This model should not present resolution issues and, therefore, it is used as our reference model.

Table 2. Experimental scenarios for the boost model.

	Case 1	Case 2	Case 3	Case 4	Case 5	Case 6
Output load	100%	100%	20%	20%	10%	10%
Starting capacitor voltage	400 V	410 V	400 V	410 V	400 V	410 V

Table 3 summarizes the theoretically minimum floating-point format needed for every load. These widths have been calculated using Equations (7) and (8) and the scenarios of Table 2. Regardless of the calculated widths, all scenarios have been simulated with significand widths between 24 and 32 bits.

Table 3. Significand length needed for optimal simulations using Equation (7).

	100% Load Case 1 & 2	20% Load Case 3 & 4	10% Load Case 5 & 6
i_L	$7 + n$	$5 + n$	$4 + n$
v_C	$26 + n$	$29 + n$	$30 + n$

In order to compare the simulation results quantitatively, some figures of merit should be defined. The Mean Absolute Error (MAE) between the state variables and their references (double-precision model) offers an overview of the precision of the model. The main drawback is that it does not take into account whether the error is spread out along the simulation or condensed in a small zone.

The RMSE (Root Mean Square Error) considers the square of the errors, so the main advantage of using RMSE is that it gives a high weight to large errors, and therefore it is much more sensitive to outliers. It is important to note that RMSE is the square root of the average squared error, so the results will be given directly in volts and amperes, and therefore will be directly compared with MAE.

Figure 4 shows the MAE and RMSE for the inductor current and capacitor voltage in every scenario, relative to the RMS current and RMS voltage respectively. It can be seen that the voltage calculation is more sensitive to the variable width, which is consistent with Table 3, as the width for the current variable is less restrictive. As Table 3 predicted, the scenarios with 100% of load improve when the capacitor voltage is stored using 26 bits or more. Likewise, scenarios with 20% and 10% of load improve over 30 bits.

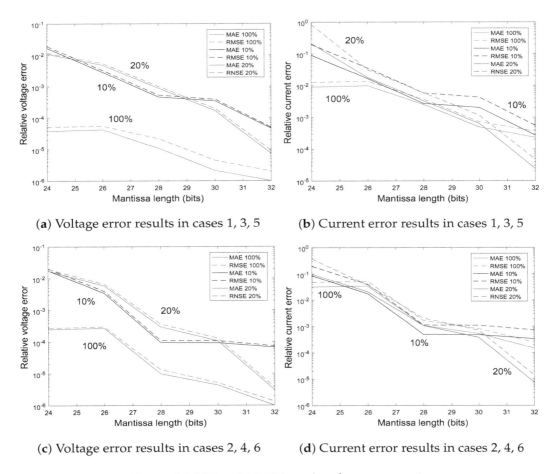

(**a**) Voltage error results in cases 1, 3, 5   (**b**) Current error results in cases 1, 3, 5

(**c**) Voltage error results in cases 2, 4, 6   (**d**) Current error results in cases 2, 4, 6

Figure 4. MAE and RMSE results of every scenario.

There are several cases that are worth mentioning. For instance, Figure 5 shows the capacitor voltage of case 6 (10% of load and voltage transient). It can be seen that, using 24 bits for the significand field, the capacitor voltage not only does not decrease but even increases. This is due to the insufficient resolution of the output voltage. As Table 3 shows that at least 30 bits are needed for the voltage state variable. The capacitor voltage increments are $(i_L - i_R) \cdot \frac{\Delta t}{C}$ when it is increasing and $-i_R \cdot \frac{\Delta t}{C}$ when it is decreasing. Taking into account that the voltage is around 400 V and the output current is around 0.1 A, the negative increments are around $-0.1 \times \frac{50 \text{ ns}}{540.5 \,\mu\text{F}} \approx -9 \times 10^{-6}$ V and $400 - 9 \times 10^{-6}$ V is rounded to 400 V when using 24 bits for the significand. When the switch is on, the current inductor reaches 1 A and, after switch-off, positive increments are around $(1 - 0.1) \times \frac{50 \text{ ns}}{540.5 \,\mu\text{F}} \approx 8 \times 10^{-5}$ V, and $400 + 8 \times 10^{-5}$ V is rounded to 400.00009 V using 24 bits for the significand. Therefore, the problem is that $-i_R \cdot \frac{\Delta t}{C}$ is so small that it is rounded to 0 when it is compared with the actual capacitor voltage. However, $(i_L - i_R) \cdot \frac{\Delta t}{C}$ is numerically bigger, so it is not rounded to 0, and the capacitor voltage only increases until the model reaches steady state.

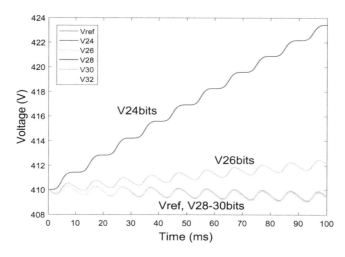

Figure 5. Simulation of 10% of load and a voltage transient (Case 6).

Case 3 is another interesting simulation (20% of load without voltage transient), which can be seen in Figure 6. In this scenario, the system is in DCM as the load is low. When using 24 bits, the output voltage decreases due to resolution problems, until it crosses the limit between DCM and CCM. However, the duty cycles are calculated to operate in DCM and the simulation is in open loop, so they are not modified. As the DCM mode needs higher duty cycles than CCM for the same values of input and output voltages, when the model enters CCM mode, the inductor suffers a short but pronounced transient. The system does not become unstable because the current transient is followed by a growth of the capacitor voltage, and the model comes back to the DCM mode.

As stated, depending on the scenarios and the variable widths, some simulations offer completely wrong waveforms. The previous statistics — MAE and RMSE — give an idea of the simulation error but do not provide clear information about the similarity of the waveforms in terms of tendency. Therefore, another statistic could be found to achieve that. The PCC (Pearson Correlation Coefficient) measures the correlation between a model and its reference, so it also offers a quick test to know if the signs of a state variable and its reference match (positive when matching and negative otherwise). The Pearson correlations for all scenarios are presented in Table 4.

As almost all the simulations present relatively similar waveforms, the PCC is around 1 in almost all cases. In fact, the case of Figure 6 presents a PCC of 0.8311, because the waveforms tendency is similar most of the time, but the current transient worsens this similarity. The case of Figure 5 gives a negative PCC because the signs of the capacitor voltage tendency are opposite. It can be seen that, only when the PCC is over 0.999, the errors of Figure 4 may be acceptable. It is also important to note that the MAE and RMSE statistics make sense only when the tendencies of the tested simulation and its reference are similar. In other words, the similarity in the tendency is reached before the error reaches acceptable values. A comparison of Tables 3 and 4 shows that the results are coherent. The theoretical widths are 26+n, 29+n and 30+n bits for 100%, 20% and 10% of load respectively, while the PCC results show that the simulations are sufficiently similar to their references (in terms of tendency) with 24, 28 and 32 bits.

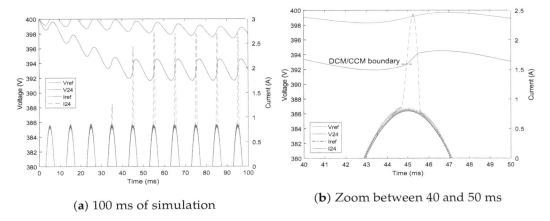

(a) 100 ms of simulation

(b) Zoom between 40 and 50 ms

Figure 6. Simulation of 20% of load (Case 3). (a) 100 ms of simulation. (b) Zoom between 40 and 50 ms.

Table 4. Pearson correlation taking the capacitor voltage.

	Case 1 100% Steady	Case 2 100% Trans	Case 3 20% Steady	Case 4 20% Trans	Case 5 10% Steady	Case 6 10% Trans
24 bits	1.0000	0.9991	0.8311	0.9832	0.8708	0.9712
26 bits	0.9999	0.9990	−0.0555	0.7582	0.9496	−0.0685
28 bits	1.0000	1.0000	1.0000	1.0000	0.9928	0.9980
30 bits	1.0000	1.0000	0.9993	0.9998	0.9976	0.9980
32 bits	1.0000	1.0000	1.0000	1.0000	1.0000	0.9994

As Table 3 shows, the necessary width for the current is much smaller than the voltage width. However, the current error of Figure 4 is similar to the voltage error using the same widths. The reason for such similarity is that both state variables depend on each other, and the accuracy issues of one variable affect the other one, so the worst case — the maximum width — should be considered.

Taking all the results into account, some conclusions can be reached regarding the necessary variable widths. First of all, the waveforms of the state variables should be similar to their references. This similarity can be measured using the PCC and selecting only the widths that obtain a PCC above 0.999, and taking the worst case – the most sensitive state variable. The previous step may choose insufficient widths. For instance, the case of 28 bits and 20% of load has a PCC of 1.0, but the MAE and RMSE errors are relatively high (around 0.5%), however, with 30 bits, the PCC is 0.993, while the MAE and RMSE errors drop to 0.1%. Therefore, once PCC has chosen a reasonable width, RMSE should also be considered. Errors below 0.1% may be sufficient for almost any application. It is possible to increase even more the width to reduce the numerical resolution error, but it is important to note that the model inherently has other error sources, such as non-idealities that have not been modeled, or tolerances in the values of C, R, etc.

For example, for this application we should look for widths that produce a PCC over 0.999 and MAE or RMSE between 0.1% and 0.5% in both state variables. These constraints would imply 28–30 bits for 100% load, 30–32 for 20% load and also 30–32 bits for 10% load. Table 3 predicts these results when using $n = 2$ or $n = 4$, verifying that the method proposed in Section 3 is a good approximation to determine the necessary width of state variables without having to run long simulations.

68

Theory, Design and Applications of Power Electronics

5. Synthesis Results

In this work, the synthesis is targeted to a device of the Arria 10 FPGA family using the Quartus Prime version 17.0 Standard Edition tool configured with the default parameters and automatic constraints except for the required clock period.

Table 5 shows area and time results for several significand widths, s. On the one hand, the area occupied in the target device for the converter is studied in terms of the logic utilization in ALMs (Adaptive Logic Module), total registers, number of DSP Blocks inferred by the synthesizer, and number of required pins. On the other hand, the maximum clock frequency for each converter configuration is evaluated. The constraint for the CLK period was set such that the synthesis and fitter (place and route) tools generate the fastest circuits.

Concerning time results, the circuit speed worsens at a rate of 1.5% per additional bit in the significand, while the area does so at the higher rate of 4%. It means that this architecture is more tolerant, in terms of speed, to resolution improvements than it is in terms of area, which is good for real-time simulations. Furthermore, the selected device, one of the largest in this family, has 427,200 ALMs.

Area and speed are closely related in digital circuits. In this case, although DSP blocks in the Arria 10 family have dedicated single-precision floating-point operators implemented in silicon, these resources are not inferred by the HDL synthesizer. Instead, the synthesizer configures the DSP block to compute the significand-part fixed-point operations. Therefore, similar results should be obtained using other FPGA families.

As the results have shown, the significand-part growth has not influenced the hardware usage or the maximum achievable frequency significantly. Therefore, the estimation method explained in Section 3, using a value between 4 and 6 for n, is valid. The simulations done in Section 4 are not necessary to estimate the state variable widths, but they were accomplished to demonstrate the validity of the method.

Table 5. Post place & route area and time results.

Significand Width	24	26	28	30	32
ALMs	4998	5413	5522	6146	6606
Regs	64	68	72	76	80
DSP	3	3	9	9	9
Pins	131	139	147	155	163
CLK const. [ns]	52	52	53	55	57
Fmax [MHz]	19.38	19.21	18.6	17.86	17.23

6. Conclusions

Thanks to the improvement in the performance of digital devices, HIL systems are starting to be used in applications that require small simulation steps (below 1 µs). The reduction of the simulation step allows more accurate simulations or make it possible to apply the technique to systems with higher natural frequencies, but the integration increments are inherently reduced. This can cause resolution problems if the arithmetics cannot handle values which are so small compared with the actual values of the state variables. In FPGA-based HIL applications, 32-bit floating-point is the most widely used arithmetic because of its simplicity from the designer point of view, along with its good performance compared with 64-bit floating-point. However, 32-bit floating-point numerical resolution is not suitable for all applications as it was observed in this work. Instead of using 64-bit arithmetics, intermediate widths can be chosen. This work has shown the limits of 32-bit floating-point for HIL simulations, and it has also provided a method to calculate the optimal width, taking into account the accuracy and the performance of the HIL system. Results have proven that the addition of few bits can dramatically improve the accuracy of the simulation but, once the numerical resolution is better than the increments, it is unproductive to increase the width.

Author Contributions: Conceptualization, A.S. and A.d.C.; methodology, A.S. and A.d.C.; software, E.T.; validation, E.T.; writing—original draft preparation and writing—review and editing, A.S., E.T. and A.d.C.

References

1.   Patella, B.J.; Prodic, A.; Zirger, A.; Maksimovic, D. High-frequency digital PWM controller IC for DC-DC converters. *IEEE Trans. Power Electron.* **2003**, *18*, 438–446. [CrossRef]

2.   Peterchev, A.V.; Xiao, J.; Sanders, S.R. Architecture and IC implementation of a digital VRM controller. *IEEE Trans. Power Electron.* **2003**, *18*, 356–364. [CrossRef]

3.   Albatran, S.; Smadi, I.A.; Ahmad, H.J.; Koran, A. Online Optimal Switching Frequency Selection for Grid-Connected Voltage Source Inverters. *Electronics* **2017**, *6*, 110. [CrossRef]

4.   Nguyen, T.D.; Hobraiche, J.; Patin, N.; Friedrich, G.; Vilain, J. A Direct Digital Technique Implementation of General Discontinuous Pulse Width Modulation Strategy. *IEEE Trans. Ind. Electron.* **2011**, *58*, 4445–4454. [CrossRef]

5.   Güvengir, U.; Çadırcı, I.; Ermiş, M. On-Line Application of SHEM by Particle Swarm Optimization to Grid-Connected, Three-Phase, Two-Level VSCs with Variable DC Link Voltage. *Electronics* **2018**, *7*, 151. [CrossRef]

6.   Champagne, R.; Dessaint, L.A.; Fortin-Blanchette, H. Real-time simulation of electric drives. *Math. Comput. Simul.* **2003**, *63*, 173–181. [CrossRef]

7.   Short, M.; Abugchem, F.; Abrar, U. Dependable Control for Wireless Distributed Control Systems. *Electronics* **2015**, *4*, 857–878. [CrossRef]

8.   Dennetière, S.; Saad, H.; Clerc, B.; Mahseredjian, J. Setup and performances of the real-time simulation platform connected to the INELFE control system. *Electr. Power Syst. Res.* **2016**, *138*, 180–187. [CrossRef]

9.   Barragan, L.A.; Urriza, I.; Navarro, D.; Artigas, J.I.; Acero, J.; Burdio, J.M. Comparing simulation alternatives of FPGA-based controllers for switching converters. In Proceedings of the 2007 IEEE International Symposium on Industrial Electronics, Vigo, Spain, 4–7 June 2007; pp. 419–424. [CrossRef]

10. Short, M.; Abugchem, F. A Microcontroller-Based Adaptive Model Predictive Control Platform for Process Control Applications. *Electronics* **2017**, *6*, 88. [CrossRef]

11. Viola, F.; Romano, P.; Miceli, R. Finite-Difference Time-Domain Simulation of Towers Cascade Under Lightning Surge Conditions. *IEEE Trans. Ind. Appl.* **2015**, *51*, 4917–4923. [CrossRef]

12. Aiello, G.; Cacciato, M.; Scarcella, G.; Scelba, G. Failure analysis of AC motor drives via FPGA-based hardware-in-the-loop simulations. *Electr. Eng.* **2017**, *99*, 1337–1347. [CrossRef]

13. Herrera, L.; Li, C.; Yao, X.; Wang, J. FPGA-Based Detailed Real-Time Simulation of Power Converters and Electric Machines for EV HIL Applications. *IEEE Trans. Ind. Appl.* **2015**, *51*, 1702–1712. [CrossRef]

14. Sandre-Hernandez, O.; Rangel-Magdaleno, J.; Morales-Caporal, R.; Bonilla-Huerta, E. HIL simulation of the DTC for a three-level inverter fed a PMSM with neutral-point balancing control based on FPGA. *Electr. Eng.* **2018**, *100*, 1441–1454. [CrossRef]

15. Morales-Caporal, M.; Rangel-Magdaleno, J.; Peregrina-Barreto, H.; Morales-Caporal, R. FPGA-in-the-loop simulation of a grid-connected photovoltaic system by using a predictive control. *Electr. Eng.* **2018**, *100*, 1327–1337. [CrossRef]

16. Waidyasooriya, H.M.; Takei, Y.; Tatsumi, S.; Hariyama, M. OpenCL-Based FPGA-Platform for Stencil Computation and Its Optimization Methodology. *IEEE Trans. Parallel Distrib. Syst.* **2017**, *28*, 1390–1402. [CrossRef]

17. Fernández-Álvarez, A.; Portela-García, M.; García-Valderas, M.; López, J.; Sanz, M. HW/SW Co-Simulation System for Enhancing Hardware-in-the-Loop of Power Converter Digital Controllers. *IEEE J. Emerg. Sel. Top. Power Electron.* **2017**, *5*, 1779–1786. [CrossRef]

18. Typhoon HIL. Available online: https://www.typhoon-hil.com (accessed on 11 December 2017).

19. dSPACE. Available online: https://www.dspace.com (accessed on 11 December 2017).

20. OPAL-RT. Available online: https://www.opal-rt.com (accessed on 11 December 2017).

21. Sanchez, A.; de Castro, A.; Garrido, J. A Comparison of Simulation and Hardware-in-the-Loop Alternatives for Digital Control of Power Converters. *IEEE Trans. Ind. Inform.* **2012**, *8*, 491–500. [CrossRef]

22. Razzaghi, R.; Mitjans, M.; Rachidi, F.; Paolone, M. An automated FPGA real-time simulator for power electronics and power systems electromagnetic transient applications. *Electr. Power Syst. Res.* **2016**, *141*, 147–156. [CrossRef]

23. MacCleery, B.; Trescases, O.; Mujagic, M.; Bohls, D.M.; Stepanov, O.; Fick, G. A new platform and methodology for system-level design of next-generation FPGA-based digital SMPS. In Proceedings of the 2012 IEEE Energy Conversion Congress and Exposition (ECCE), Raleigh, NC, USA, 15–20 September 2012; pp. 1599–1606. [CrossRef]

24. Parma, G.; Dinavahi, V. Real-Time Digital Hardware Simulation of Power Electronics and Drives. *IEEE Trans. Power Deliv.* **2007**, *22*, 1235–1246. [CrossRef]

25. Matar, M.; Iravani, R. Massively Parallel Implementation of AC Machine Models for FPGA-Based Real-Time Simulation of Electromagnetic Transients. *IEEE Trans. Power Deliv.* **2011**, *26*, 830–840. [CrossRef]

26. Myaing, A.; Dinavahi, V. FPGA-Based Real-Time Emulation of Power Electronic Systems with Detailed Representation of Device Characteristics. *IEEE Trans. Ind. Electron.* **2011**, *58*, 358–368. [CrossRef]

27. Lucia, O.; Urriza, I.; Barragan, L.A.; Navarro, D.; Jimenez, O.; Burdio, J.M. Real-Time FPGA-Based Hardware-in-the-Loop Simulation Test Bench Applied to Multiple-Output Power Converters. *IEEE Trans. Ind. Appl.* **2011**, *47*, 853–860. [CrossRef]

28. Sanchez, A.; Todorovich, E.; de Castro, A. Impact of the hardened floating-point cores on HIL technology. *Electr. Power Syst. Res.* **2018**, *165*, 53–59. [CrossRef]

29. Infineon Technologies AG. *Design Note: CCM PFC Boost Converter Design (DN 2013-01)*; Rev. 1.0.; Infineon Technologies AG: Neubiberg, Germany, 2013.

30. Butcher, J.C. *The Numerical Analysis of Ordinary Differential Equations: Runge-Kutta and General Linear Methods*; Wiley-Interscience: New York, NY, USA, 1987.

31. Ibarra, L.; Rosales, A.; Ponce, P.; Molina, A.; Ayyanar, R. Overview of Real-Time Simulation as a Supporting Effort to Smart-Grid Attainment. *Energies* **2017**, *10*, 817. [CrossRef]

32. Saralegui, R.; Sanchez, A.; Martínez-García, M.S.; Novo, J.; de Castro, A. Comparison of Numerical Methods for Hardware-In-the-Loop Simulation of Switched-Mode Power Supplies. In Proceedings of the 2018 IEEE 19th Workshop on Control and Modeling for Power Electronics (COMPEL), Padova, Italy, 25–28 June 2018; pp. 1–6. [CrossRef]

33. Sutikno, T.; Idris, N.R.N.; Jidin, A.Z.; Daud, M.Z. FPGA based high precision torque and flux estimator of direct torque control drives. In Proceedings of the 2011 IEEE Applied Power Electronics Colloquium (IAPEC), Johor Bahru, Malaysia, 18–19 April 2011; pp. 122–127. [CrossRef]

34. Goñi, O.; Sanchez, A.; Todorovich, E.; de Castro, A. Resolution Analysis of Switching Converter Models for Hardware-in-the-Loop. *IEEE Trans. Ind. Inf.* **2014**, *10*, 1162–1170. [CrossRef]

35. *IEEE Standard for Floating-Point Arithmetic*; IEEE Std 754-2008; IEEE Standards: Piscataway, NJ, USA, 2008; pp. 1–70. [CrossRef]

36. *IEEE Standard VHDL Language Reference Manual*; IEEE Std 1076-2008 (Revision of IEEE Std 1076-2002); IEEE Standards: Piscataway, NJ, USA, 2009; pp. 1–626. [CrossRef]

37. Merfert, I. Analysis and application of a new control method for continuous-mode boost converters in power factor correction circuits. In Proceedings of the PESC97, Record 28th Annual IEEE Power Electronics Specialists Conference, Formerly Power Conditioning Specialists Conference 1970-71, Power Processing and Electronic Specialists Conference 1972, Saint Louis, MO, USA, 27 June 1997; Volume 1, pp. 96–102. [CrossRef]

38. Merfert, I.W. Stored-duty-ratio control for power factor correction. In Proceedings of the Fourteenth Annual Applied Power Electronics Conference and Exposition, APEC '99, Dallas, TX, USA, 14–18 March 1999; Volume 2, pp. 1123–1129. [CrossRef]

39. Sanchez, A.; de Castro, A.; López, V.M.; Azcondo, F.J.; Garrido, J. Single ADC Digital PFC Controller Using Precalculated Duty Cycles. *IEEE Trans. Power Electron.* **2014**, *29*, 996–1005. [CrossRef]

5

Analysis of Equivalent Inductance of Three-Phase Induction Motors in the Switching Frequency Range

Milan Srndovic [1,*], Rastko Fišer [2] and Gabriele Grandi [1]

[1] Department of Electrical, Electronic, and Information Engineering, University of Bologna, 40136 Bologna, Italy; gabriele.grandi@unibo.it

[2] Department of Mechatronics, Faculty of Electrical Engineering, University of Ljubljana, 1000 Ljubljana, Slovenia; rastko.fiser@fe.uni-lj.si

* Correspondence: milan.srndovic2@unibo.it

Abstract: The equivalent inductance of three-phase induction motors is experimentally investigated in this paper, with particular reference to the frequency range from 1 kHz to 20 kHz, typical for the switching frequency in inverter-fed electrical drives. The equivalent inductance is a basic parameter when determining the inverter-motor current distortion introduced by switching modulation, such as rms of current ripple, peak-to-peak current ripple amplitude, total harmonic distortion (THD), and synthesis of the optimal PWM strategy to minimize the THD itself. In case of squirrel-cage rotors, the experimental evidence shows that the equivalent inductance cannot be considered constant in the frequency range up to 20 kHz, and it considerably differs from the value measured at 50 Hz. This frequency-dependent behaviour can be justified mainly by the skin effect in rotor bars affecting the rotor leakage inductance in the considered frequency range. Experimental results are presented for a set of squirrel-cage induction motors with different rated power and one wound-rotor motor in order to emphasize the aforesaid phenomenon. The measurements were carried out by a three-phase sinusoidal generator with the maximum operating frequency of 5 kHz and a voltage source inverter operating in the six-step mode with the frequency up to 20 kHz.

Keywords: equivalent inductance; leakage inductance; switching frequency modelling; induction motor; current switching ripple

1. Introduction

A well pronounced frequency-dependent behaviour of an equivalent inductance in a three-phase induction motor (IM) has been analysed several times so far. Generally, the equivalent inductance of an IM, i.e., leakage inductance seen from the stator side at a stand-still, is at times determined as the sum of the stator and rotor inductances, both constant [1,2]. Such a simplification is not acceptable for higher order harmonics due to the frequency-dependent character of the rotor leakage inductance as a result of a skin effect in squirrel-cage rotor bars [3–5].

There are just a few analyses where the high-frequency behaviour of three-phase IMs was carried out over the actual switching-frequency range, but using impedance meters with an insufficient voltage level, and occasionally supplying just one out of three phases. The low power supply voltage might cause a significant limitation on the validity and applicability of proposed measurements [6,7]. Furthermore, supplying just one phase is not an entirely accurate approach and it might lead to a mismatch between estimated parameters when comparing them with the ones obtained at symmetrically supplied three phases.

The phenomenon of varying the rotor leakage inductance with frequency, and consequently the equivalent inductance, contributes considerably to motors' characteristics. The estimation of inverter current harmonic distortion, current ripple rms, and peak-to-peak current ripple amplitude [1–4,8,9]

is strongly affected as well. Especially, in the case of multiphase IMs, the equivalent inductance of each α-β plane seen from the stator side may additionally vary due to the different arrangements of machine connections [4].

On the whole, variation of the rotor leakage inductance over the frequency range is firmly connected to motor parameters, such as the power range, pole-pairs number, rotor bar shape, and depth [7]. In [10], a comprehensive equation for calculating the rotor leakage inductance considering some of the aforesaid parameters was introduced. The equation was derived with respect to the DC value of the rotor leakage inductance. The rotor leakage inductance decreases following a reciprocal value of the frequency square root over the whole frequency range. On the other hand, in [7], it was noted that the rotor leakage behaviour for frequencies lower than 1 kHz is reciprocally dependent on the square of frequency, and for frequencies higher than 1 kHz, the behaviour is the same as it was given by the equation in [10]. The explanation is that for lower frequencies, the skin depth and conductor size of squirrel-cage IM are of the same order of magnitude. Considering this, the ratio of leakage inductances at 20 kHz compared to the rated motor frequency is about 0.5 to 0.6. In [7], it was experimentally noted that the equivalent inductance of one IM had an increasing behaviour above a certain given frequency value, but without a proper explanation. Since a similar phenomenon was observed also in one of the tested motors, a capacitive effect was introduced, considering it as an adequate explanation. Similar effects were analysed in [11,12].

This paper gives a comprehensive set of experimental results for three different IMs with squirrel-cage rotors and one IM with a wound-rotor. The comparison between theoretical developments and the experiments are given over the considered frequency range, resulting in good correspondence. Apart from this, an additional experiment was carried out in order to show the effect of the total inductance frequency dependence on the current ripple estimation in case of one IM supplied by a three-phase inverter. Based on this, the graphical evidence of the analytically calculated current ripple envelopes at 50 Hz and at the switching frequency of 3 kHz are presented.

2. Theoretical Background and Basic Assumptions

In order to analyse the high-frequency behaviour of IM, the basic electrical scheme at a standstill is presented in Figure 1.

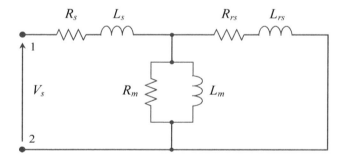

Figure 1. Basic per-phase equivalent circuit of the IM at a standstill.

The model consists of stator resistance, R_s, and stator leakage inductance, L_s, rotor resistance, R_{rs}, and rotor leakage inductance, L_{rs}, both seen from the stator side, and the magnetizing branch with parallel resistance, R_m, and inductance, L_m. For the higher order harmonics, the stator and rotor resistances can be neglected due to the fact that stator and rotor leakage inductances dominate [4]. By omitting these two parameters, the equivalent circuit is simplified to just four parameters. The parallel connection between the rotor leakage inductance and magnetizing inductance can be replaced with just one inductance with the proper equivalent value. If the magnetizing inductance is much higher than the rotor leakage inductance, which is usually the case in IMs, the parallel connection can be considered as the rotor leakage inductance per se, where $L_{rs} \mid \mid L_m \cong L_{rs}$.

Such an assumption is made in this paper. After simplifying the proposed model, the equivalent impedance seen from the terminals 1 and 2 (Figure 1) can be derived:

$$Z' = R' + j\omega\, L' = \frac{\omega^2 L_{rs}^{\,2} R_m}{R_m^{\,2} + \omega^2\, L_{rs}^{\,2}} + j\omega\left(L_s + \frac{L_{rs} R_m^{\,2}}{R_m^{\,2} + \omega^2\, L_{rs}^{\,2}}\right), \tag{1}$$

where f is the frequency, and $\omega = 2\pi f$ is the angular frequency. The frequency dependent behaviour of the rotor leakage inductance in case of rotor bars having a depth, d, is presented in [10,13]:

$$L_{rs}(f) = \frac{3\, L_{rs}^{dc}}{k\, d\, \sqrt{f}}\, \frac{\sinh\left(k\, d\sqrt{f}\right) - \sin\left(k\, d\sqrt{f}\right)}{\cosh\left(k\, d\sqrt{f}\right) - \cos\left(k\, d\sqrt{f}\right)}, \tag{2}$$

where L_{rs}^{dc} denotes the rotor leakage inductance at DC (in our case, at the lowest initial measured frequency [5–7]), and:

$$k = \sqrt{\frac{4\pi\mu_o}{\rho}}. \tag{3}$$

The relative permeability of free space in (3) is $\mu_o = 4\pi \times 10^{-7}$ H/m and the resistivity of the aluminium bars is $\rho = 2.65 \times 10^{-8}$ Ω·m. It can be noted that the behaviour of the rotor leakage inductance is inversely proportional to the square root of the frequency. In order to introduce the effect of stray capacitances for higher frequencies, the modified IM equivalent circuit at the standstill is presented in Figure 2.

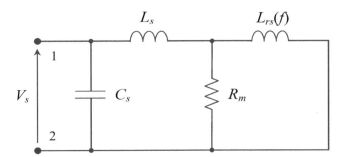

Figure 2. Proposed per-phase equivalent circuit of IM in the switching-frequency range, including stray capacitance and a simplified RL network.

The value of the capacitor, C_s, which is connected in parallel to the simplified circuit, is within the range of 0.5 nF to 5 nF [11,12]. Usually, the capacitive effect is not noticeable in the lower switching frequency range; however, in some particular cases, it has a relatively small contribution for higher frequencies. Considering the equivalent series resistance, R', and inductance, L', as defined in (1), and introducing the parallel capacitor, C_s, the overall equivalent inductance becomes:

$$L_{eq} = \frac{L'\left(1 - \omega^2 L' C_s\right) - C_s R'^2}{\left(1 - \omega^2 L' C_s\right)^2 + \left(\omega C_s R'\right)^2}. \tag{4}$$

3. Experimental Evaluation of Equivalent Inductance

Experimental analyses were carried out in order to estimate the frequency dependent behaviour of the equivalent inductance of three-phase IMs and to verify the analytical developments based on the simplified equivalent circuit presented in Figure 2. The set of motors under investigation consists of four IMs with different rated power (see Figure 3). Three out of four tested IMs have squirrel-cage rotors, where the skin effect in aluminium rotor bars must be considered. The fourth one has wound-rotor windings, mainly used experimentally to emphasise the considerably less pronounced skin effect in such rotor windings.

Figure 3. The considered set of IMs under testing.

The rated parameters of the tested IMs represented in Figure 2 are given in Table 1.

Table 1. Rated parameters of tested induction motors.

IM	Squirrel Cage Rotor			Wound Rotor	
Power (kW)	2.2	4.0	7.5	5.5	
Voltage (V)	400	400	400	Stator	Rotor
				380	186
Current (A)	5.2	9.2	15.3	14.0	19.5
Frequency (Hz)	50	50	50	50	
Speed (r/min)	1400	1425	1450	1400	

During measurements, each rotor shaft was mechanically locked to prevent its rotation. A three-phase sinusoidal power source, HP6834B (300 V, 4500 VA, 1Φ/3Φ, Keysight Technologies, Santa Rosa, CA, USA), was used for supplying the star-connected motors at a standstill. Due to its upper frequency limit of 5 kHz, in order to extend the test frequency range up to 20 kHz, a three-phase custom-made inverter operating in the six-step mode was used for the additional set of measurements, also guaranteeing almost sinusoidal motor currents. It consists of the three-phase IGBT Mitsubishi PS22A76 intelligent power module (1200 V, 25 A, Mitsubishi Electric Corporation, Tokyo, Japan) controlled by the Arduino Due microcontroller board (84 MHz Atmel, SAM3X83 Cortex-M3 CPU, Somerville, MA, USA). The Yokogawa DLM 2024 oscilloscope (Yokogawa Electric Corporation, Tokyo, Japan) with the PICO TA057 differential voltage probe (25 MHz, ±1400 V, ±2%, Pico Technology, Tyler, TX, USA) and LEM PR30 current probe (DC to 20 kHz, ±20 A, ±1%, LEM International SA, Plan-les-Ouates, Switzerland) were used to acquire motors' phase voltage and current. The whole experimental setup is shown in Figure 4.

The measurement determines the equivalent impedance for all the four IMs, calculated as the ratio between rms values of fundamental components of the voltage and current in the frequency range from 1 kHz to 20 kHz, with an equidistant step of 1 kHz. The equivalent reactance, and consequently the equivalent inductance (seen from the stator side), was calculated based on the phase displacement between the two fundamental components. In addition, the same quantities were also determined at the motor rated frequency of 50 Hz. The built-in digital filters of the oscilloscope were used to properly handle the current and voltage waveforms.

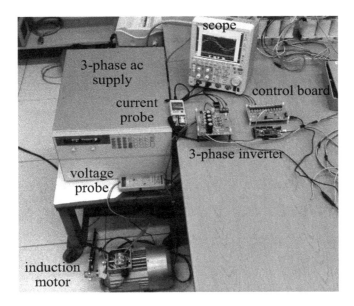

Figure 4. Experimental setup for measuring frequency dependence of L_{eq}.

To exclude the possible influence of voltage harmonics on the inductance measurements, a correlation between results acquired using the three-phase sinusoidal power source (up to 5 kHz) and the three-phase inverter with the six-step mode control was examined for all four motors. Comparing the results given in Table 2 for the 2.2 kW and 4.0 kW IMs clearly shows that the applied voltage source does not considerably influence the accuracy (within ±5%) of the determined equivalent inductance; therefore, only results for inverter supplied motors were presented to cover the whole frequency range from 1 kHz to 20 kHz. In Table 2, the parameter, $L_{eq,\sim}$ represents the equivalent inductance measured in the case of the sinusoidal power source, $L_{eq,inv}$ denotes the equivalent inductance for the inverter supplied motor, and ε is the relative difference (in percent) between them.

Table 2. Variance of equivalent inductance with two different supplies.

IM	2.2 kW			4.0 kW		
f_{sw} (kHz)	$L_{eq\sim}$ (mH)	$L_{eq,inv}$ (mH)	ε (%)	$L_{eq\sim}$ (mH)	$L_{eq,inv}$ (mH)	ε (%)
1	19.43	18.95	+2.53	13.53	13.49	+0.29
2	17.95	18.37	−2.39	12.37	12.76	−3.06
3	17.63	17.44	+1.09	11.60	11.83	−1.94
4	17.09	17.11	−0.12	11.06	11.57	−4.41
5	16.79	16.46	+2.00	10.66	11.20	−4.82

Due to the voltage limitation of the supply source, it was not possible to perform the short circuit test at the rated current for the whole frequency range up to 20 kHz (the necessary voltage would also destroy the motor isolation). Therefore, the input inductance was determined at the same highest possible % of rated current for a particular IM. To evaluate the magnetizing conditions at lower currents than the rated stator currents and to eliminate possible mistakes in the determination of the equivalent inductance, verification tests with different currents were done for two motors at the rated frequency. The results presented in Table 3 show that the equivalent inductance, L_{eq}, practically, does not change (regularly within ±5%), considering different test currents (I_s).

The equivalent inductance, experimentally determined by the procedure described above, was compared with the value calculated by the proposed method (4), considering the motors' parameters given in Table 4. The results show an acceptable agreement in the whole considered frequency range from 1 kHz to 20 kHz.

Table 3. Variance of equivalent inductance at reduced stator currents.

	IM 2.2 kW					
I_s (A)	5.2 [(*)]	4	3	2	1	0.35
L_{eq} (mH)	23.56	23.44	24.04	24.64	24.04	25.02
ε (%)	0.00	−0.51	+2.04	+4.58	+2.04	+6.19
	IM 4.0 kW					
I_s (A)	9.2 [(*)]	7	5	3	1	0.33
L_{eq} (mH)	16.20	16.56	15.83	16.49	16.41	16.30
ε (%)	0.00	+2.22	−2.28	+1.79	+1.30	+0.62

* Rated current.

Table 4. Equivalent circuit parameters of tested induction motors.

IM	Squirrel-Cage Rotor			Wound-Rotor
Power (kW)	2.2	4.0	7.5	5.5
C_S (nF)	0.25	0.1	0.1	3.5
L_S (mH)	13	8.8	4	4
L_{rs}^{dc} (mH)	12	8	4.6	3.7
R_m (Ω)	500	150	80	350
d (mm)	6	5	5.5	—

Particularly, in Figure 5, the case of IMs with rotor bars is presented, showing the expected decreasing behaviour of the equivalent inductance over the frequency due to the rotor leakage inductance frequency dependence. In the case of the 2.2 kW IM, the effect of stray capacitances is noticeable after the frequency of 15 kHz (modelled by the capacitor connected parallel to the equivalent circuit, Figure 2). The capacitive effects regarding the other two IMs (4 kW and 7.5 kW) are not visible within the considered frequency range.

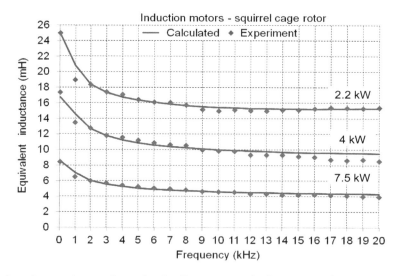

Figure 5. Model and experimental results for L_{eq} vs. supply frequency for three squirrel-cage IMs.

In the case of the IM with a wound rotor, as shown in Figure 6, the rotor leakage inductance is almost constant, with the skin effect being ineffective in such a rotor winding construction. The smooth decrease of the equivalent inductance is motivated by the resistive and reactive branches in the equivalent circuit (Figure 2), acting as a variable current divider as function of the frequency. A small increase of the L_{eq} after 15 kHz is also noticeable, well represented by the parallel capacitor.

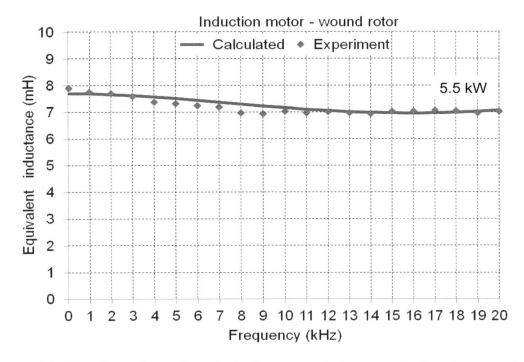

Figure 6. Model and experimental results for L_{eq} vs. supply frequency for the rotor-wound IM.

4. Application Example

When analysing the output current ripple regarding different PWM techniques and inverter configurations [9,14,15], the switching frequency and the load inductance are directly involved. In case of a squirrel-cage IM supplied by an inverter operating in the typical switching frequency range from 1 kHz to 20 kHz, and considering its equivalent inductance measured at a standstill at 50 Hz, this will result in a wrong current ripple estimation due to the frequency dependency of L_{eq}, which could generally decrease more than 50% (Figure 5).

Experimental results were carried out in order to demonstrate the influence of L_{eq} on the peak-to-peak current ripple amplitude in case of a three-phase squirrel-cage IM drive. The modulation technique used in this analysis was the space vector PWM technique, practically obtained by centering the three sinusoidal modulating signals of a carrier-based PWM with a common-mode injection (the so-called min/max injection).

The IM under test is the first given one in Table 2; Table 4 (2.2 kW). Figure 7 shows the line-to-neutral voltage (blue trace), instantaneous output current (red trace), and its fundamental component (grey trace) and ripple (orange trace), over one fundamental period (20 ms). The presented case corresponds to the DC bus voltage of $V_{dc} = 300$ V, switching frequency of 3 kHz, and modulation index, $m = 0.5$ ($m_{max} = 0.577$). The current ripple was obtained by subtracting the fundamental component from the instantaneous current [15]. All waveforms in Figure 7 are given in real scales.

In Figure 8, the following waveforms are presented: Measured current ripple (orange trace, obtained by downloading the experimental data with a high sample resolution and post-processing in Matlab/Simu-link), current ripple envelope analytically calculated by using $L_{eq} = 17.36$ mH, measured at $f_{sw} = 3$ kHz (blue trace), and the current ripple envelope, analytically calculated by using $L_{eq} = 24.98$ mH, measured at 50 Hz (red trace). The procedure of the current ripple envelope calculation is explained in detail in [9,14,15].

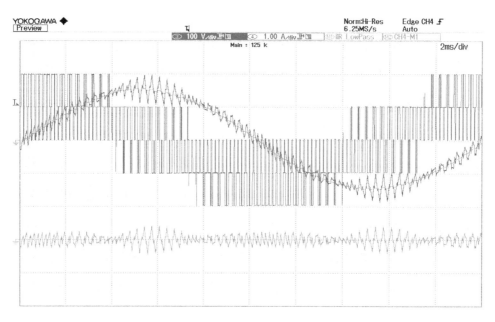

Figure 7. Details of experimentally obtained line-to-neutral voltage (blue trace), instantaneous current (red trace), fundamental current (grey trace), and current ripple (orange trace) for the 2.2 kW IM supplied by a three-phase inverter (centered PWM): $V_{dc} = 300$ V, $f_{sw} = 3$ kHz and $m = 0.5$.

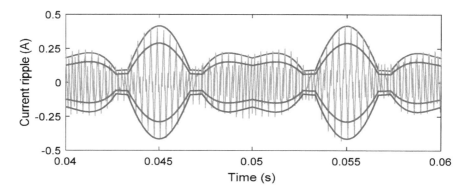

Figure 8. Measured current ripple (orange trace) and calculated current ripple envelopes (blue trace: L_{eq} measured at $f_{sw} = 3$ kHz, red trace: L_{eq} measured at 50 Hz).

The results clearly emphasise the case of an incorrect current ripple estimation when considering L_{eq} measured at the rated frequency instead of the switching frequency, giving a smaller current ripple amplitude (red envelope) compared with the correct one (blue envelope). L_{eq} further changes with the switching frequency according to Figure 5. Such an explicit discrepancy may lead to several mismatches in the procedures of current ripple minimization, modulation strategies' optimization, harmonic losses evaluation, EMC analysis, etc.

5. Conclusions

In this paper, the frequency-dependent behaviour of the IM equivalent inductance was presented in more detail compared to the existing literature. A simplified circuit model was proposed for induction motors in order to evaluate the equivalent inductance over the switching frequency range from 1 kHz to 20 kHz, which is the typical operating range of a PWM inverter in industrial ac motor drives.

Analytical results were compared with experimental ones, carried out by considering IMs with different rated powers, both squirrel-cage and wound-rotor types. The results show a good match, proving that the skin effect in rotor bars is mainly responsible for the equivalent inductance variation. There was huge variation in the frequency range up to 10 kHz, leading to a value that was 0.5 to

0.6 times more than the corresponding value measured at 50 Hz, depending on the IM design and power range. For higher frequencies, in the considered range up to 20 kHz, the equivalent inductance was almost constant. Thus, while selecting the switching frequency, it is very important to take into account the corresponding equivalent inductance, especially in the case of precise and sensitive motor control algorithms. The effect of stray capacitances was just noticeable, and it starts to be effective for frequencies higher than 20 kHz, which is already above the commonly used switching frequencies of mass-produced industrial inverters. More precise evaluation of the stray capacitance would require more complex circuits, instruments with higher resolution, and experiments in the capacitively-dominant working region; however, its influence is hardly noticeable in most cases. In the case of IM with wound rotor winding, the overall decrease of equivalent inductance was much less pronounced, and was not determined by the skin effect, and was well represented by the proposed circuit model. Apart from the presented analysis of the equivalent inductance, the evaluation of the equivalent resistance over the switching frequency range that is responsible for other relevant changes in some parameters could be performed in further research.

Author Contributions: Conceptualization, M.S. and G.G.; methodology, M.S., R.F., and G.G.; validation, M.S.; formal analysis, M.S., R.F. and G.G.; writing—original draft preparation, M.S.; writing—review and editing, R.F. and G.G..; funding acquisition, M.S.

References

1. Casadei, D.; Serra, G.; Tani, A.; Zarri, L. Theoretical and experimental analysis for the RMS current ripple minimization in induction motor drives controlled by SVM technique. *IE* **2004**, *51*, 1056–1065. [CrossRef]
2. Kubo, H.; Yamamoto, Y.; Kondo, T.; Rajashekara, K.; Zhu, B. Current ripple analysis of PWM methods for open-end winding induction motor. In Proceedings of the 2014 IEEE Energy Conversion Congress and Exposition (ECCE), Pittsburgh, PA, USA, 14–18 September 2014.
3. Dujic, D.; Jones, M.; Levi, E. Analysis of output current-ripple RMS in multiphase drives using polygon approach. *PEL* **2010**, *25*, 1838–1849. [CrossRef]
4. Jones, M.; Dujic, D.; Levi, E.; Prieto, J.; Barrero, F. Switching ripple char-acteristics of space vector PWM schemes for five-phase two-level voltage source inverters—Part 2: Current ripple. *IA* **2011**, *57*, 2799–2808.
5. Calin, M.; Rezmerita, F.; Ileana, C.; Iordache, M.; Galan, N. Performance analysis of three phase squirrel cage induction motors with deep rotor bars in transient behavior. *Electr. Electron. Eng.* **2012**, *2*, 11–17. [CrossRef]
6. Potter, B.A.; Shirsavar, S.A.; Mcculloch, M.D. Study of the variation of the input impedance of induction machines with frequency. *IET Electr. Power Appl.* **2007**, *1*, 36–42. [CrossRef]
7. Novotny, D.W.; Nasar, S.A.; Jeftenic, B.; Maly, D. Frequency depend-ence of time harmonic losses in induction machines. In Proceedings of the ICEM, Boston, FL, USA, 13–15 August 1990.
8. Jiang, D.; Wang, F. Current ripple prediction for three-phase PWM converters. *IA* **2014**, *50*, 531–538. [CrossRef]
9. Grandi, G.; Loncarski, J.; Dordevic, O. Analysis and comparison of peak-to-peak current ripple in two-level and multilevel PWM inverters. *IE* **2015**, *62*, 2721–2730. [CrossRef]
10. Kwon, Y.S.; Lee, J.H.; Moon, S.H.; Kwon, B.K. Standstill parameter identification of vector-controlled induction motors using the frequency characteristics of rotor bars. *IA* **2009**, *45*, 1610–1618.
11. Hidaka, T.; Ishida, M.; Hori, T.; Fujita, H. High-frequency equivalent cir-cuit of an induction motor driven by a PWM inverter. *Electr. Eng. Jpn.* **2001**, *135*, 65–76. [CrossRef]
12. Grandi, G.; Casadei, D.; Reggiani, U. Equivalent circuit of mush wound AC windings for high frequency analysis. In Proceedings of the ISIE, Guimarães, Portugal, 7–11 July 1997.
13. Cho, K.R.; Seok, J.K. Induction motor rotor temperature estimation based on high-frequency model of a rotor bar. *IA* **2009**, *45*, 1267–1275.
14. Grandi, G.; Loncarski, J.; Srndovic, M. Analysis and minimization of output current ripple for discontinuous pulse-width modulation techniques in three-phase inverters. *Energies* **2016**, *9*, 380. [CrossRef]
15. Loncarski, J.; Leijon, M.; Srndovic, M.; Rossi, C.; Grandi, G. Comparison of output current ripple in single and dual three-phase inverters for electric vehicle motor drives. *Energies* **2015**, *8*, 3832–3848. [CrossRef]

6

Automatic Fault Diagnostic System for Induction Motors under Transient Regime Optimized with Expert Systems

Jordi Burriel-Valencia, Ruben Puche-Panadero, Javier Martinez-Roman, Angel Sapena-Bano *, Manuel Pineda-Sanchez, Juan Perez-Cruz and Martin Riera-Guasp

Institute for Energy Engineering, Universitat Politècnica de València, 46022 Valencia, Spain;
joburva@die.upv.es (J.B.-V.); rupucpa@die.upv.es (R.P.-P.); jmroman@die.upv.es (J.M.-R.);
mpineda@die.upv.es (M.P.-S.); juperez@die.upv.es (J.P.-C.); mriera@die.upv.es (M.R.-G.)
* Correspondence: asapena@die.upv.es

Abstract: Induction machines (IMs) power most modern industrial processes (induction motors) and generate an increasing portion of our electricity (doubly fed induction generators). A continuous monitoring of the machine's condition can identify faults at an early stage, and it can avoid costly, unexpected shutdowns of production processes, with economic losses well beyond the cost of the machine itself. Machine current signature analysis (MCSA), has become a prominent technique for condition-based maintenance, because, in its basic approach, it is non-invasive, requires just a current sensor, and can process the current signal using a standard fast Fourier transform (FFT). Nevertheless, the industrial application of MCSA requires well-trained maintenance personnel, able to interpret the current spectra and to avoid false diagnostics that can appear due to electrical noise in harsh industrial environments. This task faces increasing difficulties, especially when dealing with machines that work under non-stationary conditions, such as wind generators under variable wind regime, or motors fed from variable speed drives. In these cases, the resulting spectra are no longer simple one-dimensional plots in the time domain; instead, they become two-dimensional images in the joint time-frequency domain, requiring highly specialized personnel to evaluate the machine condition. To alleviate these problems, supporting the maintenance staff in their decision process, and simplifying the correct use of fault diagnosis systems, expert systems based on neural networks have been proposed for automatic fault diagnosis. However, all these systems, up to the best knowledge of the authors, operate under steady-state conditions, and are not applicable in a transient regime. To solve this problem, this paper presents an automatic system for generating optimized expert diagnostic systems for fault detection when the machine works under transient conditions. The proposed method is first theoretically introduced, and then it is applied to the experimental diagnosis of broken bars in a commercial cage induction motor.

Keywords: fault diagnosis; condition monitoring; induction machines; support vector machines; expert systems; neural networks

---

1. Introduction

Induction machines (IMs) power most modern industrial processes (induction motors) and generate an increasing portion of our electricity (doubly fed induction generators). Therefore, fault diagnosis of IMs has become an important area of condition-based maintenance (CBM) programs, to avoid the high economic losses generated by unexpected breakdowns of IMs and sudden stoppages of the production lines that they drive. Specifically, fault diagnosis techniques based on the analysis of the MCSA [1–6] have gained a wide industrial deployment, due to their simplicity, low requirements of hardware and software, and capability for on-line simultaneous detection of a wide range of machine faults. Despite its advantages, industrial application of MCSA in harsh industrial environments, under real working conditions, is challenging. The spectral lines, whose amplitude signals the presence of a fault, can be difficult to evaluate under the myriad of spectral lines in the spectrum of the machine current, especially in case of incipient faults, where the fault harmonics have small amplitudes, or in case of low slip working conditions, where the leakage of the fundamental can bury the fault harmonics appearing at very close frequencies. To deal with these difficulties, several ongoing research works [7,8] propose the development of expert systems that can improve the diagnostic hit ratio, mostly where the fault features information obtained to detect these faults is scarce or unrepresentative. Nevertheless, developing and combining expert systems with fault diagnostic methods to improve hit ratio it is not trivial at all. An optimum combination of both elements can lead to a significant hit ratio improvement in fault detection, but an inadequate combination can even result in a misdiagnose. In the scientific literature, some works such as [9–13] are focused on the analysis, explanation, and development of recommendations, techniques, and methodologies to achieve a correct expert system implementation with optimal problem resolution. Following these recommendations, two main aspects are relevant to build an accurate expert system for the diagnosis of a faulty IM: on the one hand it is necessary to use a method able to detect and obtain features of the motor that can characterize a given type of fault; on the other hand, some algorithm or methodology able to interpret these features to discern about fault existence must be developed.

The design of an expert system for fault detection of IMs, taking into account both aspects, is a complex task, with many design variables that can influence the performance of such a diagnostic system. In this context, this paper proposes the automation of this design stage, through the development an automated system (the supra-system) which automatically generates custom fault diagnostic systems with high precision rate for fault detection. The proposed, so called supra-system is based on the exhaustive comparison of different combinations of fault diagnostic methods and optimized expert systems. It has been applied with success to the generation of an expert system for the detection of broken bars in a squirrel cage IM, both in steady-state regime and in transient state. The application to the detection of other types of fault is straightforward.

The paper is structured as follows: Section 2 introduces methods to detect faults in the induction machine under transient conditions. Section 3 describes the expert systems more commonly used in this field. Section 4 describes the development of the supra-system to generate optimized diagnostic systems. Section 5 is devoted to the experimental results and validation, and finally, in Section 6 the conclusions are presented.

2. Components of the Generator of Expert Systems for Fault Diagnosis of IMs

The three main components of an expert system for fault diagnosis of IMs, to be generated by the proposed supra- system, are the following ones:

- The quantity measured in the IM.
- The method used for extracting fault representative features from the measured quantities.
- The type of expert system used to perform the fault diagnosis from the selected IM features.

These three characteristics are analyzed in the following sections.

2.1. Quantities Measured in the IM

Fault diagnostic methods for IMs can obtain representative fault features from different motor quantities, like phase currents or voltages, acoustic, temperature, vibrations, etc. Using these features, these methods must able to detect the presence or absence of failures like broken rotor bars, winding short circuits, bearing damages, eccentricities, etc. In addition, these methods must work under any machine operation regimes (standstill, start-up, steady-state or transient regime). In this paper, the selected quantity is the machine stator current, because it is non-invasive (a Hall sensor or a current transformer placed in the line feeding the machine can acquire the current), can be acquired on-line, without disturbing the machine work, and can identify a wide range of simultaneous machine faults. Each type of fault generates a characteristic signature in the stator current, and, thus, it is possible to detect different type of faults by the on-line analysis of the machine current. Furthermore, it is easy to adapt this approach to other IM quantities.

2.2. Methods to Obtain Fault Representative Features

Under steady-state regime, the stator current analysis is performed using the frequency spectrum of the stator current, via the FFT [14]. Nevertheless, in many modern industrial processes, IMs operate under transient conditions (due to varying loads, action of controllers, etc.). The methods used for obtaining fault representative features and analyzing them in steady-state regime are usually not valid in transient state (for example, the FFT cannot be used under varying speed conditions).

On the contrary, most methods designed for transient state diagnosis are also valid for the detection of failures in steady-state condition, but at the cost of a higher computational complexity both of the set of diagnostic features and of the analysis algorithms. For these reasons, this research focuses on the fault diagnosis both in steady-state regime and in transient state.

Recently, the development of diagnostic techniques focused on machines working in transient regimes have attracted the attention of many researches, giving raise to works as [15–17] dealing with this subject. In [15] the empirical mode decomposition (EMD) is used to obtain the fault features. The proposed approach, to obtain the fault features not only needs to compute the consecutive intrinsic mode functions (IMFs) containing the fault related components, but also the average value and the zero crossing of the sum of the IMFs.

This process adds disturbances losing information about the fault evolution. In [16] the discrete wavelet transform (DWT) is used to decompose the Park vector of the stator currents in 12 levels; this approach implies a dyadic decomposition of the frequency bands and a computation of the energy contained in each range. Therefore, the information used (the energy of frequency band) is not only related to the fault features but also to other components or effects such as the spectral leakage that can significantly influence on the results. In a second step, the standard deviation of this data is used as fault features being less precise. Therefore, the use of a self-organizing map (SOM) network is proposed to detect the presence or absence of a given fault.

In [17] the time-frequency plane is used to detect the fault. Nevertheless, the proposed method is only valid for the detection of the rotor broken bar fault in induction motors and only for the start-up transient. The fault feature is a complex representation of the characteristic V-shape pattern (with specific width and angle) for this fault (rotor broken bar) and this regime (start-up transient). It requires complex image treatment (dilatations, erosions, thresholds, subtraction of images, etc.) to obtain the fault features and implies a loss of fault information and a very limited field of application. Therefore, although these techniques can be also applied to obtain the fault features in steady-state regimes theses fault features would be more complex and less precise than the specific methods for steady state regimes.

A recent development in fault diagnosis of IMs allows to apply the same tool for analyzing the spectral content of the motor current in both regimes. This technique, known as harmonic order tracking analysis (HOTA) [4,5,18] allows to design a unique expert system for both regimes. Therefore, it is chosen as the basis of the diagnostic system generator proposed in this paper.

2.3. The Harmonic Order Tracking Analysis (HOTA) Method

In this section, the HOTA method, introduced in [4,5,18], is briefly described. Unlike other methods where the frequency position of fault features (obtained from the current or its envelope) depend on other variables, in the HOTA method the fault frequencies are normalized to an integer, harmonic k-order scale, which is independent of the motor supply frequency and of the motor slip. This further simplifies and accelerates the procedure of conditioning the current signal prior to its processing by the expert system.

HOTA method for transient state is a method based on reducing the 2D time-frequency content of the fault harmonics of the motor current to a much simpler 1D harmonic k-order domain. As explained in [5], this simplification process is implemented using a Gaussian window and a short time Fourier transform (STFT), by iteratively moving the Gaussian window along the time domain and performing the frequency axis re-scaling at each step (see an example in Figure 1).

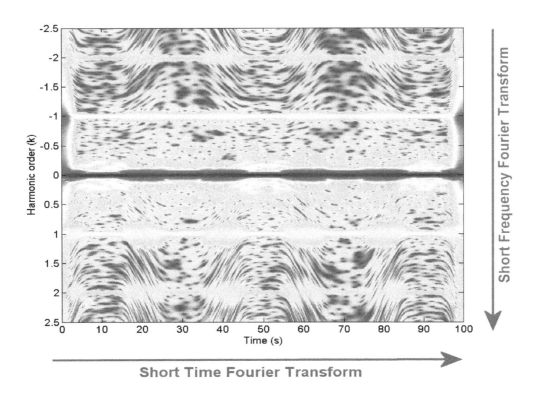

Figure 1. Example of time/harmonic k-order space generated with HOTA method for induction motors fault diagnosis in transient state.

An alternative option is to replace the STFT transform in HOTA by the short-frequency Fourier transform (SFFT), as in [6]. SFFT generates a time-frequency Gaussian window which is displaced along the frequency axis, instead of the time axis (Figure 1). Both the STFT and the SFFT transform generate the same time/frequency representations, although the SFFT has speed computing advantages when applied to fault diagnosis of IMs, as will be shown in Section 4.1.

In the last step of HOTA algorithm, once the time/frequency space is generated, a conversion into the harmonic k-order domain is made, obtaining as a result a single vector. Inside this vector each fault component is clearly shown through its k component number, as shown in Figure 2. This figure shows an example of the final result generated from the time-harmonic k-order space example shown in Figure 1. This graphic shows a fault k component within the harmonic k-order domain corresponding to a rotor broken bar fault.

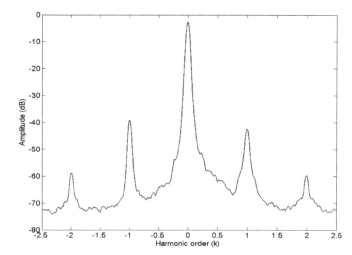

Figure 2. Final result vector in harmonic k-order domain obtained with HOTA method.

3. Expert Systems for Features Classification

Expert systems implementation to automate decision-making process are applied to interpret complex data or correlation features with some degree of uncertainty. Although diverse types of expert systems are used to resolve problems with different origins, in the field of IMs fault diagnosis the classification expert systems prevail. Classification expert systems generate discriminant results (decisions made regarding features), precluding the generation of ambiguous results.

As shown in [7,8,19], the support vector machine (SVM) expert system and artificial neural network (ANN) expert system are the most commonly used expert systems to solve features classification problems, due to their high learning coefficient of the problem. With a failure representative features database, obtained via fault diagnostic methods, these expert systems analyze and interpret the data to assess the presence or absence of an IM fault.

SVM and ANN expert systems have an input interface to the system where fault features are inserted. With these features, the expert system deduces a solution for the fault diagnosis problem. The internal part of the system is then adjusted to perform the interpretation of input data, and an output of one or more results is finally generated. Every solution is related to its respective input data.

However, in a different way about procedural programs, an expert system is not an execution of a sequence of commands that finally generate a result. Both SVM and ANN systems need to develop a previous training of the expert system to "learn" the mechanics of the problem to find the optimal solution. Adjusting the properties and parameters of the expert system to obtain a good training and good failure rates is a highly complex task, which in many cases is carried on by a manual trial and error procedure.

3.1. Support Vector Machine

SVM is an expert system based on n-dimensional spaces where each n parameter of the feature corresponds to the n dimension of the space (n feature parameters $= n$ dimensions of the space). The learning stage is based on the generation of a $(n-1)$ hyperplane that divides the n-dimensional space into two subspaces where each one represents one solution of the problem, which is sufficient to discern the occurrence or absence of fault.

In Figure 3 it is shown an example of a 2-dimensional space divided by a hyperplane generated with previous training. This SVM space has been generated with the supra-system during the experimental test described in Section 5.1.

The SVM optimization for obtaining the best system for analysis and diagnosis of failures depends on configuring the best kernel method, on the proper parameter fitting, and on the development of an optimal learning of the problem.

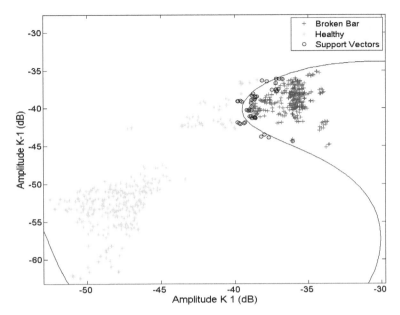

Figure 3. Space of a 2-dimensional SVM (2 parameters on each failure representative feature) where a hyperplane is splitting the space in subspaces of healthy features and failure features.

3.2. Artificial Neural Network

ANN expert systems are based on the emulation of biological neuronal networks. An ANN is composed by a set of "neurons" distributed in several interconnected layers. Although there are several types of neural networks structures and configurations, the most used for features classification is the Multilayer Perceptron. Figure 4 shows the layer structure of a Multilayer Perceptron, with an input data terminal, no or some hidden layers, and one output layer.

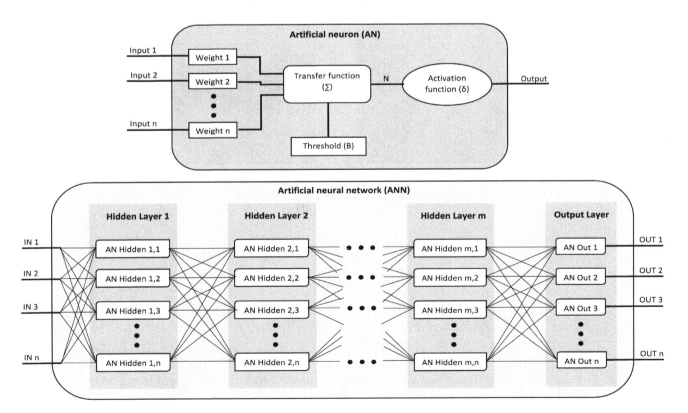

Figure 4. Scheme of Multilayer Perceptron neural network.

ANNs also must be trained before being able to correctly solve the problem. On ANN optimization, this learning is based on finding the best structure for a given problem (number of layers and number of neurons per layer) as well as the correct weight for each neuron within this network.

4. Development of the Supra-System, a System Able to Generate and Optimize Failure Diagnostic Systems Improved with Expert Systems

As presented in the previous sections, fault representative features finding methods and expert systems need to be optimized to obtain an acceptable diagnostic system. Finding the best fitting is not a trivial task since the accuracy of the expert system depends critically on the selection of its parameters. Therefore, to perform these tasks, in this paper, an autonomous system (the supra-system) is proposed. It is able not only of generating fault diagnostic systems, but also to optimize them through a set of fault representative features finding methods (Figure 5).

Figure 5. General scheme of the supra-system, the system generator of optimized fault diagnostic system.

4.1. Optimization of HOTA Method for IMs Fault Diagnosis

In HOTA method, as in the other methods that develop time/frequency spaces, it is essential to adjust the proportion of window filter to maximize the resolution of the space and the desired frequencies. Depending on the proportion applied to the Gaussian window filter, the space resolution can be improved (Figure 6a,c) or be worsened (Figure 6b,d), hindering the fault detection.

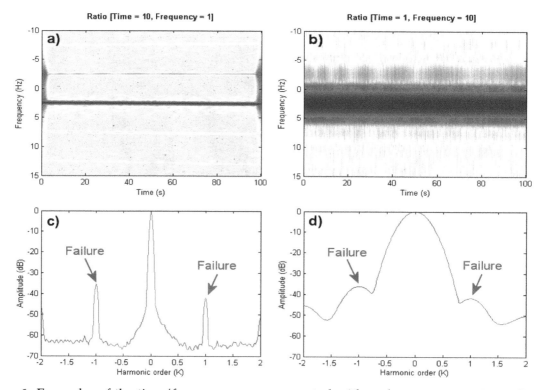

Figure 6. Examples of the time/frequency space generated with a phase current using a Gaussian window of ratio 10/1 (**a**) and its HOTA solution (**c**). The time/frequency space obtained with the same current but using a Gaussian window of ratio 1/10 is shown in (**b**), and its HOTA solution is shown in (**d**).

Figure 6 shows an example of a motor with a broken bar fault, with the fault components at $k = 1$ and $k = -1$; To detect properly these components in the joint time-frequency domain (t-f domain) is a challenging task, because their amplitude is much smaller than the main component amplitude. The performance of the fault diagnostic system depends critically on the analyzing window used for obtaining the t-f spectrum of the stator current. In this work, the Gaussian window has been selected, because it achieves the highest power density in the t-f domain. Nevertheless, different width to height ratios of a Gaussian window with the same minimum area in the t-f domain can alter substantially the shape of the fault harmonics. For each fault component, it can be observed that exists a specific window whose width to height ratio gives the sharpest fault components in the t-f spectrum [20], and consequently the fault components on HOTA result are also maximal (most representative fault features). Moving away from this ideal window size ratio, either in time or frequency, the resulting fault components on HOTA are less representatives for the fault. One of the goals of the supra-system presented in this paper is precisely to obtain this optimal window ratio, for generating the best t-f spectrum and the best fault representatives features in HOTA result.

The sharpness of the fault components is calculated with the mean value of the amplitude in the frequency range between the fault components ($k = 1$ and $k = -1$) and the main component. In the proposed system, the search for the best window size ratio has been optimized using a binary search algorithm [21], which is more specialized than a linear search algorithm.

As discussed above, the t-f spectrum is generated with a STFT transform. Nevertheless, as explained in [6], a SFFT transform can be used instead STFT, obtaining the same t-f spectrum. It is remarkable that the use of the SFFT in diagnostic applications reduces greatly the required computational power, regarding the STFT. This is due to the fact that the fault frequencies to be analyzed are known and the SFFT is able to generate the t-f spectrum by moving the Gaussian window in the frequency domain within a limited range close to the desired frequencies. This can be seen in Figure 7, where the results of the SFFT are shown in color, whereas the gray zone shows the results of the t-f spectrogram as obtained with the traditional STFT.

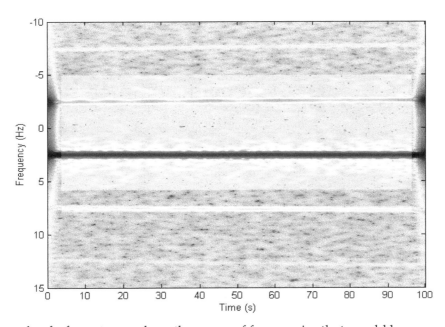

Figure 7. Example of t-f spectrum where the range of frequencies that would be necessary to obtain broken bar fault features has been marked.

4.2. Optimization of the Learning Process of the SVM Expert System for IMs Fault Diagnosis

The SVM expert system is quite optimal in solving problems that show a poorly linear distribution in its solutions. Nevertheless, its flexibility implies that it is necessary to adjust several coefficients during the training process to obtain a satisfactory learning of the expert system. The goal of the supra-system presented in this paper is also to optimize the value of such coefficients in an automatic way.

SVM allows different kernels to transform the space form to improve the discrimination of different features with the use of a hyperplane. In [10] several common kernels are discussed. In particular, with the Polynomial Kernel (1) and Radial Basis Function (RBF) kernel (2), it is possible to solve almost all cases where the solution distribution of the problem goes from totally linear to scarcely linear.

$$K(x_i, x_j) = (1 + x_i^T x_j)^d \tag{1}$$

$$K(x_i, x_j) = exp\left[- \left(\frac{\|x_i - x_j\|}{2\sigma} \right)^2 \right] \tag{2}$$

To use one of these two kernels one must set two parameters in order to obtain an optimum learning SVM. If a Polynomial Kernel is used, it is necessary to set the pair of parameters $\delta(C, d)$, where C is the box limit parameter for the search box during training and d is the polynomial order. If a RBF kernel is used, the pair of parameters that must be set are $\delta(C, \sigma)$, where C is the box limit parameter and σ is the scale factor of the Gaussian function.

In [10] it is explained how to fit these parameters by a mesh search method. Although this fitting method is less optimal than others with lower execution time, it guarantees a good approximation to the global solution avoiding local solutions that can be found by other heuristic fitting methods.

Like the scheme in Figure 8, the mesh search method divides each parameter reachable range into subsets (similar to a mesh). On each subset an SVM is trained with the parameters of this subset, obtaining an estimation of its hit rate. Finally, the subset with the trained SVM with highest hit rate is selected and the described process is iterated in a depth search until the subset with the highest hit rate (without over-training) is obtained.

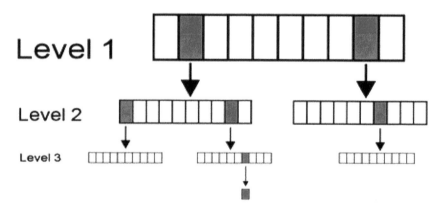

Figure 8. Schematic example of setting a parameter with the mesh search method.

It is relevant that all these trained SVM, generated during the mesh search, have had an optimal training to obtain a good approximation to the real hit rate. Achieving an optimal training for each SVM is another issue that is addressed in the supra-system presented in this paper.

An expert system cannot be trained directly with the whole training set of fault features to obtain its hit ratio. Otherwise, the hit ratio would be erroneous. There are several methods to train and validate an SVM, but the most reliable for obtaining the best hit ratio approximation for SVM is the *'leave one out cross validation'* method. With this method, the number of SVMs trained are equal to the number of fault features. In the particular field of IMs fault diagnosis, the number of fault features are not very high and therefore the computational complexity is acceptable.

As described the scheme for the *'leave one out cross validation'* (Figure 9), for each feature k, a trained SVM is generated with the remaining features, that is, excluding k feature. Once the SVM has been trained, then the k feature is validated. Repeating this process for the whole set of fault features, it yields finally an approximate hit ratio quite close to the actual hit ratio.

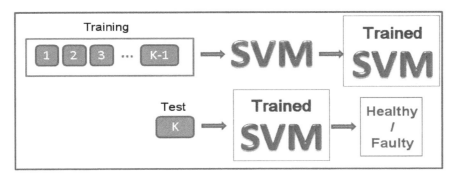

Figure 9. Scheme about training and classification for the iteration of k feature in the *'leave one out cross validation'* training algorithm.

4.3. Optimization of the Learning Process of ANN Expert System for IMs Fault Diagnosis

ANN expert systems are quite optimal in problem solving and may even be better than SVM when the distribution of problem solutions is highly non-linear. In contrast, the complexity to train and optimize an ANN expert system that solves the problem is greater than in the case of an SVM system.

The optimization of the structure of an ANN expert system implies the selection of the number of neurons per layer and the number of layers. These choices are relevant, since they influence the quality of learning and the accuracy of the results. For each concrete problem to solve, there is an optimal structural configuration. Training an unoptimized structure may lead to under-training or over-training the problem.

As of today, there is not a well defined method to find an optimal structural network in a direct way. However, in the case of neural networks for classification, several methodological rules have been published that discuss the structural limits of the network [10,11]. For classification problems these researches recommend a hidden layer range in the network between one and two layers, a heuristic search algorithm for fitting optimal network structure, and the best internal configuration for each neuron [9].

A stochastic search with heuristic optimization using the pyramid rule has been chosen to search the optimal structural network for IMs fault diagnosis. In the pyramid rule it is assumed that the best initial structure for searching optimal network has a trapezoidal pyramid shape where the network base are the inputs and the top are the outputs.

In the case of classification problems for ANN developed in the artificial intelligence field, as shown in Table 1, for each hidden layer's neuron is assigned a "hyperbolic tangent Sigmoid" transfer function. On the other hand, on the output layer, for each neuron a '*Competitive SoftMax*' transfer function is used, whose output acquires the maximum value while the other outputs are cancelled.

Table 1. Transfer functions used to solve classification problems with ANN expert systems for IMs fault diagnosis.

Layer Type	Transfer Function	In(i)/Out(o) Tie
Hidden Layer	Hyperbolic Tangent Sigmoid	$o = \frac{i^n - i^{-n}}{i^n + i^{-n}}$
Output Layer	Competitive Softmax	$o = 1$, n max $o = 0$, the others

It must be emphasized that both optimizing methods for ANN network structure and configuration used in this work are only valid and optimal when the ANN is used to solve classification problems. Therefore, these methods are valid for IMs fault diagnosis since it is a classification problem (healthy/faulty).

In each step of this ANN structural optimization algorithm, a specific network structure is generated (specific number of layers and number of neurons per layer). For each generated structure an optimal training and validation must be carried out to obtain a valid approximation hit rate for this ANN structure. The highest hit rate when the ANN structural optimization algorithm has finished is declared as the best approximation hit rate and its structure is the optimum one to solve the IMs classification problem.

For ANN, the *'leave one out cross validation'* method is not optimal to obtain a good approximation of the hit ratio, due to the operative of the algorithms used to train ANN expert systems. Accordingly, in this research, it has been implemented for training the ANN a *'training/validation/test cross validation'* method. This method, shown in the scheme of Figure 10, iterates in a loop of 100 iterations (100 basic trainings) where the fault characteristics are distributed in each iteration between the training, validation, and test sets. Between 60 and 80 percent of features are assigned to the training set, between 10 and 20 percent features for the test set, and the rest is assigned to the validation set.

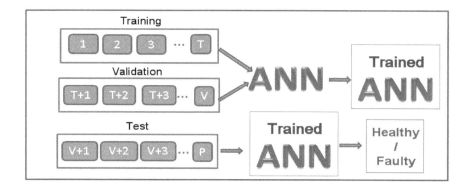

Figure 10. Scheme about training and classification for X iteration in the *'training/validation/test cross validation'* training algorithm.

In the *'training/validation/test cross validation'* method, it is important that the features selected on each set are representative of the whole range of values that the fault features can reach. In this research, this issue has been solved with the generation of a features space and its segmentation in a random sequence.

As analyzed in [12], there are several learning algorithms based on a gradient back-propagation, where each one has its advantages and disadvantages. The scaled conjugate gradient back-propagation algorithm has been chosen in this work as it guarantees a global optimal learning for classification problems with a moderate computational complexity, and without over-training locally (for the specific ANN structure trained).

The cost function used to minimize the gradient error is the *'cross entropy'* formulated as:

$$FC = -\frac{1}{n} \cdot \sum_{i=1}^{n} [y_i \cdot ln(a_i) + (1 - y_i) \cdot ln(1 - y_i)] \tag{3}$$

where the variable n denotes the total number of features for training, variable a stands for actual outputs and variable y stands for desired outputs.

5. Experimental Validation

The experimental validation of the supra-system has been carried out to detect rotor broken bars as in [1,3,22]. Nonetheless, this supra-system can be easily adapted to detect other types of faults such as eccentricity, stator inter-turn short circuits and bearing faults.

5.1. Test Bed

Figure 11 shows the test bed used in this paper. Two squirrel cage IMs whose main characteristics are given in Appendix A were tested, one in healthy conditions the other with a rotor broken bar fault. The broken bar fault was created by drilling a hole in one bar, in the junction of the bar with the end ring, as can be seen in Figure 12. This way to produce the bar breakage avoids damaging the magnetic circuit of the rotor and enables an easy verification of the complete disconnection between the bar and the end ring. To cover a wide scenario of industrial situation, the motor under test has been feed, alternatively, through variable speed drives (VSDs) of two different brands (ABB model ACS800 and Siemens model M440) with up to four different control strategies (scalar, scalar with slip compensation, field oriented control (FOC) and direct torque control (DTC)) and direct on-line (DOL) through an autotransformer. A permanent magnet synchronous machine (PMSM) is used as a mechanical load controlled by a drive ABB model ACSM1-04AS-024A-4. The test bed is controlled using a programmable logic controller (PLC) and a system control and data acquisition (SCADA) system which allows to perform the test in an automatic way and to repeat accurately the same conditions with different motors under test. A digital oscilloscope model Yokogawa DL750 has been used to acquire the currents during the different performed tests.

In this case, the test bed has been used for detecting broken bars fault, as in [1,3,22], but it can be easily adapted to detect other faults such as inter-turn short circuits, bearing damages or eccentricities. Indeed, other IMs with different faults can be coupled in the test bed and the tests set can be performed automatically.

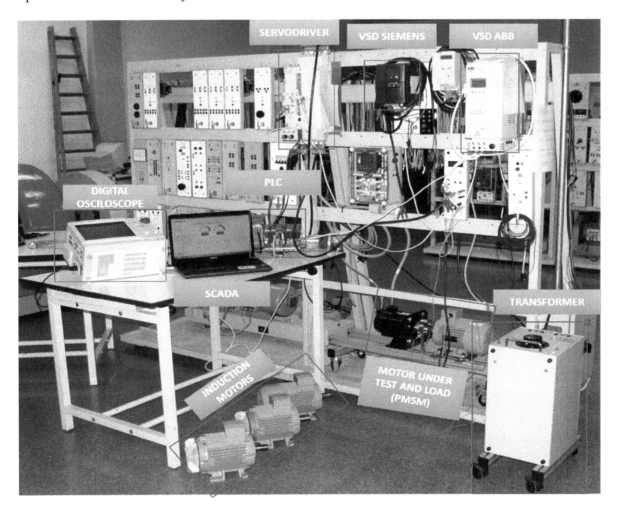

Figure 11. Test Bed used to obtain the experimental signals.

Figure 12. Top rotor with an artificially rotor broken bar fault and bottom rotor in healthy conditions.

5.2. Experimental Results

The supra-system generator has been implemented in MATLAB R2014 platform for evaluating its feasibility, the supra-system generator is applied to a large set of experimental current samples, obtained through the test bed described in the previous section, testing healthy and faulty IMs under different conditions (supply conditions, load conditions).

The samples database generated in this research comprises 726 phase current samples, with 369 healthy rotor samples and 357 faulty rotor samples. During the sampling process the frequency and load torque has been changed to obtain also transient regime samples. Table 2 summarizes the conditions under which the tests were conducted.

Table 2. Summary of the tests performed for the experimental validation.

Regime	Supply		Load	Tests
Steady	DOL	50 Hz	Constant	12
	VSD	25 Hz	Constant	24
		50 Hz	Constant	24
Transient	DOL	50 Hz	Pulse	8
			Ramps	11
	VSD	25 Hz	Pulse	15
			Ramps	19
		50 Hz	Pulse	15
			Ramps	19
		Ramps	Constant	60
			Pulse	33
			Ramps	12

The supra-system developed in this work has explored and optimized a search space given by all the possible combinations of the following diagnostic systems options:

- two HOTA implementations (with STFT or SFFT).
- two SVM expert system variants (with Polynomial or RBF Kernel).
- two ANN expert system variants (one or two hidden layers)

By an optimized search in this search space, applying the techniques introduced in Section 4.1, the global optimum result (the best fault diagnostic system) is obtained, that is, the optimized expert system for IMs fault diagnosis which has the highest diagnosis hit rate.

5.3. Time Required by the Supra-System to Generate Locals and Global Optimized IMs Fault Diagnostic Systems

In this experimental test, 286044 SVM expert systems and 252000 ANN expert systems have been analyzed in the process of finding the final optimum fault diagnostic system by the supra-system. The total time used for training (Table 3) and for classification of the IM condition (Table 4) shows that the fastest approach is to use the SFFT technique for implementing the HOTA method.

Table 3. Time needed by the supra-system to generate each local optimal diagnostic system (hours).

Optimization Time (hours)	HOTA + STFT	HOTA + SFFT
SVM with Polynomial Kernel	13.81	2.70
SVM with RBF Kernel	12.57	1.83
ANN with 1 hidden layer	13.58	3.03
ANN with 2 hidden layers	34.02	20.16

Table 4. Time used by the diagnostic system generated by the supra-system to classify the IM condition (seconds).

Classification Time (seconds)	HOTA + STFT	HOTA + SFFT
SVM with Polynomial Kernel	36.64	5.60
SVM with RBF Kernel	36.64	5.60
ANN with 1 hidden layer	36.65	5.60
ANN with 2 hidden layers	36.65	5.60

5.4. Hit Rates Obtained by the Expert Systems for IMs Fault Diagnosis Generated by the Supra-System

The hit rates obtained by the expert systems generated by the supra-system Table 5, are very high, giving a very efficient IM fault diagnostic system. It is relevant that, although SVM and ANN expert systems have different implementations, in both cases the hit rates are very similar. This similarity means that:

- In both expert systems the maximum hits ratio has been reached (the optimal diagnostic system).
- None of these local diagnostic system shows over-training. Otherwise it would show a higher hit ratio much closer to 100% and hit ratios values would be more different between local diagnosis systems.

Table 5. Hit rates obtained in each of the local optimal diagnostic systems generated by the supra-system.

Hit Rate (%)	HOTA + STFT	HOTA + SFFT
SVM with Polynomial Kernel	98.62	98.89
SVM with RBF Kernel	97.24	97.38
ANN with 1 hidden layer	97.10	97.38
ANN with 2 hidden layers	98.89	98.89

HOTA implementing SFFT with SVM with Polynomial Kernel and HOTA implementing SFFT with ANN with 2 hidden layers lead to the same result. In both cases a global optimum diagnostic system is obtained, with the best hit ratio and with the lowest diagnosis time. Therefore the use of one or another for diagnosis would be only a user decision.

Even so, the others diagnostic systems (optimal local but not optimal global) also show an excellent hit ratio with a low diagnostic time. All of them are very well optimized and could be used alike.

6. Conclusions

In this research a supra-system implementation has been proposed to generate an optimized fault diagnostic system. This supra-system has shown the following advantages regarding to traditional expert systems:

- All the diagnostic systems generated are suitable to be used in transient regime operation.
- The generation process is totally automated. That is, starting on the samples input until finishing the generation of optimum diagnostic system no user intervention is required at all.
- The generation process is totally autonomous. That is, it is not necessary to carry out any control or adjustment task on the supra-system for a successful development of the optimum diagnostic system.

This supra-system has been experimentally tested and validated, confirming that it achieves the proposed goals. The supra-system approach solves a practical industrial problem in the field of IMs fault diagnosis area, especially in transient regime, where the generation of a conventional expert diagnostic system must be manually customized for each specific motor structure.

The application of the supra-system approach to the design of IMs fault diagnostic systems addressing other types of faults (bearings faults, eccentricity, etc.) is straightforward by using the fault features of this types of fault which can be extracted with the same methods as those proposed in this paper.

Author Contributions: Conceptualization, J.M.-R. and M.R.-G.; Data curation, J.B.-V. and J.M.-R.; Formal analysis, J.B.-V. and R.P.-P.; Investigation, A.S.-B. and J.P.-C.; Methodology, J.B.-V., R.P.-P. and J.P.-C.; Project administration, M.P.-S.; Resources, M.P.-S.; Supervision, R.P.-P., M.P.-S. and M.R.-G.; Validation, J.B.-V., J.M.-R., A.S.-B., J.P.-C. and M.R.-G.; Writing—original draft, J.B.-V. and M.P.-S.; Writing—review and editing, A.S.-B., M.P.-S. and M.R.-G. These authors contributed equally to this work.

References

1. Puche-Panadero, R.; Pineda-Sanchez, M.; Riera-Guasp, M.; Roger-Folch, J.; Hurtado-Perez, E.; Perez-Cruz, J. Improved resolution of the MCSA method via Hilbert transform, enabling the diagnosis of rotor asymmetries at very low slip. *IEEE Trans. Energy Convers.* **2009**, *24*, 52–59. [CrossRef]
2. Abd-el Malek, M.; Abdelsalam, A.K.; Hassan, O.E. Induction motor broken rotor bar fault location detection through envelope analysis of start-up current using Hilbert transform. *Mech. Syst. Signal Process.* **2017**, *93*, 332–350. [CrossRef]
3. Martinez, J.; Belahcen, A.; Muetze, A. Analysis of the vibration magnitude of an induction motor with different numbers of broken bars. *IEEE Trans. Ind. Appl.* **2017**, *53*, 2711–2720. [CrossRef]
4. Sapena-Bano, A.; Pineda-Sanchez, M.; Puche-Panadero, R.; Perez-Cruz, J.; Roger-Folch, J.; Riera-Guasp, M.; Martinez-Roman, J. Harmonic order tracking analysis: A novel method for fault diagnosis in induction machines. *IEEE Trans. Energy Convers.* **2015**, *30*, 833–841. [CrossRef]
5. Sapena-Bano, A.; Burriel-Valencia, J.; Pineda-Sanchez, M.; Puche-Panadero, R.; Riera-Guasp, M. The Harmonic Order Tracking Analysis Method for the Fault Diagnosis in Induction Motors Under Time-Varying Conditions. *IEEE Trans. Energy Convers.* **2017**, *32*, 244–256. [CrossRef]
6. Burriel-Valencia, J.; Puche-Panadero, R.; Martinez-Roman, J.; Sapena-Bano, A.; Pineda-Sanchez, M. Short-Frequency Fourier Transform for Fault Diagnosis of Induction Machines Working in Transient Regime. *IEEE Trans. Instrum. Meas.* **2017**, *66*, 432–440. [CrossRef]
7. Yin, Z.; Hou, J. Recent advances on SVM-based fault diagnosis and process monitoring in complicated industrial processes. *Neurocomputing* **2016**, *174*, 643–650. [CrossRef]
8. Bazan, G.H.; Scalassara, P.R.; Endo, W.; Goedtel, A.; Godoy, W.F.; Palácios, R.H.C. Stator fault analysis of three-phase induction motors using information measures and artificial neural networks. *Electr. Power Syst. Res.* **2017**, *143*, 347–356. [CrossRef]

9. Beale, M.H.; Hagan, M.T.; Demuth, H.B. Neural network toolbox 7. In *User's Guide*; MathWorks: Natick, MA, USA, 2010.

10. Hsu, C.W.; Chang, C.-C.; Lin, C.-J. *A Practical Guide to Support Vector Classification*; Technical Report; Department of Computer Science, National Taiwan University: Taipei City, Taiwan, 2013.

11. Bishop, C.M. *Neural Networks for Pattern Recognition*; Oxford University Press: Oxford, UK, 1995.

12. Mustafidah, H.; Hartati, S.; Wardoyo, R.; Harjoko, A. Selection of Most Appropriate Backpropagation Training Algorithm in Data Pattern Recognition. *Int. J. Comput. Trends Technol.* **2014**, *2*, 92–95. [CrossRef]

13. Godoy, W.F.; da Silva, I.N.; Goedtel, A.; Palácios, R.H.C.; Lopes, T.D. Application of intelligent tools to detect and classify broken rotor bars in three-phase induction motors fed by an inverter. *IET Electr. Power Appl.* **2016**, *10*, 430–439. [CrossRef]

14. Ghorbanian, V.; Faiz, J. A survey on time and frequency characteristics of induction motors with broken rotor bars in line-start and inverter-fed modes. *Mech. Syst. Signal Process.* **2015**, *54*, 427–456. [CrossRef]

15. Valles-Novo, R.; de Jesus Rangel-Magdaleno, J.; Ramirez-Cortes, J.M.; Peregrina-Barreto, H.; Morales-Caporal, R. Empirical mode decomposition analysis for broken-bar detection on squirrel cage induction motors. *IEEE Trans. Instrum. Meas.* **2015**, *64*, 1118–1128. [CrossRef]

16. Vitor, A.L.; Scalassara, P.R.; Endo, W.; Goedtel, A. Induction motor fault diagnosis using wavelets and coordinate transformations. In Proceedings of the 2016 12th IEEE International Conference on Industry Applications (INDUSCON), Curitiba, Brazil, 20–23 November 2016; pp. 1–8.

17. De Santiago-Perez, J.J.; Rivera-Guillen, J.R.; Amezquita-Sanchez, J.P.; Valtierra-Rodriguez, M.; Romero-Troncoso, R.J.; Dominguez-Gonzalez, A. Fourier transform and image processing for automatic detection of broken rotor bars in induction motors. *Meas. Sci. Technol.* **2018**, *29*, 095008. [CrossRef]

18. Perez-Cruz, J.; Perez-Vazquez, M.; Pineda-Sanchez, M.; Puche-Panadero, R.; Sapena-Bano, A. The Harmonic Order Tracking Analysis (HOTA) for the Diagnosis of Induction Generators Working Under Steady State Regime. In Proceedings of the 2017 Asia-Pacific Engineering and Technology Conference (APETC 2017), Kuala Lumpur, Malaysia, 25–27 May 2017; pp. 1864–1869.

19. Merabet, H.; Bahi, T.; Drici, D.; Halam, N.; Bedoud, K. Diagnosis of rotor fault using neuro-fuzzy inference system. *J. Fundam. Appl. Sci.* **2017**, *9*, 170–182. [CrossRef]

20. Riera-Guasp, M.; Pineda-Sanchez, M.; Pérez-Cruz, J.; Puche-Panadero, R.; Roger-Folch, J.; Antonino-Daviu, J.A. Diagnosis of induction motor faults via Gabor analysis of the current in transient regime. *IEEE Trans. Instrum. Measur.* **2012**, *61*, 1583–1596. [CrossRef]

21. Gambhir, A.; Vijarania, M.; Gupta, S. Implementation and Application of Binary Search in 2-D Array. *Int. J. Inst. Ind. Res.* **2016**, *1*, 30–31.

22. Gyftakis, K.N.; Cardoso, A.J.M.; Antonino-Daviu, J.A. Introducing the Filtered Park's and Filtered Extended Park's Vector Approach to detect broken rotor bars in induction motors independently from the rotor slots number. *Mech. Syst. Signal Process.* **2017**, *93*, 30–50. [CrossRef]

7

Improving Performance of Three-Phase Slim DC-Link Drives Utilizing Virtual Positive Impedance-Based Active Damping Control

Ahmet Aksoz [1,*], **Yipeng Song** [2], **Ali Saygin** [1], **Frede Blaabjerg** [2] and **Pooya Davari** [2,*]

[1] Department of Electrical and Electronics Engineering, Faculty of Technology, Gazi University, 06500 Ankara, Turkey; asaygin@gazi.edu.tr (A.S.)

[2] Department of Energy Technology, Aalborg University, AAU 9220 Aalborg East, Denmark; yis@et.aau.dk (Y.S.); fbl@et.aau.dk (F.B.)

* Correspondence: ahmetaksoz@gazi.edu.tr (A.A.); pda@et.aau.dk (P.D.)

Abstract: In this paper, a virtual positive impedance (VPI) based active damping control for a slim DC-link motor drive with 24 section space vector pulse width modulation (SVPWM) is proposed. Utilizing the proposed control and modulation strategy can improve the input of current total harmonic distortion (THD) while maintaining the cogging torque of the motor. The proposed system is expected to reduce the front-end current THD according to international standards, as per IEC 61000 and IEEE-519. It is also expected to achieve lower cost, longer lifetime, and fewer losses. A permanent magnet synchronous motor (PMSM) is fed by the inverter, which adopts the 24 section SVPWM technique. The VPI based active damping control for the slim DC-link drive with/without the 24 section SVPWM are compared to confirm the performance of the proposed method. The simulation results based on MATLAB are provided to validate the proposed control strategy.

Keywords: slim DC-link drive; VPI active damping control; total harmonic distortion; cogging torque

1. Introduction

In many industrial applications, slim DC-link drives have become increasingly favored day by day. A classical driver consists of a 6-pulse diode bridge rectifier, an intermediate circuit with a big capacitor, an inductor, and an inverter. To maintain stable DC-link voltage, the DC-link capacitor needs to be carefully selected. Although the big size capacitor with large capacitance is at a higher cost and shorter lifetime, it has strong robustness against the stability problem. However, cost, lifetime, and loss must be taken into consideration for industrial applications.

Thus, using a film capacitor as the slim DC-link capacitor in drivers is preferred, in spite of the stability problem. A diode rectifier based slim DC-link drive is shown in Figure 1. This grid-connected driver has a diode rectifier, a slim DC capacitor, and a 6-switches three-phase inverter. Additionally, point of common coupling (PCC) phase currents for stiff grid and weak grid are given in Figure 2 at different operation speeds.

The i_pcc simulated waveforms show that stiff grid and high operation speed (c) is the best current waveform. However, it can be improved with control methods, especially for weak grid conditions.

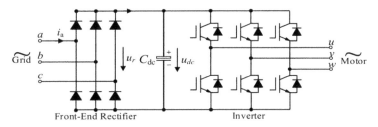

Figure 1. Diode rectified slim DC-link drive.

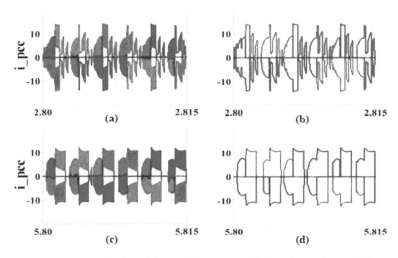

Figure 2. PCC phase current. (**a**) Stiff grid at 1500 rpm. (**b**) Weak grid at 1500 rpm. (**c**) Stiff grid at 3000 rpm. (**d**) Weak grid at 3000 rpm.

To solve the stability problem, Virtual Positive Impedance (VPI) based active damping control has been effectively implemented [1,2]. Active damping control for slim DC-link drive ensures that no extra passive damping component is needed. The negative impedance instability is described as:

$$\frac{-P_L}{v_{dc}^2} = \frac{-i_{dc}v_{dc0}}{v_{dc}(v_{dc0} - \widetilde{v})} = \frac{-(i_{dc0} + \widetilde{i_{dc}})v_{dc0}}{v_{dc}(v_{dc0} - \widetilde{v})} \tag{1}$$

where P_L is the load power, v_{dc0} is the DC component of the DC-link voltage, $\widetilde{v}$ is the AC part of the DC-link voltage, and v_{dc} is the DC-link voltage [1]. i_{dc} is the DC-link current, $\widetilde{i_{dc}}$ is the AC part of the DC-link current, and i_{dc0} is the DC component of the DC-link current. In contrast to the case using the big capacitor [3], the constant power load behavior of the motor with a slim-DC-link capacitor causes the larger ripple on the DC voltage. Both ripples on the DC-link voltage and the front-end current harmonics are higher when using a small capacitor [4,5]. In order to reduce the input current harmonics, VPI based active damping control decreases the ripple on the DC-link voltage [1–4]. A virtual positive impedance block diagram is illustrated in Figure 3 [1]. This model can be used for the generation of the DC-link voltage reference. In order to control the AC component of the DC-link voltage, the 1st-order High Pass Filter is represented by a high pass filter (HPF) block. g_v is the gain on the accompaniment of the DC-link voltage. In spite of the fact that VPI contains an HPF block, it resembles the 1st-order Low Pass Filter and harmonic detection block. Additionally, this block diagram can be used for detecting the harmonic [5]. The harmonic detection is also ensured by the VPI method, providing the lower ripples on the DC-link voltage. Harmonic mitigation, harmonic cancellation, or generally a harmonic problem is an important issue for motor drivers [6,7]. This problem can deteriorate grid voltage quality, as well as the performance of both the driver and the load. Although it cannot be completely removed, it needs to be mitigated as much as possible. For this purpose, several control techniques and PWM techniques have been studied [8–10].

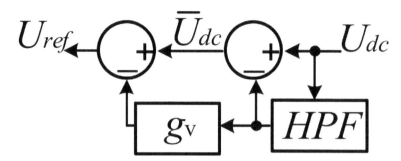

Figure 3. Virtual positive impedance block diagram [1].

At the same time, not only ripple on the DC-link voltage on a small capacitor, but also a motor cogging torque due to interactions between core and magnet result in both grid input current harmonics and motor current harmonics. Owing to the harmonic problem, active damping control (ADC) and VPI can be used to decrease THD. Even through these methods achieve harmonics suppression, the cogging torque also needs to be solved by harmonic effect, because motor current harmonics cause the higher cogging torque [11–16]. When the 3DSVPWM aims to optimize switching waveforms, it can achieve a lower cogging torque. The algorithm of the 3DSVPWM was based on four steps. Firstly, the reference vector was transformed into 2D. In addition, the length of the reference vector was described according to the length of the basis vectors. Secondly, the closest three vectors were found. When they were detected, finding high–low values of the reference vector coordinates could be facilitated. Duty cycles were calculated in the third step. Lastly, the best switching states were selected when 2D coordinates are transformed to 3D coordinates [9,12].

In this study, the DC-link voltage, the grid input current THD, the VPI bode results, the cogging torque, and the THD of motor currents were simulated in MATLAB (R2016b, MathWorks, Natick, MA, USA), where four simulation models were developed: (1) Weak grid without VPI without 3DSVPWM (wOVPIwO3D), (2) weak grid with VPI without 3DSVPWM (wVPIwO3D), (3) weak grid with VPI with 3DSVPWM (wVPIw3D), and (4) stiff grid with VPI with 3DSVPWM (stiffwVPIw3D).

Simulation results of these models are compared and discussed. Section 2 analyzes the interaction between the cogging torque and the current harmonic. The 3DSVPWM, the PMSM model, and the input admittance are described. Then, virtual positive impedance based active damping control is given in Section 3. In Section 4, the performance analysis of the DC-link current THD and the motor current THD of the slim DC-link capacitor is shown. In addition, the stability analysis, the control structure, and the control impedance Y_{ctrl} are explained in the same section. Additionally, the simulation validation of the grid input current THD_i and the cogging torque are obtained. Lastly, the study is summarized in Section 5.

2. Interaction between Cogging Torque and Harmonic

2.1. The Cogging Torque Reduction Methods

In order to decrease the cogging torque, some methods are used. These are mainly:

- Skewing stator stack or magnets;
- Modulation drive current waveform;
- Using fractional slots per pole;
- Optimizing the magnet pole arc or width [9,17].

The schema of the cogging torque reduction is illustrated in Figure 4. In order to obtain a better modulation drive current waveform, there are three main methods. They are decreasing harmonics, switching at high frequency, and using advanced PWM techniques [9,17].

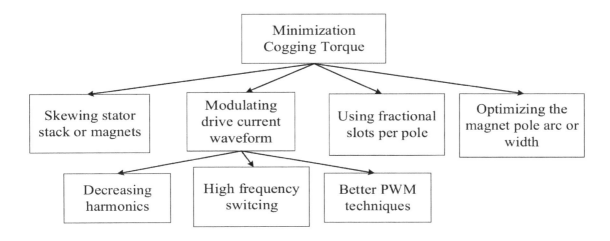

Figure 4. Reducing cogging torque schema [17].

On the other hand, the kth harmonic is related to the cogging torque, as expressed in (2). T_{ck} is the amplitude of the kth harmonic component of the cogging torque, θ is the angle of rotation, and k is the order of cogging harmonics.

$$T_{ck}(\theta) = \sum_{n=-\infty}^{\infty} T_n e^{2ni(\theta - k\theta_s)} \tag{2}$$

where T_n is the Fourier series coefficient and θ_s is the electrical angle slot pitch. It is expressed as:

$$\theta_s = \pi N_m / N_s \tag{3}$$

where N_m is the number of the magnet pole and N_s is the number of the slot. Accordingly, the cogging torque T_{cog} can be written as the Fourier series as:

$$T_{cog}(\theta) = \sum_{k=0}^{N_s-1} T_{ck}(\theta) \tag{4}$$

The proposed models are applied not only for achieving a reduced cogging torque, but also improving system stability thanks to decreased harmonics. Given the fact that the T_{ck} is decreased, the cogging torque can be reduced.

2.2. 3DSVPWM Technique

In the proposed modulation technique, there are 24 sectors, including zero voltage vectors [9]. The modulation space is divided into 6 sections (S1–S6), each section consisting of 4 delta sectors ($\Delta 1, \Delta 2, \Delta 3, \Delta 4$).

This proposed modulation technique is adopted in the 3-phase 3-level or multilevel inverter. However, it is used in this study in the 3-phase 2-level inverter. Thus, modulation angles are made smaller and the number of the sector is increased in sections. Here, using the definition of the vector norm, the vectors of the inverter are defined in a plane as given in (5):

$$v_{ab} + v_{bc} + v_{ca} = 0 \tag{5}$$

where v_{ab}, v_{bc}, and v_{ca} are the vectors of the inverter in the 3D coordinate system. The switching state vectors are shown in Figure 5 [18,19]. Additionally, the numbers of 0, 1, and 2 in Figure 5 represent V_{ab}, V_{bc}, or V_{ca}/V_{dc}.

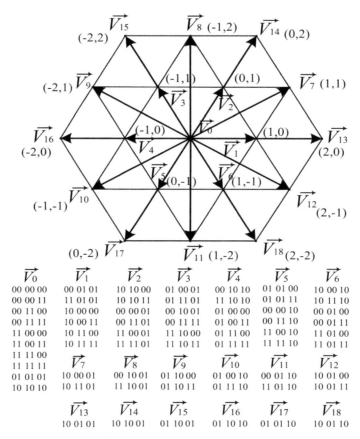

$\vec{V}_0$	$\vec{V}_1$	$\vec{V}_2$	$\vec{V}_3$	$\vec{V}_4$	$\vec{V}_5$	$\vec{V}_6$
00 00 00	00 01 01	10 10 00	01 00 01	00 10 10	01 01 00	10 00 10
00 00 11	11 01 01	10 10 11	01 11 01	11 10 10	01 01 11	10 11 10
00 11 00	10 00 00	00 00 01	00 10 01	01 00 00	00 00 10	00 01 00
00 11 11	10 00 11	00 11 01	00 11 11	01 00 11	00 11 10	00 01 11
11 00 00	10 11 00	11 00 01	11 10 00	01 11 00	11 00 10	11 01 00
11 00 11	10 11 11	11 11 01	11 10 11	01 11 11	11 11 10	11 01 11
11 11 00						
11 11 11	$\vec{V}_7$	$\vec{V}_8$	$\vec{V}_9$	$\vec{V}_{10}$	$\vec{V}_{11}$	$\vec{V}_{12}$
01 01 01	10 00 01	00 10 01	01 10 00	01 00 10	00 01 10	10 01 00
10 10 10	10 11 01	11 10 01	01 10 11	01 11 10	11 01 10	10 01 11

$\vec{V}_{13}$	$\vec{V}_{14}$	$\vec{V}_{15}$	$\vec{V}_{16}$	$\vec{V}_{17}$	$\vec{V}_{18}$
10 01 01	10 10 01	01 10 01	01 10 10	01 01 10	10 01 10

Figure 5. Space vector diagram of 3DSVPWM in 2D and 3D vectors after 3D transformation.

According to (5) and Figure 5, the delta vectors can be expressed in the 2D coordinate system as:

$$\vec{V}_{1(v_{ab},v_{bc},v_{ca})},\ \vec{V}_{2(v_{ab},v_{bc},v_{ca})} = \begin{bmatrix} V_{dc} \\ 0 \\ -V_{dc} \end{bmatrix},\ \begin{bmatrix} 0 \\ V_{dc} \\ -V_{dc} \end{bmatrix} \qquad (6)$$

where, $\vec{V}_{1(vab,\ vbc,\ vbc)}$ and $\vec{V}_{2(vab,\ vbc,\ vbc)}$ are the delta vectors. They are the transformed vectors from 3D to 2D. The reference vector can be placed in a sector and the switching-state vector is shown at the corner of each sector. It is able to produce switching-state vectors in 2D, which are then transformed into 3D switching-state vectors.

This 24 sectors SVPWM technique can eliminate the need for dead-time protection and allow the upper and lower switches to switch at the same time. In this case, the dead-time effect is removed, an additional midpoint voltage is generated, and the effective output switching frequency is doubled. Thus, the current harmonics in the output current waveform are significantly suppressed by applying the three-level voltage output and doubling the effective switching frequency. In order to reduce the current harmonics, the adjacent three vectors and the reference vector must be defined in the best way. V_x, V_y, and V_z are the adjacent three vectors as follow:

$$T_s = d_x + d_y + d_z \qquad (7)$$

$$\vec{V}_{ref} = d_x\vec{V}_x + d_y\vec{V}_y + d_z\vec{V}_y \qquad (8)$$

where the dwell time of vectors are d_x, d_y, and d_z, respectively. The reference vector is determined in the hexagon to state which triangle will be used. The biggest difference between 3DSVPWM and classical SVPWM is dwell times: The 3DSVPWM provides better dwell times for switching angles.

2.3. PMSM Model

The PMSM motor is modeled in the dq reference frame, which relies on the field oriented control (FOC), and the mathematical equations are given below:

$$\begin{bmatrix} v_{sd} \\ v_{sq} \end{bmatrix} = \begin{bmatrix} R + sL_{sd} & -\omega_r L_{sq} \\ \omega_r L_{sd} & R + sL_{sq} \end{bmatrix} \begin{bmatrix} i_{sd} \\ i_{sq} \end{bmatrix} + \begin{bmatrix} 0 \\ \omega_r \lambda \end{bmatrix} \tag{9}$$

$$T_e = \frac{3}{2} P(\lambda i_{sq} + (L_{sd} - L_{sq}) i_{sd} i_{sq}) \tag{10}$$

where the R, λ, P, L_{sd}, L_{sq}, T_e, and ω_r represent the stator resistor, the flux produced by the permanent magnets, the number of pole pairs, the stator inductances in the dq-frame, the electrical torque, and the rotor speed individually. In addition, v_{sd}, v_{sq}, i_{sd}, and i_{sq} represent the stator voltages and the stator currents in the dq-frame, respectively. Due to the fact that the speed and current loops force the stator current i_{sd} to be 0, (9) can be rewritten as:

$$V_{sd} = -\omega_r L_{sq} i_{sq} \tag{11}$$

$$V_{sq} = R i_{sq} + s L_{sq} i_{sq} + \omega_r \lambda \tag{12}$$

2.4. Input Admittance

Using the above equations, the input admittance of the control block and the constant power load are specified as follow:

$$G_{iq} = \frac{1}{Z_q + F_{iq}} \frac{V_q}{V_{dc}} (1 + g_v DA) \tag{13}$$

$$G_{id} = \frac{1}{Z_d + F_{id}} \frac{V_q}{V_{dc}} (1 + g_v DA) \tag{14}$$

$$G_{vd} = \frac{3}{2} (Z_d I_d + V_d + \omega_r L_d I_q \\ + \frac{3}{2} \frac{(L_d - L_q)^2 I_d I_q^2 N_{pp}^2}{Js}) \tag{15}$$

$$G_{vq} = \frac{3}{2} (Z_q I_q + V_q - \omega_r L_q I_d \\ + \frac{3}{2} \frac{(L_d - L_q)^2 I_q I_d^2 N_{pp}^2}{Js}) \tag{16}$$

$$Y_{in} = \frac{1}{Z_{in}} = \frac{G_{vd} G_{id} + G_{vq} G_{iq}}{V_{dc}} + \frac{-P_L}{V_{dc}^2} \tag{17}$$

$$1/Y_{ctrl} = \frac{V_{dc}}{G_{vd} G_{id} + G_{vq} G_{iq}} \tag{18}$$

Y_{ctrl} is the admittance of the control part with VPI based active damping control and Y_{cpl} $(-P_L/V_{dc}^2)$ is the admittance of the constant power load behavior. D, A, N_{pp}, J, Z_{dq}, F_{id}, and F_{iq} relate the PWM delay, the 1st-order HPF, the pole-pairs, the inertia, the dq-axis impedance of the PMSM, and the current controller of the dq-axis separately. Two input admittances in (17) are illustrated with an equivalent DC-link circuit in Figure 6, which is on simplified equivalent circuit model of the diode rectified based slim DC-link drive.

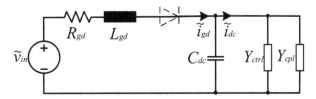

Figure 6. Simplified equivalent DC-link in a drive unit.

3. Virtual Positive Impedance Based Active Damping Control

The parameters of the drive and the PMSM are given in Table 1. Moreover, the sample period is Ts, the reference torque of 3 ph trapezoidal motor is Tm, C_{dc} is the slim DC-link capacitor value, SCR is the short circuit ratio, and the stator resistive and inductive values are R and $Ld–Lq$. In addition, the current loop, the speed loop, and PWM block are shown in Figure 7.

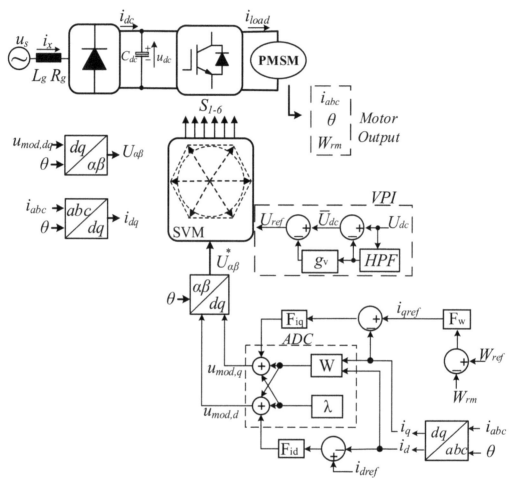

Figure 7. Control system block diagram with Virtual Positive Impedance (VPI) based active damping control (ADC).

Table 1. The parameters of the drive and the PMSM.

Parameters	Values
V_{rms}	400 V/3 ph
C_{dc}	14 uF
SCR (weak grid-stiff grid)	35–350
T_s	50 us
T_m	4 Nm
R	150 mΩ
$L_d–L_q$	8.5–8.8 mH
ω_r	1500–3000 rpm

The control diagram includes the Space Vector Modulation (SVM) block, together with a VPI and an ADC block, the Park and the inverse Park transformation, the speed controller F_ω, the current controller F_{iq} and F_{id} together with the decoupling block W. All the components are assumed as ideal. The power loss and the saturation effects in the drive system are neglected. The VPI is illustrated in Figure 3; U_{dc} is the voltage feedback from the slim DC-link capacitor, which is used for voltage ripple elimination. Then, U_{ref} is the control reference of the V_{dc}, which is calculated based on the reference voltages is 400 V, and the sum of the reversed voltage is $-U_{dc}$.

Furthermore, the speed closed-loop control and the current closed-loop control are shown in Figure 7 in the dq-frame. W is the decoupling function block presented in Equation (19) and ω_r is the rotor speed. The current control equation is shown in Equation (20) and the speed control equation is shown in Equation (21) as follows:

$$W = \begin{bmatrix} 0 & -\omega_r L_q \\ \omega_r L_d & 0 \end{bmatrix} \tag{19}$$

$$u_{\text{mod},dq}(t) = K_p\left[i_{dq,ref}(t) - i_{dq}(t)\right] \\ +K_i\left[i_{dq,ref}(t) - i_{dq}(t)\right] + 2\lambda + i_{dq}(t)W \tag{20}$$

$$i_q(t) = K_p\left[(\omega_{ref}(t) - \omega_r(t))\right] + K_i T_s(\omega_{ref}(t) - \omega_r(t)) \tag{21}$$

4. Performance Analysis of The DC-Link Voltage THD and the Motor Current THD

According to the Equation (18), bode diagrams of the $1/Y_{ctrl}$ at 1500 rpm and 3000 rpm are given in Figures 8 and 9.

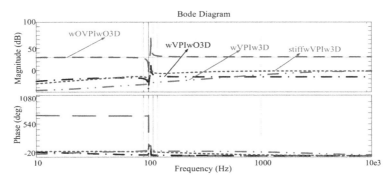

Figure 8. Bode plot of $1/Y_{ctrl}$ at 1500 rpm (Blue: wOVPIwO3D. Black: wVPIwO3D. Red: wVPIw3D. Purple: stiffwVPIwO3D).

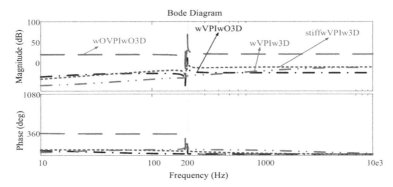

Figure 9. Bode plot of $1/Y_{ctrl}$ at 3000 rpm (Blue: wOVPIwO3D. Black: wVPIwO3D. Red: wVPIw3D. Purple: stiffwVPIwO3D).

The Bode diagrams of the control impedance of the four cases can be seen at 1500 rpm and 3000 rpm. In the case of not using VPI based ADC and 3DSVPWM that is wOVPIwO3D (blue), the impedance magnitude behaves flat, but it does not reach zero. By contrast, in the case of the wVPIwO3D (black), wVPIw3D (red), and stiff wVPIw3D (purple), the magnitude of the control impedance becomes lower than that without active damping in the frequency range. The resonance is named as Negative-Impedance (NI) resonance due to the frequency character decided by NI at the constant power load (CPL) situation. Its impedance characteristic behaves as an inductive plus negative-resistive impedance during [10, 5000] Hz. This is helpful in suppressing the harmonics, caused by the resonance between L_{gd} and C_{dc}. In order to improve the THD (lower ripple on magnitude), impedance is increased at the current controller bandwidth. Increased impedance with the bandwidth of the current controller is helpful for suppressing the current harmonics (100 Hz and 200 Hz). Additionally, control impedance always behaves as positive-resistive plus inductive at high frequency while capacitive at low frequency. This positive-resistive characteristic helps to damp the system into a stable state.

4.1. The Performance Analysis of the DC-Link Voltage THD and the Motor Current THD

According to the VPI based ADC, the performance analysis of the DC-link voltage THD and the motor current THD is presented. Owing to the fact that the big size capacitor or RLC components have a higher cost and shorter lifetime, using the film capacitor as the slim DC-link capacitor in drivers can be a good alternative [1]. In spite of the stability problem, the DC-link voltage of the slim capacitor is controlled well with VPI based ADC and 3DSVPWM. The DC-link voltage performances of the four cases are given below. The DC-link voltage when rotor speed is 1500 rpm is shown in Figure 10 and the DC-link voltage when rotor speed is 3000 rpm is shown in Figure 11. Firstly, the motor is operated at 1500 rpm from 0 s to 3 s, and then it is operated at 3000 rpm from 3 s to 6 s. However, time periods of the simulation are only 2.8–2.815 s and 5.8–5.815 s, because the results of the simulation are the same during 0s to 3 s and 3 s to 6 s. Thus, 2.8–2.815 s as TP1 (time period 1) and 5.8–5.815 as TP2 (time period 2) are used.

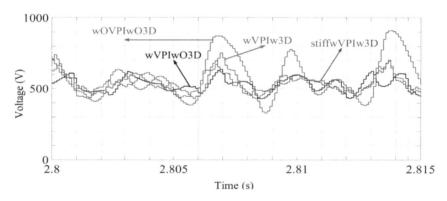

Figure 10. DC-link voltage when rotor speed is 1500 rpm.

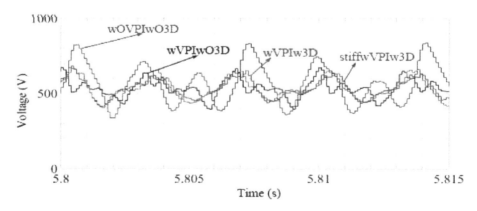

Figure 11. DC-link voltage when rotor speed is 3000 rpm.

The ripples on the DC-link voltage when rotor speed is 1500 rpm and 3000 rpm are given in Table 2.

Table 2. The ripples of the DC-link voltage.

Cases	Voltage (TP1)	Voltage (TP2)
wOVPIwO3D	240.3	190.8
wVPIwO3D	158.9	144.2
wVPIw3D	225.0	169.6
stiffwVPIwO3D	82.1	76.79

In addition, the fast Fourier transform (FFT) results of the DC-link voltage when rotor speed is 1500 rpm and 3000 rpm are given in Table 3.

Table 3. The FFT results of the DC-link voltage.

Cases	THD% (TP1)	THD% (TP2)
wOVPIwO3D	34.34	32.84
wVPIwO3D	13.15	15.80
wVPIw3D	15.93	17.01
stiffwVPIwO3D	15.93	17.02

As shown in Table 2, the case of stiffwVPIwO3D has the best performance, as expected, with the lowest ripples on the DC-link voltage for both operation speeds as 82.1 V and 76.79 V. However, wVPIwO3D with a weak grid has the best THD results for both operation speeds according to Table 3, as 13.15% at TP1 and 15.80% at TP2. Additionally, the ripples on DC-link voltage wVPIwO3D are obviously better than those on wVPIw3D (158.9–225.0 V and 144.2–169.6 V). The ripple on the DC-link voltage of the VPI based ADC with traditional SVPWM can oscillate. Lower ripples on DC-link voltage are obtained as 158.9 V at TP1 and 144.2 V at TP2. This means that using both the 3DSVPWM and the VPI based ADC does not provide better results of the ripple on DC-link voltage. Moreover, the FFT results of the motor current harmonics are displayed in Table 4.

Table 4. The FFT results of the motor current.

Cases	THD% (TP1)	THD% (TP2)
wOVPIwO3D	41.36%	29.82%
wVPIwO3D	27.83%	20.46%
wVPIw3D	10.74%	12.07%
stiffwVPIw3D	10.84%	11.95%

From Table 4, the motor current harmonics of wVPIw3D with a weak grid or stiff grid are acceptable. When the 3DSVPWM is enabled, the motor current harmonics are suppressed effectively as 10.74% at TP1 and 12.07% at TP2 for Case 3 and 10.84% at TP1 and 11.95% at TP2 for Case 4.

The motor current (i_abc) waveforms of 4 cases are given in Figure 12. As seen there, Case 4 supplies the best results (g and h), thanks to virtual positive impedance-based active damping control and 3DSVPWM under the stiff grid. Then, Case 3 gives good results (e and f), thanks to virtual positive impedance-based active damping control and 3DSVPWM under the weak grid. Motor currents without 3DSVPWM means are seen in Case 2 (c and d). Lastly, motor currents with classical SVPWM without VPI based ADC means (a and b) are given in Case 1.

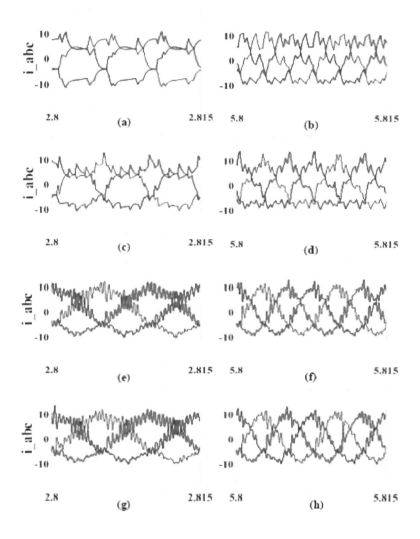

Figure 12. Motor current waveforms. (**a**) wOVPIwO3D at TP1. (**b**) wOVPIwO3D at TP2. (**c**) wVPIwO3D at TP1. (**d**) wVPIwO3D at TP2. (**e**) wVPIw3D at TP1. (**f**) wVPIw3D at TP2. (**g**) stiffwVPIw3D at TP1. (**h**) stiffwVPIw3D at TP2.

4.2. The Performance Analysis of the Grid Current THD and the Cogging Torque

Table 5 shows the analysis of the grid current THD according to the four cases of simulation results.

Table 5. The grid input current FFT results.

Cases	THD% (TP1)	THD% (TP2)
gridCurrent wOVPIwO3D	110.71%	98.45%
gridCurrent wVPIwO3D	54.53%	51.60%
gridCurrent wVPIw3D	47.74%	43.88%
gridCurrent stiffwVPIwO3D	48.60%	43.48%

As shown in Table 5, the grid input current FFT results are shown when the drive load is 3 kW. The FFT results of the grid input current of Case 1 are not as expected. When the 3DSVPWM is enabled, the THD_i decreases from 51.60% to 43.88% at TP2, and it also decreases from 54.53% to 47.74% at TP1. Although Case 4 has a better result than Case 3 at 3000 rpm, the result of Case 4 gives worse THD_i than Case 3 at 1500 rpm. The grid input current harmonics with the VPI based ADC and with the 3DSVPWM in Case 3 or Case 4 (with a weak grid or stiff grid) are acceptable. When the four cases are compared, the THD_i results of Case 3 and Case 4 are rather desirable for both operation speeds.

The cogging torque results are given in Table 6. As seen in Table 6, the cogging torque clearly decreases when adopting the 3DSVPWM. When Case 2 and Case 3 are compared, the cogging torque values get lower, from 0.25170 Nm to 0.15931 Nm at 3000 rpm and from 0.33270 Nm to 0.17871 Nm. Moreover, since the stiff grid is used, these results are 0.15285 Nm at 3000 rpm and 0.16720 Nm at 1500 rpm. In addition, the worst results are 0.28560 Nm at 3000 rpm and 0.44510 Nm at 1500 rpm from Case 1. These results are also seen in Figures 13 and 14.

Table 6. The cogging torque results.

	Cogging Torque (Nm) (TP1)	Cogging Torque (Nm) (TP2)
wOVPIwO3D	0.44510	0.29560
wVPIwO3D	0.33270	0.25170
wVPIw3D	0.17871	0.15931
stiffwVPIwO3D	0.16720	0.15285

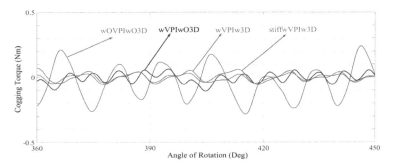

Figure 13. Cogging torque at 1500 rpm.

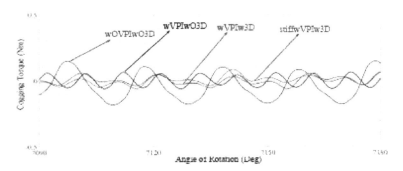

Figure 14. Cogging torque at 3000 rpm.

It can be seen that the cogging torque results are higher at a lower operation speed. The cogging torque results of Case 3 and Case 4 are more preferable than those of Case 1 and Case 2. Because the cogging torque is an important problem at low speed, the performance of the cogging torque in Case 3 and Case 4 are desired, especially at lower speed. At the same time, the cogging torque results of Case 3 and Case 4 at higher speed are better than those of the other cases. When both results in the tables and the figures are compared under either a weak or stiff grid, the adoption of the VPI based ADC and the 3DSVPWM together gives better results. The VPI based ADC ensures better harmonics, using a more advanced modulation technique, like 3DSVPWM (0.17871–0.15931 Nm and 0.16720–0.15285 Nm). Although the ripples on the DC-link voltage of the wVPIwO3D (Case 2) are lower without 3DSVPWM, the wVPIw3D (Case 3) and the stiffwVPIw3D (Case 4) are able to better suppress the grid current THD, the motor current THD, and the cogging torque.

After the results are obtained, all of them are given in Figure 15.

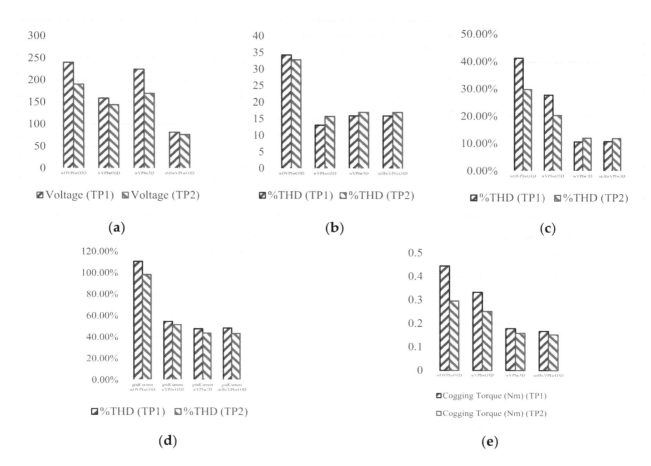

Figure 15. The performance difference between cases. (**a**) The ripples of the DC-link voltage. (**b**) The FFT results of the DC-link voltage. (**c**) The FFT results of the motor current. (**d**) The grid input current FFT results. (**e**) Cogging torque results.

The performance difference between cases is given in Figure 15. In this figure, the ripples of the DC-link voltage, the FFT results of the DC-link voltage, the FFT results of the motor current, the grid input current FFT results, and the cogging torque results are illustrated as bar graphics.

5. Conclusions

The motor drives equipped with the slim DC capacitor using the VPI based ADC and the 3DSVPWM are able to achieve lower input current harmonics and a lower cogging torque. In spite of the decreased ripples on DC-link voltages, the grid current harmonics and the motor current harmonics must be suppressed for the stable systems, because these harmonics result in a shorter lifetime for the driver and the cogging torque in the PMSM. However, a slim DC-link drive may not reach the expected performance, due to the weak grid condition. In this paper, the virtual positive impedance based active damping control and the 3DSVPWM are employed. Under either weak or stiff grid conditions, the four cases are clearly simulated. According to the simulation results, the performance of the VPI based ADC and the 3DSVPWM are investigated. In spite of the ripples on DC-link voltage, the 3DSVPWM with the VPI based ADC achieves harmonic suppression and a decreased cogging torque. Thus, a more stable and longer lifetime driver is obtained when the VPI based ADC and the 3DSVPWM are used.

Author Contributions: Conceptualization, F.B. and A.S.; Methodology, P.D; Software, A.A.; Validation, A.A., Y.S. and P.D.; Formal Analysis, A.A.; Investigation, P.D.; Resources, F.B. and P.D.; Data Curation, A.A.; Writing-Original Draft Preparation, A.A.; Writing-Review & Editing, P.D. and Y.S.; Visualization, Y.S.; Supervision, A.S. and F.B.; Project Administration, F. B. and A.S.; Funding Acquisition, A.S.

References

1. Yang, F.; Mathe, L.; Lu, K.; Blaabjerg, F.; Wang, X.; Davari, P. Analysis of harmonics suppression by active damping control on multi slim dc-link drives. In Proceedings of the IEEE IECON 2016, Florence, Italy, 24–27 October 2016; pp. 5001–5006.

2. Feng, Y.; Wang, D.; Blaabjerg, F.; Wang, X.; Davari, P.; Lu, K. Active damping control methods for three-phase slim DC-link drive system. In Proceedings of the IEEE IFEEC-ECCE 2017, Kaohsiung Taiwan, 3–7 June 2017; pp. 2165–2170.

3. Soltani, H.; Davari, P.; Kumar, D.; Zare, F.; Blaabjerg, F. Effects of DC-link filter on harmonic and interharmonic generation in three-phase adjustable speed drive systems. In Proceedings of the IEEE ECCE 2017, Cincinnati, OH, USA, 1–5 October 2017; pp. 675–681.

4. Maheshwari, R.; Munk-Nielsen, S.; Lu, K. An active damping technique for small dc-link capacitor based drive system. *IEEE Trans. Ind. Inf.* **2013**, *9*, 848–858. [CrossRef]

5. Wang, D.; Lu, K.; Rasmussen, P.O.; Mathe, L.; Feng, Y. Analysis of voltage modulation based active damping techniques for small dc-link drive system. In Proceedings of the IEEE ECCE 2015, Montreal, CA, USA, 20–24 September 2015; pp. 20–24.

6. Davari, P.; Yang, Y.; Zare, F.; Blaabjerg, F. A multi-pulse pattern modulation scheme for harmonic mitigation in three-phase multi-motor drives. *IEEE J. Emerg. Sel. Top. Power Electron.* **2016**, *4*, 174–185. [CrossRef]

7. Hansen, S.; Nielsen, P.; Blaabjerg, F. Harmonic cancellation by mixing nonlinear single-phase and three-phase loads. *IEEE Trans. Ind. Appl.* **2000**, *36*, 152–159. [CrossRef]

8. Malinowski, M.; Kazmierkowski, M.P.; Hansen, S.; Blaabjerg, F.; Marques, G.D. Virtual-flux-based direct power control of three-phase PWM rectifiers. *IEEE Trans. Ind. Appl.* **2001**, *37*, 1019–1027. [CrossRef]

9. Kerem, A.; Aksoz, A.; Saygin, A.; Yilmaz, E.N. Smart grid integration of micro hybrid power system using 6-switched 3-level inverter. In Proceedings of the IEEE ICSG 2017, Istanbul, Turkey, 19–21 April 2017; pp. 161–165.

10. Erol, C.; Sayan, H.H. A novel SSPWM controlling inverter running nonlinear device. *Electr. Eng.* **2018**, *100*, 39–46.

11. Vafakhah, B.; Salmon, J.; Knight, A.M. A New Space-Vector PWM with Optimal Switching Selection for Multilevel Coupled Inductor Inverters. *IEEE Trans. Ind. Electron.* **2010**, *57*, 2354–2364. [CrossRef]

12. Wang, Z.; Wang, Y.; Chen, J.; Hu, Y. Decoupled Vector Space Decomposition Based Space Vector Modulation for Dual Three-Phase Three-Level Motor Drives. *IEEE Trans. Power Electron.* **2018**, *33*, 10683–10697. [CrossRef]

13. Song, J.Y.; Kang, K.J.; Kang, C.H.; Jang, G.H. Cogging Torque and Unbalanced Magnetic Pull Due to Simultaneous Existence of Dynamic and Static Eccentricities and Uneven Magnetization in Permanent Magnet Motors. *IEEE Trans. Mag.* **2017**, *53*, 1–9. [CrossRef]

14. Yang, Y.; Zhou, K.; Wang, H.; Blaabjerg, F. Analysis and Mitigation of Dead Time Harmonics in the Single-Phase Full-Bridge PWM Converters with Repetitive Controllers. *IEEE Trans. Ind. Appl.* **2018**. [CrossRef]

15. Perez-Basante, A.; Ceballos, S.; Konstantinou, G.; Pou, J.; Kortabarria, I.; de Alegria, I.M. A Universal Formulation for Multilevel Selective Harmonic Elimination—PWM with Half-Wave Symmetry. *IEEE Trans. Power Electron.* **2018**, 1. [CrossRef]

16. Lee, J.S.; Kwak, R.; Lee, K.B. Novel Discontinuous PWM Method for a Single-Phase Three-Level Neutral Point Clamped Inverter with Efficiency Improvement and Harmonic Reduction. *IEEE Trans. Power Electron.* **2018**, *33*, 9253–9266. [CrossRef]

17. Flankl, M.; Tüysüz, A.; Kolar, J.W. Cogging Torque Shape Optimization of an Integrated Generator for Electromechanical Energy Harvesting. *IEEE Trans. Ind. Electron.* **2017**, *64*, 9806–9814. [CrossRef]

18. Holmes, D.G.; Lipo, T.A. *Pulse Width Modulation for Power Converters*; IEEE Press: Piscataway, NJ, USA, 2003.

19. Celanovic, N.; Boroyevich, D. A fast space-vector modulation algorithm for multilevel three-phase converters. *IEEE Trans. Ind. Appl.* **2001**, *37*, 637–641. [CrossRef]

8

Optimized Design of Modular Multilevel DC De-Icer for High Voltage Transmission Lines

Jiazheng Lu, Qingjun Huang *, Xinguo Mao, Yanjun Tan, Siguo Zhu and Yuan Zhu

State Key Laboratory of Disaster Prevention and Reduction for Power Grid Transmission and Distribution Equipment, State Grid Hunan Electric Company Limited Disaster Prevention and Reduction Center, Changsha 410129, China; lujz1969@163.com (J.L.); huangqj@hust.edu.cn; maoxg_0@163.com (X.M.); zhengyuan2017307@126.com (Y.T.); zhusiguo2005@ 163.com (S.Z.); zhuyuan1278@163.com (Y.Z.)
* Correspondence: huangqj@hust.edu.cn

Abstract: Ice covering on overhead transmission lines would cause damage to transmission system and long-term power outage. Among various de-icing devices, a modular multilevel converter based direct-current (DC)de-icer (MMC-DDI) is recognized as a promising solution due to its excellent technical performance. Its principle feasibility has been well studied, but only a small amount of literature discusses its economy or hardware optimization. To fill this gap, this paper presents a quantitative analysis and calculation on the converter characteristics of MMC-DDI. It reveals that, for a given DC de-icing requirement, the converter rating varies greatly with its alternating-current (AC) -side voltage, and it sometimes far exceeds the melting power. To reduce converter rating and improve its economy, an optimized configuration is proposed in which a proper transformer should be configured on the input AC-side of converter under certain conditions. This configuration is verified in an MMC-DDI for a 500 kV transmission line as a case study. The result shows, in the case of outputting the same de-icing characteristics, the optimized converter is reduced from 151 MVA to 68 MVA, and the total cost of the MMC-DDI system is reduced by 48%. This conclusion is conducive to the design optimization of multilevel DC de-icer and then to its engineering application.

Keywords: converter; ice melting; modular multilevel converter (MMC); optimization design; transmission line; static var generator (SVG)

1. Introduction

Ice covering on overhead transmission lines is a serious threat to the safe operation of power grids. Overweight ice would break wires or collapse towers, and then cause disruption of power transmission and large-scale outage [1,2]. The ice storms in North America 1998 [3], Germany 2005 [4], and China 2008 [5] are good examples of such consequences. In order to protect the grid from ice disaster, dozens of anti-icing or de-icing methods have been proposed [1,3,5,6], such as thermal de-icing, mechanical de-icing, passive icephobic coatings, etc.

Among various de-icing methods, heating of ice-covered line conductors by electrical current is recognized as the most efficient engineering approach to minimize the catastrophic consequences of ice events [5–8] because it can eliminate the ice covered on hundreds of kilometers of line within an hour, without damaging the grid structure or polluting the environment. Both alternating-current (AC) and direct-current (DC) can be used to melt ice, but AC ice-melting is usually used for transmission lines up to 110 kV, while DC ice-melting is more recommended for high voltage lines up to 500 kV [3,4]. In a DC de-icing system, the most critical part is the DC de-icer (DDI), which generates the required DC voltage and current.

Nowadays, the most widely adopted de-icer is the thyristor-based line-commutated converter (LCC) [9–11], derived from the conventional high voltage direct current transmission (HVDC) technology. It can output a wide range of DC voltages by regulating the thyristor phase shift angle to meet the de-icing requirement of various lines; moreover, it can operate as a static var compensator (SVC) when there is no de-icing requirement. Thus, it has been widely used in Russia, Canada, China [5,12,13], etc. However, due to the inherent characteristics of thyristors, LCCabsorbs much reactive power and generate a lot of harmonics. Thus, it has to deploy an extra series of harmonic filters and many shunt capacitors to meet the grid requirements. Thus, it is bulky, inflexible, and costly. In order to overcome these shortcomings, some proposed constructing the DC de-icer using a voltage source converter (VSC). In [14], a multiple phase shift de-icer was proposed, but it needs a complex multi-winding transformer. In [15], a concept of DC de-icer constructing with a static synchronous compensator (STATCOM) was proposed, but it didn't give specific solutions. In [16], a 3-level STATCOM scheme was proposed for the de-icer application. It presents excellent harmonic and reactive power features, but it requires high-power 3-level converters up to 100 MVA, and such a high-power 3-level converter is difficult to manufacture. Moreover, its DC voltage has to exceed its AC voltage, thus it has a limited DC voltage range.

In the last few decades, modular multilevel converter (MMC) topology has been rapidly developed [17,18]. Since it was presented for the first time by Lesnicar and Marquardt in 2003 [19], it has been widely used in many high-voltage and medium-voltage applications [20–22]. It can output a smooth and nearly ideal sinusoidal voltage with little filters, and it has modularity and scalability, and is facile and flexible. The main application of MMC is VSC-based HVDC transmission [22,23]. In the last five years, dozens of large-capacity MMC-based HVDC systems have been built [22], their rated DC voltage is up to ±500 kV and their rated power is hundreds of MW or even 2000 MW. Another typical application of MMC is the STATCOM [24]. In recent years, most of the STATCOM above 10 Mvar have adopted the MMC structure.

For the de-icer application, an MMC-based DC de-icer (MMC-DDI) with full-bridge submodules (SM) was firstly presented in 2013 [25]. Its structure is similar to a pair of parallel star-configured static var generators (SVGs), and their neutral points are respectively led out as the DC positive and negative poles of DC de-icer. It inherits all the advantages of MMC topology. Moreover, since it employs the full-bridge SMs, it can provide both the buck and boost functions for the DC-link voltage [26]. Thus, it has a wide DC output voltage range to satisfy the de-icing requirements of different lines. In addition, it can be operated as SVG to provide reactive power compensation for the grid. Due to these advantages, the MMC-DDI is recognized as a promising de-icing solution [27]. Since MMC-DDI was first proposed in 2013 [25], its operation principle and control optimization have been further studied in [27–29]. In [28], the hardware selection of MMC-DDI was studied, and a quantitative comparison with an LCC-based de-icer was given. As is shown, both the electrical characteristics and the land occupation have more advantages. In [29], the control and modulation algorithms of MMC-DDI are described. In [27], the dynamic model of MMC-DDI and its harmonic features under phase-shifted carrier modulation are analyzed, and then a detailed control scheme is developed, and the MMC-DDI topology was experimentally verified by utilizing a downscaled prototype. The literature above mainly focus on the technical feasibility of MMC-DDI and pay little attention to its economy optimization.

Like most STATCOMs, the existing MMC-DDI is recommended to be directly connected to the substation distribution network without a transformer. This is considered as a major advantage of the MMC-DDI scheme because the absence of transformer is believed to make the whole device small, light, and compact. Under this configuration, the arm voltage and current of MMC are substantially determined by the grid-connected voltage in addition to the required DC melting voltage and current. For the common high-voltage transmission lines up to 500 kV, their DC melting current is 4000–5000 A or even higher, while their DC melting voltage is usually no more than 10 kV. When the distribution network voltage is unsuitable—for example, 35 kV for most 500 kV substations—the MMC in this configuration simultaneously withstands higher arm voltage and larger arm current.

Thus, the converter rating of MMC-DDI far exceeds its output ice-melting power, resulting in a poor economy to engineering apply.

To address this issue, this paper presents a quantitative analysis on the converter characteristics of MMC-DDI, and then calculates the required converter rating and its influencing factors. It reveals that, for a certain DC de-icing requirement, converter rating varies greatly with its AC-side voltage, and then an optimized design method is proposed to improve the economy of MMC-DDI. Finally, a design example and its corresponding simulation results are given. As this case shows, under the same de-icing outputting characteristics, the optimized converter rating is reduced from 151 MVA to 68 MVA, and the total cost of MMC-DDI system is reduced by 48%.

2. Circuit Configuration and Operation Principle

The circuit configuration of the MMC-DDI is shown in Figure 1. It contains two sets of star-configured arms and each arm has several full-bridge SMs along with a connection reactance. Structurally speaking, it can be viewed as a pair of three-phase star-configured SVGs. The AC terminals of these two SVGs are in parallel and connected to the grid, whereas their neutral points are respectively led out as the DC positive and negative poles of MMC-DDI, and then connected to the ice-covered overhead lines through a set of de-icing disconnectors.

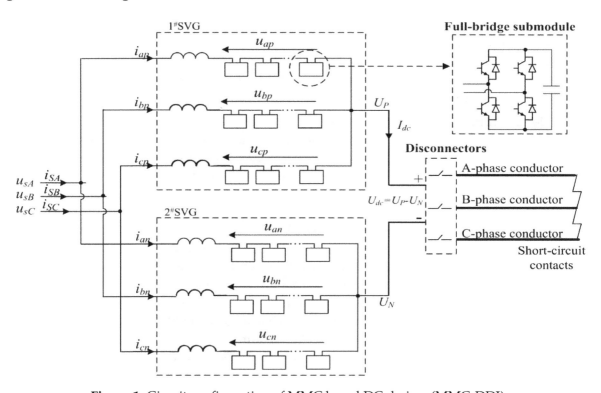

Figure 1. Circuit configuration of MMC based DC de-icer (MMC-DDI).

Since MMC-DDI can provide both buck and boost functions for DC-link voltage, it theoretically does not require a transformer to supply a wide and adjustable DC output voltage. In the existing literature, the AC terminal of MMC-DDI is directly connected to the distribution network with no transformer. This is considered as a major advantage of the MMC-DDI scheme because it can save the cost and floor area of a transformer, making the device small, light, and compact.

According to the grid requirements, MMC-DDI can have two different operation modes:

- Ice-melting Mode. When there is an ice-covered line to melt in the winter, the disconnectors are close to connect the MMC-DDI and the transmission line together, and the other terminal of the transmission line is artificially three-phase short-circuited to form a DC current loop. Then, the MMC-DDI provides a controlled DC voltage to generate the required current through the ice-covered line. At that time, the operation mode of MMC-DDI is similar to the MMC rectifier station in the VSC-HVDC transmission system, except that the DC-side output voltage almost remains unchanged in the VSC-HVDC system while it may vary with the line parameters in the MMC-DDI system. In addition, the typical control methods for the common MMC system are also applicable to MMC-DDI system, such as the capacitor voltage control, the active and reactive current control, the capacitor voltage balancing control, the circulating current control, etc.

- SVG Mode. When there is no icing line, the de-icing disconnectors can be open circuit. Then, the upper three arms and the lower three arms can operate as two parallel conventional SVGs, and provide reactive power compensation or alleviate other power quality problems.

3. Converter Characteristic of MMC-DDI

3.1. Arm Voltage and Current

Take the A-phase as an example, the dynamic equations of MMC-DDI can be expressed as:

$$u_{sA} = Ri_{ap} + L\frac{d}{dt}i_{ap} + u_{ap} + U_p \tag{1}$$

$$u_{sA} = Ri_{an} + L\frac{d}{dt}i_{an} + u_{an} + U_n \tag{2}$$

$$i_{sa} = i_{ap} + i_{an} \tag{3}$$

where u_{sA}, i_{sA} are the AC-side input phase voltage and current of converter. u_{ap}, u_{an} are respectively the output voltage of the upper arm and lower arm. i_{ap}, i_{an} are respectively the arm current of the upper and lower arms. U_p is the electric potential of the neutral point of 1#SVG, relative to the grid neutral point. U_n is the electric potential of the neutral point of 2#SVG. R and L represent the equivalent resistance and inductance of the connection reactance in each arm:

$$\begin{cases} I_{dc} = i_{ap} + i_{bp} + i_{cp} = -(i_{an} + i_{bn} + i_{cn}) \\ U_{dc} = U_p - U_n \end{cases} \tag{4}$$

where U_{dc} and I_{dc} are the DC-side output de-icing voltage and current of MMC-DDI.

Generally, the voltage and current of each arm are symmetrical, and the circulation current can be effectively suppressed with proper circulation current control method, and the voltage drop across the connection reactance is far less than other items in Equations (1) and (2). As a result, the A-phase arm voltages and currents can be expressed as:

$$u_{ap} = \sqrt{2}U_m \sin(\omega t) - 0.5U_{dc} \tag{5}$$

$$u_{an} = \sqrt{2}U_m \sin(\omega t) + 0.5U_{dc} \tag{6}$$

$$i_{ap} = \frac{\sqrt{2}}{2}I_m \sin(\omega t + \varphi) + \frac{I_{dc}}{3} \tag{7}$$

$$i_{an} = \frac{\sqrt{2}}{2}I_m \sin(\omega t + \varphi) - \frac{I_{dc}}{3} \tag{8}$$

where U_m, I_m are the root mean square (RMS) values of the AC-side input phase voltage and current of MMC converter. ω is the angular frequency of gird voltage while ϕ presents the AC-side power factor angle.

Similarly, the B-phase and C-phase arm voltage/current can also be expressed. As shown in Equations (5)–(8), the voltage/current of each arm contains both AC and DC components. Moreover, their peak values are the same for each arm, and can be expressed as

$$\begin{cases} I_{arm\_peak} = \frac{\sqrt{2}}{2} I_m + \frac{1}{3} I_{dc} \\ U_{arm\_peak} = \sqrt{2} U_m + 0.5 U_{dc} \end{cases} \tag{9}$$

where I_{arm_peak}, U_{arm_peak} present the peak values of arm current and arm voltage.

According to Equations (5)–(8), the RMS values of arm voltage and current can be expressed as

$$\begin{cases} I_{arm\_RMS} = \sqrt{\frac{1}{2} I_m{}^2 + \frac{1}{9} I_{dc}{}^2} \\ U_{arm\_RMS} = \sqrt{2 U_m{}^2 + \frac{1}{4} U_{dc}{}^2} \end{cases} \tag{10}$$

where I_{arm_RMS}, U_{arm_RMS} present the RMS values of the arm current and arm voltage.

Compared with that of common SVGs, the converter voltage/current of the MMC-DDI has different characteristics:

(1)   The arm voltage/current of MMC-DDI contains both DC and AC components, while in the conventional SVG, there is only AC component.
(2)   The arm voltage/current no longer equals the AC-side input voltage/current in MMC-DDI.
(3)   The peak value of the arm voltage/current is no longer than $\sqrt{2}$ times of its RMS value.
(4)   Due to these differences, although the MMC-DDI is structurally similar to a pair of common star-connected SVGs, their inner converter characteristics are quite different.

3.2. Influence of AC Side Input Voltage

Under normal operating conditions, the AC side input active power of the MMC converter is substantially equal to its DC side output power (neglecting tiny converter loss). According to the power balance between the AC and DC sides, the output DC ice-melting power can be obtained:

$$P_{dc} = U_{dc} I_{dc} = 3 I_m U_m \cos \varphi \tag{11}$$

where P_{dc} is the output ice-melting power, $\cos\varphi$ is AC-side power factor and generally $\cos\varphi = 1.0$.

With (11), the AC-side input current of converter can be expressed as

$$I_m = \frac{U_{dc}}{3 U_m \cos \varphi} I_{dc}. \tag{12}$$

Substituting (12) into (9), the peak values of arm voltage and arm current can be expressed as

$$\begin{cases} I_{arm\_peak} = \left(\frac{\sqrt{2}}{6 \cos \varphi} \frac{U_{dc}}{U_m} + \frac{1}{3} \right) I_{dc} \\ U_{arm\_peak} = \left(\sqrt{2} \frac{U_m}{U_{dc}} + 0.5 \right) U_{dc} \end{cases} . \tag{13}$$

According to (13), the influence of AC side input voltage on the arm voltage and current peaks can be plotted and shown in Figure 2. As it shown, for a certain DC ice-melting requirement, with the increasing of AC-side voltage, arm voltage peak increases linearly (but not proportionally) while arm current peak decreases and tends to $1/3\ I_{dc}$. This is quite different from common SVG. In an SVG, in the case of a certain output reactive power, with the increasing of the AC-side voltage, the arm voltage peak increases proportionally while the arm current peak decreases and tends to 0.

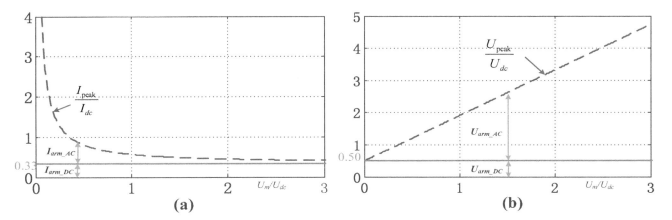

Figure 2. Influence of the AC side input voltage on the peaks of arm voltage and current (**a**) on the current (**b**) on arm voltage.

3.3. Converter Rating of MMC-DDI

In a power electronics system, the converter rating is an important technical indicator because device cost is closely related with the converter rating. For an MMC converter, its converter rating is mainly determined by the arm voltage peak and arm current peak because they largely determine the size and quantity of submodules, and then determines the main hardware of the converter. Therefore, the converter rating of the MMC-based devices can be collectively defined as

$$S_c = \sum_1^n \frac{U_{pi} I_{pi}}{2} \tag{14}$$

where S_c presents the converter rating. n presents the total number of arms. U_{pi}, I_{pi} are the output voltage and current peak of the i-th arm.

For a conventional star-connected SVG, there are three arms, and the current peak of each arm is approximately equal to the AC side phase current while arm voltage peak is approximately equal to the AC-side phase voltage (ignoring the voltage drop across the connection reactance). Then, its converter rating can be expressed as

$$S_c = 3\frac{U_p I_p}{2} = 3\frac{\sqrt{2}U_{sP} \times \sqrt{2}I_{sp}}{2} = 3U_{sp}I_{sp} = S_{out} \tag{15}$$

where U_{sp}, I_{sp} are respectively the RMS values of AC-side phase voltage and phase current, S_{out} presents the output apparent power of SVG.

Indeed, Equation (15) also applies to the delta-connected SVGs or an SVG group composed of several converters. In summary, for any SVG, the converter rating can be directly characterized by its rated output power.

For the MMC-DDI, the six arms share the same voltage and current peaks. Substituting Equation (9) into Equation (16), then the converter rating can be expressed as

$$S_c = 6\frac{U_{\text{arm\_peak}} I_{\text{arm\_peak}}}{2} = 3U_m I_m + \sqrt{2}U_m I_{dc} + \frac{3\sqrt{2}}{4}I_m U_{dc} + 0.5U_{dc}I_{dc} \tag{16}$$

Compared with equation (15), there are three other items in Equation (16), thus the converter rating characteristics of MMC-DDI are significantly different from that of common SVG.

Substituting Equation (13) into Equation (16) and considering $\cos\phi = 1.0$, the converter rating can be simplified as

$$S_c = 3\left(\frac{\sqrt{2}}{6\cos\varphi}\frac{U_{dc}}{U_m} + \frac{1}{3}\right)I_{dc} \cdot \left(\sqrt{2}\frac{U_m}{U_{dc}} + 0.5\right)U_{dc} = \left(1.5 + \frac{\sqrt{2}}{4}\frac{U_{dc}}{U_m} + \sqrt{2}\frac{U_m}{U_{dc}}\right)P_{dc} \quad (17)$$

With Equation (17), the relationship of the converter rating of MMC-DDI with its AC-side voltage can be calculated and shown as Figure 3. As it shown, under a certain DC-side output voltage and power requirement, the converter rating varies greatly with its AC input voltage. It can be analytically solved that when and only when $U_m = 0.5\,U_{dc}$, the converter rating gets its minimum value, and the minimum rating is 2.91 times the output ice-melting power. This conclusion can be expressed as

$$S_{c\_min} = \left(1.5 + \sqrt{2}\right)P_{dc} \quad \text{when} \quad U_m = 0.5U_{dc} \quad (18)$$

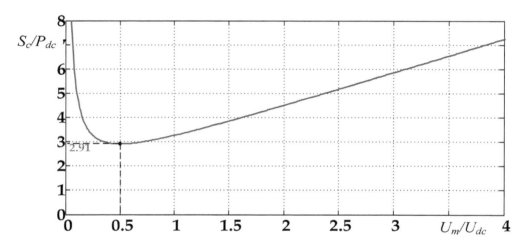

Figure 3. Relationship of the converter rating of MMC-DDI with its AC-side voltage.

4. The Proposed Optimization Design Method

4.1. General Design Process of IMD

For any type of DC ice melting device, its design process generally follows these steps:

Step 1: According to the line parameters and meteorological conditions of the transmission lines to be melted, calculate the required DC-side output de-icing current, voltage and power, and then determine the rated DC-side output parameters of IMD.

For a given transmission line, its required de-icing current depends on many parameters, such as conductor type, ambient temperature, wind velocity, ice thickness and de-icing duration, etc. The thermal behavior of overhead conductors has been well studied, and some formulas are given to calculate the de-icing current in many standards—for example, IEEE standard [30] and CIGRE standard [31]. Generally, the de-icing current should be greater than the minimum de-icing current and no more than the maximum endure current of the line conductor. For some typical conductor types used in China, the minimum de-icing current and the maximum endure current are shown as Table A1 (see Appendix A). In actual ice melting system, it generally tries to choose the intermediate value of the maximum and minimum values as the rated de-icing current.

After determining the de-icing current, the required de-icing DC voltage can be calculated as

$$U_{dc} = k_{icing}R_{line}I_{icing} \quad (19)$$

where I_{icing} is the required de-icing current and R_{line} is the phase resistance of transmission line. k_{icing} corresponds to the ice-melting mode, $k_{icing} = 2$ when the de-icing current is passed down one phase conductor and back along another, and $k_{icing} = 1.5$ when down one and back along the other two [16].

When there are several lines to be melted, the de-icing DC current and voltage of each line can be calculated one by one, and then the rated DC-side output parameters of the IMD are determined by the output DC voltage range, the maximum de-icing current, and the maximum de-icing power.

Step 2: According to the optional voltage levels of the power substation as well as the rated IMD output power, select the proper access voltage of the IMD.

For typical transmission lines, their DC ice-melting power is generally among several MW and hundreds of MW. Within this range, the IMD is usually connected to the low-voltage distribution network of the substation, generally 10 kV or 35 kV in China.

Step 3: According to the DC-side output parameter requirements and the grid access voltage, design the internal structure and parameters of the IMD.

In the process of designing the internal IMD parameters, it is usually necessary to consider both the technical feasibility and the economy.

4.2. The Proposed Circuit Configuration and Its Economic Analysis

According to the above calculation, for a certain ice-melting requirement, the converter rating of MMC-DDI varies greatly with its AC-side voltage. Traditionally, MMC-DDI is directly connected to the grid, thus its AC-side input voltage always equals the grid voltage. This may correspond to a very high converter rating, resulting in poor economy. To solve such a problem, this paper proposes an optimization MMC-DDI configuration structure as shown in Figure 4, i.e., a transformer should be inserted between the grid and the converter under certain conditions. In order to realize this idea, there are two main questions:

(A)   When should the transformer be desired and when is it undesired?
(B)   If a transformer is inserted, what are the specifications and parameters of the transformer?

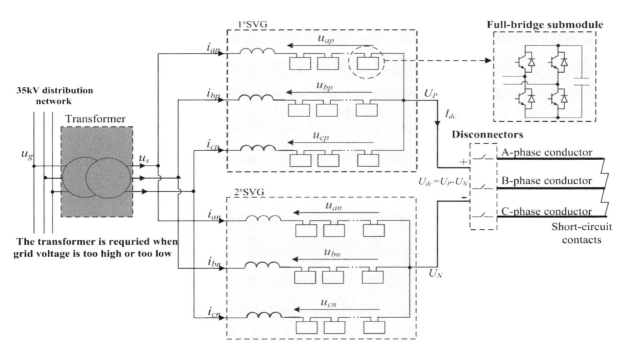

Figure 4. The proposed configuration structure of MMC-DDI.

According to (11), when the power factor is controlled as $\cos\phi = 1$, the AC-side input apparent power of MMC-DDI always equals its DC side output power regardless of the AC-side voltage. Therefore, if a transformer is inserted, its rating only needs to equal the output de-icing power rather than the converter rating. In order to get the minimum converter rating as shown in (18), the output phase voltage of the transformer can be set as $U_{\mathrm{m}} = 0.5\,U_{\mathrm{dc}}$, corresponding to a line voltage $\sqrt{3} \times 0.5U_{dc}$. In summary, the specification of the transformer can be determined as

$$\begin{cases} S_{Tran} = P_{dc} \\ T_r = U_g/(\sqrt{3} \times 0.5U_{dc}) \end{cases} \tag{20}$$

where S_{Tran} is the transformer rating, and T_r is the transformer rating voltage radio.

In order to get the timing of transformer insertion, the cost of converter and transformer should be compared. Since the MMC-DDI is rarely applied, it is difficult to obtain its market cost; here, its cost is estimated by referring to that of SVGs. This is due to three reasons: (1) MMC-DDI is structurally equivalent to a pair of star-connected SVGs, (2) SVG has been widely used and its cost is transparent, and (3) the rating range of common SVGs is wide enough to cover the potential MMC-DDI. Table A2 shows the deal prices of several high capacity SVGs built in China from 2013 to 2018.

As (15) shows, the converter rating of SVG is approximately equal to its rated output power, so the converter cost can be directly evaluated with the SVG deal price. As Table A2 shows, SVG cost is basically proportional to the rating, and its unit cost is around 15,000 $/Mvar. For some SVGs over 60 Mvar, the unit cost is 40% higher. This is because there are only a few applications for such high-power SVGs, thus their R&D cost is higher. Moreover, such high-power SVG usually require higher reliability and larger configuration margin, and this also increases the device cost. For simplicity, here the MMC converter cost is estimated with the average unit price 15,000 $/Mvar.

When a transformer is inserted as Figure 4, the transformer would bring a cost itself. Table A3 shows the deal prices of several 10 MVA-class rectifier transformers built in China. As is shown, the cost of 10 MVA rectifier transformer is about $86,000, about half of the same rating SVG. With the rating growth of transformer, its unit cost decreases rapidly. For a 56 MVA transformer, its unit cost is 4400 $/Mvar and about 1/3 of a similar rating SVG. For a 100 MVA transformer, its unit cost reduces to 3300 $/Mvar and about 1/6 of the same rating SVG.

Based on these cost data, it can be obtained that the cost of a common transformer is much lower than that of the same-rating MMC converter.

In the proposed configuration of MMC-DDI, it can get a minimum MMC converter rating at the cost of an extra transformer. In order to quantitatively compare the economics of the proposed configuration, the costs of MMC-DDI with and without the transformer can be expressed as

$$\begin{cases} P_{no} = P_{con}(u_s = u_g) \\ P_{with} = P_{trans} + P_{con}(u_s = \sqrt{3} \times 0.5U_{dc}) \end{cases} \tag{21}$$

where P_{no} presents the cost of MMC-DDI with no transformer, and $P_{con}(u_s = u_g)$ presents the cost of the MMC converter when its AC-side voltage is equal to the grid voltage. P_{with} presents the cost of the MMC-DDI with a transformer; P_{trans} presents the transformer cost. $P_{con}(u_s = \sqrt{3} \times 0.5U_{dc})$ presents the cost of the MMC converter with an AC-side input voltage of $u_s = \sqrt{3} \times 0.5U_{dc}$.

As long as the cost of MMC-DDI with transformer is lower than that without a transformer, i.e., the reduced converter cost is greater than transformer cost, the proposed configuration structure is cost-effective. At this point, a transformer can be inserted on the AC side of converter to improve the system economy. Otherwise, this is no need to plug in the transformer.

4.3. Applicable Scope of the Proposed Configuration

Compared with the traditional MMC-DDI structure, the proposed MMC-DDI configuration structure requires an extra transformer. It seems that this would increase the cost of the total system, and partially offset the advantages of the MMC topology. However, in fact, the converter rating of traditional MMC-DDI varies greatly with its AC-side voltage, thus the insertion of transformer can sometimes reduce the converter rating and its cost. As long as the reduction of the converter cost is sufficient to offset the transformer cost, the proposed MMC-DDI structure is cost-effective.

According to the cost comparison data of the converter and transformer in the previous section, the unit cost of an MMC converter is generally much higher than that of a conventional transformer, especially for large-capacity converters above 50 MVA. Moreover, the reduced converter rating caused by an introduction of transformer is sometimes much higher than the transformer rating.

In order to obtain quantitative guidance, here an assumption is made of the cost of converter and SVG:

A   The converter cost is approximately considered to be proportional to the converter rating.
B   The transformer cost is a quarter of the same rating MMC converter cost.

Based on the above quantitative assumption, we can get the following conclusions:

a.   When the ratio of the grid line voltage to DC-side output voltage exceeds 2.0 or falls below 0.25, the overall cost of MMC-DDI with a transformer is less than that without transformer, i.e., a transformer can be inserted on the AC side of a converter to improve the system economy.
b.   When the ratio is between 0.25–2.0, the cost of the transformer exceeds its revenue. In that case, no transformer is required.

Indeed, for the common high-voltage transmission lines up to 500 kV, the required ice-melting voltage is generally less than 15 kV. Under such DC voltage range, if the MMC-DDI is connected to a 35 kV network, the grid voltage is more than two times the ice-melting DC voltage. In that case, the proposed MMC-DDI configuration is more applicable than the traditional one. However, if MMC-DDI is connected to a 10 kV distribution network, the grid voltage is usually among 0.25–2.0 times DC voltage, thus the traditional configuration is more applicable. In China, almost all of the distribution network voltage of 500 kV substations is 35 kV. Thus, at least for 500 kV transmission lines, the proposed MMC-DDI configuration is superior to the traditional configuration in most cases.

5. Design Example and Simulation Results

5.1. A Typical Design Example

In order to verify the above analysis and the proposed configuration, a design example of MMC-DDI is given here. For a 500 kV transmission line, the wire type is 4 × LGJ-400, the line length is 40 km, and its single-phase resistance is 0.72 Ω. The minimum ambient temperature along the line is −5 °C, and the maximum wind speed in winter is about 5 m/s. In the 500 kV substation at one end of the transmission line, the distribution grid voltage is 35 kV, corresponding a 20.2 kV phase voltage.

With the data shown in Table A1, the required de-icing current of the above transmission line should be between 3475–4768 A. Within this range, the smaller the current, the longer the de-icing process lasts. Considering a balance between ice-melting rapidity and IMD economics, the rated DC de-icing current can be set as 4.0 kA. Then, with (19), the required de-icing voltage can be calculated as 5.76 kV (2 × 4.0 kA × 0.72 Ω). Thus, the rated de-icing output power is 23.2 MW (= 5.76 kV × 4.0 kA).

With the formulas in Chapter 3, the detailed electrical parameters of above MMC-DDI can be calculated and then listed in Table 1. The voltage and current peaks of the six arms are respectively 31.5 kV and 1.6 kV, thus the converter is equivalent to two conventional star-connected SVGs and each SVG has a 38.5 kV rated line voltage ($31.5 \text{kV}/\sqrt{2} \times \sqrt{3}$), a 1.13 kA rated current ($1.6 \text{kA}/\sqrt{2}$), and a 75.4 Mvar rating ($\sqrt{3} \times 38.5$ kV × 1.13kA). Under the above total arm voltage and arm current, the specifications and numbers of MMC submodules can be freely selected within a certain range. As the 1700 V-level insulated gate bipolar transistor (IGBT) module is widely used in many medium-voltage engineering applications, here the submodule is construed with such IGBT, so the rated capacitor voltage of is set as 900 V and each arm contains 39 submodules. Referring to the SVG price list in Table A2, the converter cost can be estimated as 2.26 million dollar (15,000 \$/Mvar × 75.4 × 2 = 2.26 million). With respect to its 23.2 MW output de-icing power, such high cost is too high to be acceptable.

Table 1. Electrical parameter comparison of the MMC-DDI under conventional configuration and optimized configuration.

Parameter	Symbol	Conventional Configuration (with No Transformer)	Optimized Configuration (with Transformer)
Rated DC voltage	U_{dc}	5.8 kV	5.8 kV
Rated DC current	I_{dc}	4.0 kA	4.0 kA
Rated output DC power	P_{dc}	23.2 MW	23.2 MW
AC-side phase voltage	U_m	20.2 kV	2.9 kV
AC-side phase current	I_m	0.38 A	4.6 kA
Arm voltage peak	U_{arm_peak}	31.5 kV	7.0 kV
Arm current peak	I_{arm_peak}	1.6 kA	3.2 kA
Converter rating	S_c	151 MVA	68 MVA
Transformer		None	23 MVA–35 kV/5 kV
Submodule number in each arm	N	39	9
Submodule capacitor voltage	U_{c0}	900 V	900 V
Submodule capacitance	C_c	10 mF	20 mF

If the proposed optimization method is adopted, a 23 MVA–35 kV/5 kV transformer should be inserted between the MMC converter and the 35 kV grid. At this time, the optimized MMC-DDI is mainly composed of an MMC converter and a transformer, and the detailed electrical parameters of MMC-DDI are also listed in Table 1. As Table 1shown, the voltage and current peaks of the six arms are 7.0 kV and 3.2 kV, thus the converter is equivalent to two common SVGs and each SVG has a rated line voltage 8.57 kV ($7.0\ kV/\sqrt{2} \times \sqrt{3}$), 2.26 kA rated current ($3.2\ kA/\sqrt{2}$) and 33.5 Mvar rating ($\sqrt{3} \times 8.57\ kV \times 2.26\ kA$). Considering the approximate SVG unit cost (15,000 \$/Mvar), the converter cost can be estimated as 1.01 million dollar (\$15,000 /Mvar × 33.5 Mvar × 2). In addition, in Table A3, the cost of a 24 MVA transformer is \$166,000. Then, the total cost of the optimized MMC-DDI can be estimated as 1.18 million dollars.

The above cost comparison results are listed in Table 2. Compared with the cost of the original MMC-DDI with no transformer, the optimized cost of the IMD device has dropped by 48%.

Table 2. Cost comparison of the MMC-DDI under conventional configuration and optimized configuration.

Component	Original Cost (Million Dollar)	Optimized Cost (Million Dollar)
Converter	2.26	1.01
Transformer	-	0.17
Total	2.26	1.18

Besides the cost, the size and weight of the de-icer are also concerned in engineering applications. In practical projects, a complete MMC-DDI system contains not only the connection reactance, the converter valves and the disconnectors as shown in Figure 1, but also inlet cabinet, startup cabinet, control system, cooling subsystem, power distribution cabinet, cable and other auxiliary equipment. The equipment footprint not only includes the size of these devices, but also the insulation distance and other factors. Considering the fact that the main difference of the two MMC-DDI configurations is the converter and transformer, here only the converter chain and transformer are carefully compared.

For simplicity, here refers to a 100 Mvar SVG project built in Hunan in 2016 as a benchmark to compare the size and weight of the two topologies. This project consists of two Y-connected SVGs based on IGBT, and each SVG is 50 Mvar with a 20 kV rated voltage. Its arm current peak is 2125A, and 1.2 times that of that the conventional MMC-DDI configuration. Each SVG contains 63 power submodules, packed in 11 power cabinets. The converter hall is arranged on the first floor, while the cooling system is arranged on the roof. The floorplan of the SVG room is shown in Figure A1 and the main installation parameters of the SVGs are shown in Table A3. As Figure A1 shows, the SVG room covers an area 280 m², wherein the converter chain occupies 163 m² (17.6 m × 9.25 m).

The submodule current peak of the above SVG is about 1.2 times that of the conventional MMC-DDI. Here, we adopt the same submodule to form MMC-DDI. Considering that the arm current peak of the optimized MMC-DDI configuration is just twice that of the conventional one, the submodules of the optimized MMC-DDI configuration can be constructed with two parallel SVG submodules. Based on the above ideas, the conventional MMC-DDI requires 234 power modules while the optimized one requires 108 modules, and then their size and weight parameter can be calculated and shown in Table 3. The size and weight of the transformer are based on a 24 MVA rectifier produced for another project, the body size of the transformer is 5.4 m × 4.7 m, but its actual land occupation is set as 8 m × 9 m while considering the insulation distance and ancillary facilities.

Table 3. Size and weight comparisons of the MMC-DDI under conventional and optimized configuration.

Items	100 Mvar SVG	Conventional MMC-DDI	Optimized MMC-DDI
Main components	Converter (2 × 50 Mvar)	Converter (151 MVA)	Converter + Transformer (68 MVA) (24 MVA)
Number of power units	2 × 63	2 × 117	2 × 54
Number of power cabinets	2 × 11	2 × 20	2 × 10
Submodule capacitor voltage	900 V	900 V	900 V
Size of each submodule	0.7 m × 0.7 m × 0.8 m	0.7 m × 0.7 m × 0.8 m	0.7 m × 0.7 m × 0.8 m
Weight of each submodule	250 kg	250 kg	250 kg
Total weight of submodules	31.5 t	59 t	26 t
Transformer weight	-	-	38 t
Converter area	163 m²	296 m²	133 m²
Transformer area	None	None	72 m²
Other floor area	117 m²	117 m²	117 m²
Total floor area	280 m²	413 m²	322 m²
Total weight	31.5 t	59t	64 t

Compared with the conventional MMC-DDI, the optimized scheme required additional 72 m² to place the transformer, but the converter area is reduced from 296 m² to 133 m², namely a reduction of 163 m². As a result, the overall footprint of MMC-DDI system is reduced by 91 m², corresponding to a ratio of 22%. It shows that the optimized scheme also has an advantage in the land occupation. On the other hand, the optimized scheme requires a transformer with weight of 38 Ton, but its converter weight is reduced by 35 Ton, thus the total weight was slightly increased by 5 Ton. It shows that the optimized scheme have no advantage in weight. However, the DC de-icer built for high voltage transmission lines up to 500 kV is generally installed in the substations, so this weight disadvantage is still acceptable.

5.2. Simulation Results

To verify the above analysis and calculation on the converter characteristic, a corresponding MMC-DDI system is built in Matlab/Simulink (MathWorks, Natick City, MA, USA), and the simulation parameters are listed in Table A4.

For comparison, a dual-SVGs system (2 × 11.6 Mvar), which is similar to Figure 1 but has a zero DC-side output current reference, is also simulated. Since this article focuses on the converter rating characteristics, the number of submodules in each arm was set as $n = 4$ to speed up the simulation. This is also sufficient to compare the converter characteristics of the two schemes. The circuit image of the MMC-DDI simulation model built in Matlab/Simulink is shown in Figure A2, and its control block is shown in Figure A3. The simulation results of the dual-SVGs, the conventional MMC-DDI and the optimized MMC-DDI are shown in Figures 5–7, respectively.

As Figure 5a–c shows, in the dual-SVGs system, both the arm current and voltage are positive-negative symmetrical. The arm current equals half of the AC-side input current, while arm voltage is slightly higher than AC-side phase voltage due to the voltage drop across arm reactance. Their peaks are respectively 0.28 kA and 32.0 kV. With (15), the corresponding converter rating can be calculated as 26.9 MVA, just slightly higher than its output reactive power.

As Figure 5d,f shows, in the dual-SVGs system, the center point voltages of two SVGs only have tiny low-frequency component although they have obvious high frequency ripples. These ripples are mainly caused by the separate phase control method adopted in this simulation.

As Figure 5e shows, in the dual-SVGs system, the submodule capacitor voltages are around their set reference 8000 V and have tiny second harmonic fluctuations. The fluctuation amplitude is about 150 V, corresponding to 1% ripple factor.

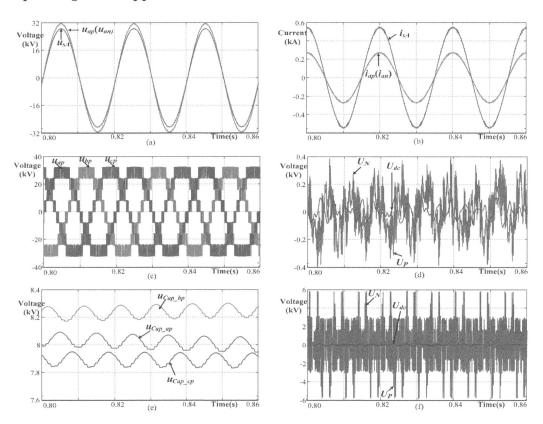

Figure 5. Simulation results of dual-SVGs system. (**a**) arm voltage (fileted high frequency ripple), (**b**) arm current, (**c**) arm voltage (unfiltered), (**d**) DC-side voltage (fileted high frequency ripple), (**e**) capacitor voltage of the first submodule in upper three arms, (**f**) original DC-side voltage (unfiltered).

As Figures 6g and 7g show, the DC-side output voltage and current in the conventional MMC-DDI and the optimized MMC-DDI are almost the same, and they rise gradually to their rated values during the melting-ice startup process. Correspondingly, both the DC and AC components in arm current rise slowly to the expected value. This indicates that the DC-side output voltage of MMC-DDI can be freely regulated within a range not exceeding its rated value, so that it can adapt to the different melting requirement of multiple transmission lines. As Figures 6d and 7d show, the DC-side output voltages are all around their expected value in both the conventional MMC-DDI and optimized MMC-DDI. The only difference is that the voltage ripple of the optimized MMC-DDI is smaller. Since the ice-melting process is mainly based on the Joule heat of the line current, this difference has little effect on the melting results.

As Figure 6a–c shows, in the conventional MMC-DDI that has no transformer, there is a visible DC component in the arm voltage and arm current. Especially in the arm current, the DC component far exceeds AC component. The arm voltage peak is 31.5 kV, which is slightly higher than AC phase voltage, while the arm current peak is 1.6 kA and far higher than the amplitude of its AC component. With (15), the corresponding converter rating can be calculated as 151 MVA, about 6.5 times the DC-side output power.

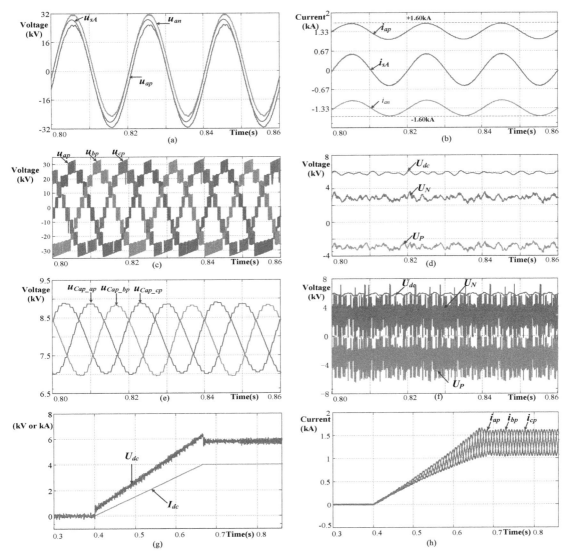

Figure 6. Simulation results of the conventional MMC-DDI. (**a**) arm voltage (fileted high-frequency ripple), (**b**) arm current, (**c**) arm voltage (unfiltered), (**d**) DC-side voltage (fileted high frequency ripple), (**e**) submodule capacitor voltage, (**f**) DC-side voltage (unfiltered), (**g**) DC-side output voltage and current during melting-ice startup process, (**h**) arm current during melting-ice startup process.

As Figure 7a–c shows, in the optimized MMC-DDI system that has a transformer, there is an obvious DC component in the arm voltage and current. The arm voltage and current peaks are respectively 7.0 kV and 3.2 kA, corresponding to a 67.2 MVA converter rating. Compared with the original MMC-DDI without a transformer, the arm current peak increases by 100% while the arm voltage peak reduces by 78%, thus the converter rating is only 44% of its original value.

The converter characteristics in such simulation results are consistent with the above analysis and calculation. In addition, the values of the converter voltage and current are also consistent with the theoretical results listed in Table 1. This proves the accuracy of the analysis and calculation on the MMC converter rating present in the paper.

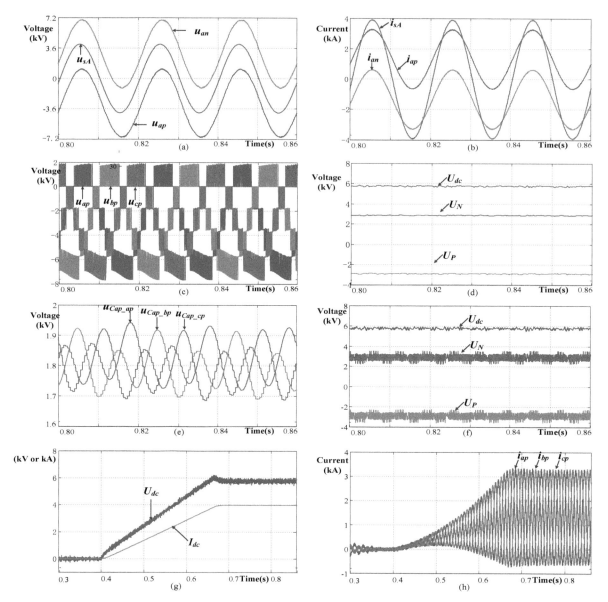

Figure 7. Simulation results of the optimized MMC-DDI. (**a**) arm voltage (fileted high-frequency ripple), (**b**) arm current, (**c**) arm voltage (unfiltered), (**d**) DC-side voltage (fileted high frequency ripple), (**e**) submodule capacitor voltage, (**f**) DC-side voltage (unfiltered), (**g**) DC-side output voltage and current during melting-ice startup process, (**h**) arm current during melting-ice startup process.

6. Discussion

Concerning the converter rating of MMC-DDI presented in this paper, the goal is to improve the economics of MMC-DDI while maintaining the same output de-icing characteristics. It turns out that, for a given DC ice-melting requirement, the converter rating of MMC-DDI varies greatly with its AC-side input voltage. Then, it is proposed to insert a transformer on the AC side of the MMC converter so that the converter rating as well as its cost can be significantly reduced, and then the economics of MMC-DDI can be improved.

It seems that this proposed configuration scheme is contradictory to traditional understanding of the MMC structure. Conventionally, in the common MMC system such as SVG, the AC side input transformers are expected to be avoided as much as possible.

This difference can be explained due to the converter characteristic of MMC-DDI having significant differences with that of the common MMC system:

(1) In an SVG, both the arm voltage and current contain only an AC component. As a result, in the case of a certain output power, the arm voltage is inversely proportional to arm current, thus the converter rating remains basically constant under any AC-side voltage. In that case, if a transformer was configured on the AC side of MMC converter, it has little influence on the converter rating while increasing a transformer. Therefore, in the common SVG, it tries to avoid a transformer.

(2) In the MMC-DDI, the arm voltage and arm current of converter contain both DC and AC components. As a result of the crossover between the DC and AC components, the converter rating of MMC-DDI varies greatly with its AC-side voltage. Due to such converter characteristics, a transformer can affect the converter rating. In this case, although the introduction of transformer will increase transformer cost, it can cause a cost increment or reduction of the converter. As long as the reduction of the converter cost is sufficient to offset the transformer cost, the introduction of the transformer is cost-effective. In addition, because the unit cost of MMC converter is generally much higher than that of the transformer, the above condition is easy to satisfy under the typical DC ice melting system parameters. Therefore, the optimized configuration scheme proposed in this paper is cost-effective in many cases.

It should be noted that the MMC-DDI can have two operation modes: ice-melting mode and SVG mode. This paper only considers the requirement of the ice melting mode, while not analyzing the operating characteristics of the SVG mode. In the optimization design process, the requirements of SVG mode have not been taken into account. This requirement can be further studied to get more comprehensive optimization results.

7. Conclusions

An MMC-based DC de-icer has been recognized as a promising de-icing solution. Conventionally, the MMC-DDI is recommended to be directly connected to the grid without a transformer.

In this paper, the converter rating of MMC-DDI was quantitatively analyzed. For a given DC ice-melting requirement, the converter rating varies greatly with its AC-side input voltage, and its minimum is 2.9 times the output ice-melting power. When the grid access point voltage is far more than DC de-icing voltage, the conventional MMC-DDI structure requires a far higher converter rating than its output de-icing power, thus the economy of MMC-DDI is very poor.

In order to improve the economy of MMC-DDI, this paper proposes an optimized MMC-DDI configuration structure in which a common two-winding transformer should be inserted at the AC-side of converter in some cases. Thus, the converter rating can be greatly reduced at the cost of an extra transformer. Since the cost of transformer is much lower than the same rating MMC converter, the introduction of transformer is cost-effective in many cases.Actually, for most 500 kV transmission lines, the optimized MMC-DDI configuration is superior to the transformerless MMC-DDI.

A design example and simulation results are given in this paper. In the case of outputting the same de-icing characteristics, the optimized converter rating is reduced from 151 MVA to 68 MVA, and the saved cost on the converter is much higher than the cost of the transformer, thus the total cost of MMC-DDI system is reduced by 48%. At the same time, the total floor space of MMC-DDI system is also greatly reduced by 22%, while, in total, the weight has a small increase.

This analysis and case show that, although the transformer is not technically necessary in an MMC-DDI, it can actually bring considerable benefits related to the total cost and space of MMC-DDI.

This conclusion is conducive to the optimized configuration of modular multilevel DC de-icer, and then to its engineering application for high voltage transmission lines.

Author Contributions: Conceptualization, J.L. and Q.H.; Formal Analysis, Q.H.; Data Curation, X.M., and Y.Z; Writing—Review and Editing, S.Z.; Supervision, Y.T.

Appendix A

Table A1. The minimum de-icing current and maximum endure current for typical power lines [5].

Conductor Type	Min. De-Icing Current (A) ($-5\,^{\circ}$C, 5 m/s, 10 mm, 1 h)	Max. Endure Current(A) ($5\,^{\circ}$C, 0.5 m/s, No Icing)
LGJ-4 × 400/50	3475	4764
LGJ-2 × 500/45	1989	2698
LGJ-2 × 240/40	1218	1716
LGJ-1 × 240/40	609	858
LGJ-1 × 185/45	515	733
LGJ-1 × 150/35	441	633
LGJ-1 × 95/55	345	500

Table A2. Deal prices of several typical SVG projects in China from 2013 to 2018.

No.	Project Location	Rated Voltage (kV)	Rating (MVA)	Deal Price [1] ($1000)	Unit Cost (1000 $/MVA)
1	Kunming, Yunnan	35	10	154	15.4
2	Zhangjiakou, Hebei	35	12	175	14.6
3	Huimin, Shandong	35	15	215	14.4
4	Huangpi, Hubei	35	16	251	15.7
5	Tongyu, Gansu	35	20	269	13.5
6	Hua County, Henan	35	20	257	12.8
7	Chenzhou, Hunan	10	20	330	16.5
8	Qiaojia, Yunnan	35	30	385	12.8
9	Linwu, Ningxia	35	40	458	11.5
10	Dabancheng, Xinjiang	35	50	615	12.3
11	Yinan, Shandong	35	60	1023	17.1
12	Haixi, Xinjiang	35	60	1154	19.2
13	Hami, Xinjiang	35	80	1508	18.8
14	Huaping, Yunnan	35	100	2109	21.1
15	Xiangtan, Hunan	35	120	2615	21.8

[1] The deal price covers a complete set of SVG equipment (including the converter chain, connection reactance, startup circuit, cooling system, control system and other ancillary facilities) and its technical service.

Table A3. Deal prices of several 10 MVA-class rectifier transformers in China.

No.	Project Location	Rated Voltage (kV)	Rating (MVA)	Deal Price ($1000)	Unit Cost (1000 $/MVA)
1	Baoding, Hebei	10/5	10	86	8.6
2	Changsha, Hunan	10/7	14	110	7.8
3	Changsha, Hunan	35/6	24	166	6.9
4	Xinyu, JiangXi	35/12	56	246	4.4
5	Chongqing	35/15	86	284	3.3
6	Zhuzhou, Gansu	35/17	100	323	3.2
7	Hengyang, Hunan	35/19	120	361	3.0

Table A4. Simulation parameters of the two MMC-DDI and dual-SVG systems.

Parameter	Symbol	Dual-SVG	Conventional MMC-DDI	Optimized MMC-DDI
AC-side rated voltage	U_S	35 kV	35 kV	5 kV
AC-side rated current	I_M	(0.38 kA) [2]	(0.38 kA)	(2.68 kA)
AC-side rated power		+23.2 Mvar	(23.2 MW)	(23.2 MW)
Arm inductance	L	35 mH	35 mH	1 mH
Arm equivalent resistance	R	0.1 Ω	0.1 Ω	0.02 Ω
DC-side output de-icing voltage	U_{dc}	0	(5.8 kV)	(5.8 kV)
DC-side output de-icing current	I_{dc}	0	4.0 kA	4.0 kA
Resistance of de-icing line	R_{dc}	-	1.45 Ω	1.45 Ω
Inductance of de-icing line	L_{line}	-	32 mH	32 mH
Submodule number of each arm	N	4	4	4
Submodule capacitance	C_{cap}	4 mF	4 mF	10 mF
Submodule capacitor voltage	U_{cap}	9.0 kV	8.0 kV	1.8 kV
Switching frequency		500 Hz	500 Hz	500 Hz

[2] The parameters in parentheses indicate the calculated value, while the parameters in parentheses indicate the values directly set in the simulation.

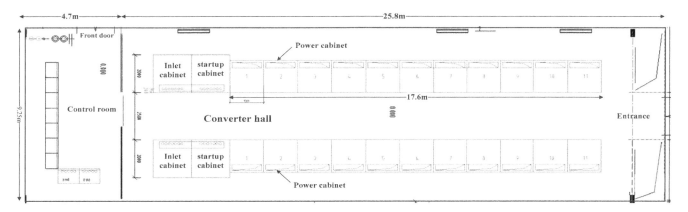

Figure A1. Floorplan of 100 Mvar SVG room in 500 kV Chuanshan substation.

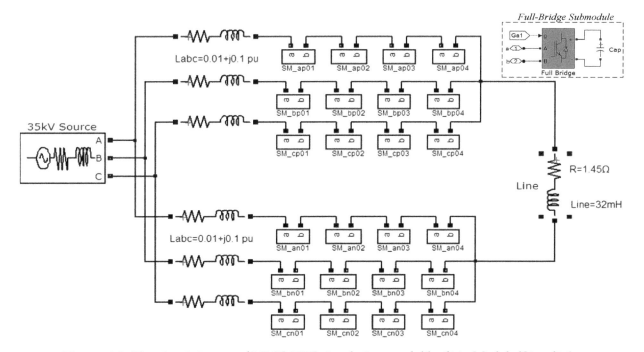

Figure A2. The circuit image of MMC-DDI simulation model built in Matlab/Simulink.

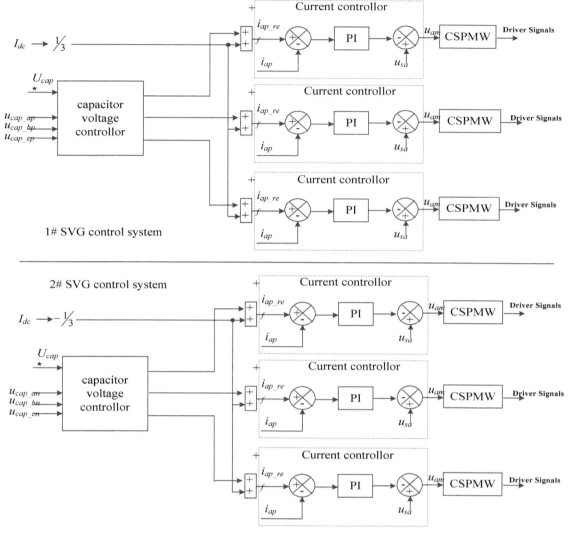

Figure A3. The control block of MMC-DDI simulation model built in Matlab/Simulink.

References

1. Joe, C.P.; Phillip, L. Present State-of-the-Art of Transmission Line Icing. *IEEE Trans. Power Appar. Syst.* **1982**, *8*, 2443–2450.
2. Farzaneh, M.; Savadjiev, K. Statistical analysis of field data for precipitation icing accretion on overhead power lines. *IEEE Trans. Power Deliv.* **2005**, *2*, 1080–1087. [CrossRef]
3. Volat, C.; Farzaneh, M.; Leblond, A. De-icing/Anti-icing Techniques for Power Lines: Current Methods and Future Direction. In Proceedings of the 11th International Workshop on Atmospheric Icing of Structures, Montreal, QC, Canada, 13–16 June 2005.
4. Brostrom, E.; Ahlberg, J.; Soder, L. Modelling of Ice Storms and their Impact Applied to a Part of the Swedish Transmission Network. In Proceedings of the 2007 IEEE Lausanne Power Tech 2007, Lausanne, Switzerland, 1–5 July 2007; pp. 1593–1598.
5. Wang, J.; Fu, C.; Chen, Y.; Rao, H.; Xu, S.; Yu, T.; Li, L. Research and application of DC de-icing technology in China southern power grid. *IEEE Trans. Power Deliv.* **2012**, *3*, 1234–1242. [CrossRef]
6. Laforte, J.L.; Allaire, M.A.; Laflamme, J. State-of-the-art on power line de-icing. *Atmos. Res.* **1998**, *46*, 143–158. [CrossRef]
7. Motlis, Y. Melting ice on overhead-line conductors by electrical current. In Proceedings of the CIGRE SC22/WG12, Paris, France, 26–30 August 2002.
8. Farzaneh, M.; Jakl, F.; Arabani, M.P.; Eliasson, A.J.; Fikke, S.M.; Gallego, A.; Haldar, A.; Isozaki, M.; Lake, R.; Leblond, L.; et al. *Systems for Prediction and Monitoring of Ice Shedding, Anti-Icing and De-Icing for Power Line Conductors And Ground Wires*; CIGRE: Paris, France, 2010.

9. Davidson, C.C.; Horwill, C.; Granger, M.; Dery, A. A power-electronics-based transmission line de-icing system. In Proceedings of the 8th IEE International Conference on AC and DC Power Transmission, London, UK, 28–31 March 2006; pp. 135–139.
10. Dery, A.; Granger, M.; Davidson, C.C.; Horwill, C.; Dery, A.; Granger, M.; Davidson, C.C.; Horwill, C. An Application of HVDC to the de-icing of Transmission Lines. In Proceedings of the 2005/2006 IEEE PES Transmission and Distribution Conference and Exhibition, Dallas, TX, USA, 21–24 May 2006; pp. 529–534.
11. Fu, C.; Rao, H.; Li, X.; Chao, J.; Tian, J.; Chen, S.; Zhao, L.; Xu, S.; Ma, X. Development and application of DC deicer. *Autom. Electr. Power Syst.* **2009**, *63*, 118–119.
12. Rao, H.; Li, L.; Li, X.; Fu, C. Study of DC based De-icing Technology in China Southern Power Grid. *South. Power Syst. Technol.* **2008**, *2*, 7–12.
13. Davidson, C.C.; Horwill, C.; Granger, M.; Dery, A. Thaw point. *Power Eng.* **2007**, *21*, 26–31. [CrossRef]
14. Jing, H.; Nian, X.; Fan, R.; Liu, D.; Deng, M. Control and switchover strategy of full-controlled ice-melting DC power for ice-covered power lines. *Autom. Electr. Power Syst.* **2012**, *36*, 86–91.
15. Zhao, G.S.; Li, X.-Y.; Fu, C.; Li, X.-L.; Wang, Y.H.; Xia, W. Overview of de-icing technology for transmission lines. *Power Syst. Prot. Control* **2011**, *39*, 148–154.
16. Bhattacharya, S.; Xi, Z.; Fardenesh, B.; Uzunovic, E. Control reconfiguration of VSC based STATCOM for de-icer application. In Proceedings of the 2008 IEEE Power and Energy Society General Meeting, Pittsburgh, PA, USA, 20–24 July 2008; pp. 1–7.
17. Hagiwara, M.; Akagi, H. Control and Experiment of Pulsewidth-Modulated Modular Multilevel Converters. *IEEE Trans. Power Electron.* **2009**, *24*, 1737–1746. [CrossRef]
18. Rohner, S.; Bernet, S.; Hiller, M.; Sommer, R. Modulation, Losses, and Semiconductor Requirements of Modular Multilevel Converters. *IEEE Trans. Ind. Electron.* **2010**, *57*, 2633–2642. [CrossRef]
19. Lesnicar, A.; Marquardt, R. An innovative modular multilevel converter topology suitable for a wide power range. In Proceedings of the 2003 IEEE Bologna Power Tech Conference Proceedings, Bologna, Italy, 23–26 June 2003.
20. Martinez-Rodrigo, F.; Ramirez, D.; Rey-Boue, A.B.; de Pablo, S.; Lucas, L.C.H. Modular Multilevel Converters: Control and Applications. *Energies* **2017**, *11*, 1709. [CrossRef]
21. Mehrasa, M.; Pouresmaeil, E.; Zabihi, S.; Mehrasa, M.; Pouresmaeil, E.; Zabihi, S.; Caballero, J.C.T.; Catalão, J.P.S. A Novel Modulation Function-Based Control of Modular Multilevel Converters for High Voltage Direct Current Transmission Systems. *Energies* **2016**, *9*, 867. [CrossRef]
22. Perez, M.A.; Bernet, S.; Rodriguez, J.; Kouro, S.; Lizana, R. Circuit topologies, modeling, control schemes, and applications of modular mul-tilevel converters. *IEEE Trans. Power Electron.* **2015**, *1*, 4–14. [CrossRef]
23. Flourentzou, N.; Agelidis, V.G.; Demetriades, G.D. VSC-Based HVDC Power Transmission Systems: An Overview. *IEEE Trans. Power Electron.* **2009**, *24*, 592–602. [CrossRef]
24. Mohammadi, H.P.; Bina, M.T. A Transformerless Medium-Voltage STATCOM Topology Based on Extended Modular Multilevel Converters. *IEEE Trans. Power Electron.* **2011**, *26*, 1534–1545.
25. Mei, H.; Liu, J. A Novel DC Ice-melting Equipment Based on Modular Multilevel Cascade Converter. *Autom. Power Syst.* **2013**, *37*, 96–102.
26. Thitichaiworakorn, N.; Hagiwara, M.; Akagi, H. Experimental verification of a modular multilevel cascade inverter based on double-star bridge cells. *IEEE Trans. Ind. Appl.* **2014**, *50*, 509–519. [CrossRef]
27. Li, B.; Shi, S.; Xu, D.; Wang, W. Control and Analysis of the Modular Multilevel DC De-Icer with STATCOM Functionality. *IEEE Trans. Ind. Electron.* **2016**, *9*, 5465–5476. [CrossRef]
28. Ning, Y.; Zheng, J.; Chen, Z. Application of star-connected cascaded STATCOM in DC ice-melting system. *Autom. Electr. Power Syst.* **2015**, *27*, 1920–1922.
29. Guo, Y.; Xu, J.; Guo, C.; Zhao, C.; Fu, C.; Zhou, Y.; Xu, S. Control of full-bridge modular multilevel converter for dc ice-melting application. In Proceedings of the 11th IET International Conference on AC and DC Power Transmission, Birmingham, UK, 10–12 February 2015; pp. 1–8.
30. *IEEE Standard for Calculating the Current-Temperature Relationship of Bare Overhead Conductors*; IEEE Std 738-2012; IEEE Standards Association: Piscataway, NJ, USA, 2013; pp. 1–72.
31. CIGRE. Thermal Behavior of Overhead Conductors. *Electra* **1992**, *144*, 107–125.

9

Review on Health Management System for Lithium-Ion Batteries of Electric Vehicles

2

Zachary Bosire Omariba [1,2], **Lijun Zhang** [1,*] **and Dongbai Sun** [1]

[1]   National Center for Materials Service Safety, University of Science and Technology Beijing,
      Beijing 100083, China; zomariba@egerton.ac.ke (Z.B.O.); dbsun@ustb.edu.cn (D.S.)
[2]   Computer Science Department, Egerton University, Egerton 20115, Kenya
*   Correspondence: ljzhang@ustb.edu.cn

Abstract: The battery is the most ideal power source of the twenty-first century, and has a bright future in many applications, such as portable consumer electronics, electric vehicles (EVs), military and aerospace systems, and power storage for renewable energy sources, because of its many advantages that make it the most promising technology. EVs are viewed as one of the novel solutions to land transport systems, as they reduce overdependence on fossil energy. With the current growth of EVs, it calls for innovative ways of supplementing EVs power, as overdependence on electric power may add to expensive loads on the power grid. However lithium-ion batteries (LIBs) for EVs have high capacity, and large serial/parallel numbers, when coupled with problems like safety, durability, uniformity, and cost imposes limitations on the wide application of lithium-ion batteries in EVs. These LIBs face a major challenge of battery life, which research has shown can be extended by cell balancing. The common areas under which these batteries operate with safety and reliability require the effective control and management of battery health systems. A great deal of research is being carried out to see that this technology does not lead to failure in the applications, as its failure may lead to catastrophes or lessen performance. This paper, through an analytical review of the literature, gives a brief introduction to battery management system (BMS), opportunities, and challenges, and provides a future research agenda on battery health management. With issues raised in this review paper, further exploration is essential.

Keywords: lithium-ion batteries; electric vehicles; battery management system; electric power

1. Introduction

Lithium-ion batteries (LIBs) are one of the most promising technologies due to advantages like high efficiency, lower volume, small weight, temperature sensitivity, and maintenance [1–3]. They are the most ideal power source of the twenty-first century and have a bright future in many applications, such as portable electronic devices, electric vehicles (EVs) [4], aerospace systems, and power storage for renewable energy sources, like solar and wind turbines. However, there are many shortfalls, such as lack of safety, fragility, and aging, which may restrict the extensive use of LIBs. The consequences of battery failure can lead to catastrophes and inconveniences, which have turned to be popular and challenging issues [5], as reliability of LIBs is yet to be improved. However, determining the remaining useful life (RUL) of LIBs can aid to some level in curbing this problem [6,7].

There are many different techniques proposed in the literature that capture these crucial parameters to determine the battery state, to ensure that the battery delivers its specified output while optimizing the charge/discharge processes, and must be communicated to on-board systems. The battery management system (BMS) plays a significant role in the prediction of RUL for LIBs, as it acts as a connector between the battery and the EVs. The main goal of the BMS is three-fold: to protect the battery system from damage by detecting malfunctions, such as overcharge, excessive rise in temperature, and electric leak; to predict and increase the battery life; and to maintain the battery system in an accurate and reliable operational condition [8,9]. The BMS is a combination of sensors, controllers, communication, and computation hardware, with software algorithms designed to decide the maximum charge/discharge current and duration from the estimation of state-of-charge (SOC) and state-of-health (SOH) of the battery pack [10]. From this definition BMS performs the two main roles of monitoring the battery to determine information, such as SOH, SOC, and RUL, as well as to operate the battery in a safe, efficient, and non-damaging way [11–13].

In many industrial applications that make use of LIBs as one of the main power sources, the BMS has proven useful. The BMS contains a set of activities that monitor and perform SOH, SOC, state-of-life (SOL), end-of-life (EOL), and state-of-power (SOP) estimation throughout the battery's entire life, and make a suitable decision to predict RUL. Thus, the BMS for LIBs is a decision process to intelligently perform maintenance, logistics, and system configuration activities on the basis of diagnostic and/or prognostics with the aim of producing actionable information to enable timely decisions [14] on maintenance optimization support, and reduce the costs of maintenance [7]. The BMS, therefore, implements state monitoring and evaluation, charge control, and cell balancing functionalities in order to maintain the safety and reliability of batteries [15]. Failure to perform these functionalities can result in battery failure, which can lead to reduced performance, operational impairment, and even catastrophic failure [16], making the performance of accurate prediction of RUL essential. RUL has attracted a great deal of interest from researchers and funding agencies around the world to mitigate the challenges associated with LIB use, in many high-impact applications, while protecting the environment.

There is a growing increase of EVs according to Bruen et al. [17], dependent on LIBs due to their numerous advantages, as compared to other batteries. This is further accelerated with the climate change concerns having a focus on a spotlight to EVs, and LIBs are believed to be the future to widespread EV adoption. However EVs are also faced with a number of drawbacks, as illustrated in Table 1, although technology is advancing fast to curb these challenges.

Table 1. Advantages and disadvantages of EVs from selected review papers.

Advantages	Disadvantages
- Highly efficient	- Electricity storage is still expensive
- Reduced emissions	- Battery charging is time consuming
- High performance and low maintenance	- Primary resource depletion for some elements of the LIB
- Very responsive and have very good torque	- Range anxiety
- EV motors are quiet and smooth	- Battery degradation costs
- Are more digitally connected than conventional vehicles	- Sufficient public charging infrastructure is still lacking
- Simplified powertrain	- Causes indirect pollution
- Low electricity consumption	- Lacks the power to accelerate and climb quickly
- Good acceleration	- Are heavy due to overloaded batteries
- Can be charged overnight on low cost electricity produces by any type of power station, including renewables	

(Source: [11,18–28]).

The BMS contains a portion responsible to monitor and control the SOH of a battery pack, and it is also referred to as the battery health management system (BHMS). However, according to Saha et al. [29], the BMS is a hardware designed to be a low-cost analog-to-digital data acquisition system. This hardware has three components: the signal conditioning board, the data acquisition board, and the embedded processor board. However, the BMS' main function is to monitor, control, and report the SOH of a battery. This review work will be a comprehensive collation of existing prognostic methods, and will provide convenience and inspiration for scholars to study and conduct further research. This review paper is organized into five sections. Section 2 talks about the BMS. Section 3 is about opportunities and challenges with respect to battery health and prognostics. Finally, Section 4 is about the critical future battery prognostic research work, and the conclusion are provided in Section 5.

2. Health Management Systems for Batteries

There is continuous increase in EV stock, but annual growth rates have been reducing consistently since 2011. In 2016 the EV stock growth was 59%, down from 76% in 2015, and 84% in 2014, but statistics shows that battery electric vehicles (BEVs) still account for the majority of the electric car stock, at 60%, as per the "Global EV Outlook 2017: Two Million and Counting" report [30]. According to this report the number of EVs increased from the previous report of a projection of one million EVs in 2016, to two million, projected in 2017. This trend shows that the number of EVs has been doubling over the years and this will put more pressure in the demand for LIBs, which has proved to be the main source of efficient power.

The current trend clearly demonstrates that with proper a BMS the number of BEVs will continue to rise, and LIBs are the main source of energy. This is because there are many aspects of the reliability process, such as requirement analysis, modelling and simulation, control strategy research, and online hardware testing of developing a BMS which requires a model to identify the characteristics of LIBs [31]. In recent years, a tremendous growth in sales of battery electric cars has been experienced and this puts more pressure on the battery technology. Table 2 shows the battery electric car stock by country, 2005–2016.

Table 2. Battery electric car stock by country, 2005–2016 (thousands).

	2005	2006	2007	2008	2009	2010	2011	2012	2013	2014	2015	2016
Canada							0.22	0.84	2.48	5.31	9.69	14.91
China					0.48	1.57	6.32	15.96	30.57	79.48	226.19	483.19
France	0.01	0.01	0.01	0.01	0.12	0.30	2.93	8.60	17.38	27.94	45.21	66.97
Germany	0.02	0.02	0.02	0.09	0.10	0.25	1.65	3.86	9.18	17.52	29.60	40.92
India				0.37	0.53	0.88	1.33	2.76	2.95	3.35	4.35	4.80
Japan					1.08	3.52	16.13	29.60	44.35	60.46	70.93	86.39
Korea						0.06	0.34	0.85	1.45	2.76	5.67	10.77
Netherlands				0.01	0.15	0.27	1.12	1.91	4.16	6.83	9.37	13.11
Norway			0.01	0.26	0.40	3.35	5.38	9.55	19.68	41.80	72.04	98.88
Sweden							0.18	0.45	0.88	2.12	5.08	8.03
United Kingdom	0.22	0.55	1.00	1.22	1.40	1.65	2.87	4.57	7.25	14.06	20.95	31.46
United States	1.12	1.12	1.12	2.58	2.58	3.77	13.52	28.17	75.86	139.28	210.33	297.06
Others					0.64	0.80	3.17	5.83	10.60	19.43	36.20	52.41
Total	**1.37**	**1.70**	**2.16**	**4.54**	**7.48**	**16.42**	**55.16**	**112.95**	**226.79**	**420.34**	**745.61**	**1208.90**

(Source: Global EV Outlook 2017 report: two million and counting [30]).

Due to the promising growth in sales of battery-powered cars, there is an increased research interest towards the BMS of LIBs. This is attributed to the need of models and technologies for accurate estimation of a battery's RUL for different high-impact applications, including mobility applications in EVs [32]. Additionally, LIBs are the most promising power source for EVs due to their numerous benefits, like being lightweight, their high energy density, and relatively low self-discharge compared to nickel-cadmium (NI-cad) and NiMH batteries [33,34]. Battery health management and RUL estimation are performed in order to ascertain the current and previous battery states and to predict the future state and RUL.

The BMS is a hardware and software system that is in charge of battery protection and SOH, SOC, and state-of-function (SOF) estimation [6,35]. The key performance parameters tradeoffs, like safety, life span, performance, charging time, and cost, are managed by the BMS. Among those, an accurate quantification of the battery state is one of the most critical tasks for the BMS, along with the task of supervising lithium-ion cells when they are used in large battery packs [10]. In LIB systems, the BMS is used to maintain safety specifications of the battery system by ensuring that each cell is equally charged and voltage balance exists in the battery pack [36]. This means that a reliable BMS is crucial; otherwise the LIBs can move into the danger zone below the threshold area, which can be catastrophic, or lead to reduced performance.

2.1. Battery Terminologies

RUL: RUL is the remaining time or number of load cycles that the battery has during which it will be able to meet its operating requirements [6,7,37], or it is simply the length of time from the present time to the end of life [38]. RUL has attracted major emphasis in research and manufacturing vehicles' BMS so as to meet the requirement of reduced costs, increased accuracy and reliability, and avoidance of catastrophic failure. RUL can be computed as:

$$RUL = T_f - T_c \tag{1}$$

where T_f is a random variable of time of failure when degradation is detected, and T_C is the current time when the predicted signal passes the failure time with some confidence to show uncertainty of the prediction [14]. The different sources of uncertainty, however, must be propagated together with the confidence of prediction and RUL estimation, since inherent uncertainties of the degradation process, measurements, environmental/operational conditions, and modeling errors exist.

SOC: SOC represents the available capacity and is one of the most important states that needs to be monitored to optimize the performance and extend the lifetime of the batteries [6]. The battery SOC is an expression of the present battery capacity as a percentage of the maximum capacity. SOC is estimated according to such conditions as working current, temperature, and voltage [39]. The SOC is generally calculated using the current integration to determine the change in battery capacity over time. If we consider a completely discharged battery, with $I_b(\tau)$ as the charging current, the charge delivered to the battery is $\int_{t_0}^{t} I_b(\tau)d\tau$. The SOC of the battery is simply expressed as:

$$SOC(t) = \frac{\int_{t_0}^{t} I_b(\tau)d\tau}{Q_0} \times 100\% \tag{2}$$

as the charging current, the charge where Q_0 is the battery capacity at time t. According to Saxena et al. [40] estimation of SOC is, by far, the most popular approach where charge counting or current integration is used in different ways to estimate battery capacity. This makes SOC estimation the most important approach in battery management since it represents the available battery capacity, which enables performance optimization and extension of battery lifetime [6]. BMS prevents the battery from discharging below a certain SOC and charging when it is full [41]. From specifications of EV batteries, as shown in Table 3, the safety range for charging and discharging is about $-20\,°C$ to $60\,°C$.

DOD: The depth-of-discharge (DOD) is the percentage (%) of the battery capacity that has been discharged, expressed as a percentage of maximum capacity. A discharge of at least 80% DOD [42], is referred to as deep discharge, and many studies assume a fixed cycle lifetime. This is a strong simplification of reality as a traction battery will not be fully discharged every single time until the allowed minimum SOC of 20%. The battery DOD is given by the equation:

$$\text{DOD}(t) = 1 - \text{SOC}(t) = \frac{Q_0 - \int\limits_0^t I_b(\tau)d\tau}{Q_0} \times 100\% \qquad (3)$$

It is common that when the DOD is higher the shorter the cycle life. To achieve this higher cycle life, a larger battery can be used for a lower DOD during normal operations [10].

SOH: The SOH is a function of depth-of-charge, and is defined as the ratio of the maximum charge capacity of an aged battery to the maximum charge capacity when the battery is new [10]. The indicator that the battery capability to store energy is deteriorating, and decreases over the battery lifetime, is the measured SOH [6]. The BMS of EVs ensures that the battery cells charge within the safety ranges. SOH is tracked by measuring the internal resistance, since the internal resistance increases as the capacity increases. The SOH is computed as:

$$\text{SOH} = \frac{Q_{act}}{Q_R} \times 100\% \qquad (4)$$

where Q_R is the rated capacity and Q_{act} is the actual battery capacity.

EOL: Prognostics are focused on predicting when a fault, damage, or wear of a component, subsystem, or system will progress to a point that is deemed unsafe, or which a system will not function as specified [43,44]. This time point is called EOL, and the time remaining until that point is reached is known as RUL [45]. At the EOL the system value of performance is deemed to be unreliable, or can lead to failure of the EVs. When the battery usage cycle reaches the EOL, the prediction accuracy increases, and prediction variance gradually decreases [32].

Table 3. Specifications of EVs LIB.

Energy density (W/Kg)	72 to 200 (chargeable electric energy per weight of battery pack)
Nominal voltage	3.7 V
Power density	1800 (proportion of dischargeable electric energy to charged energy)
Overcharge tolerance	Very low
Cycle life	500 to 1000 (Number of charge/discharge cycles in battery's entire life)
Operating rate of temperature	$-20\,°C$ to $60\,°C$
Energy efficiency	85 to 98%
Energy cost	500–2500 $/kWh
Lifetime	5 to 15 years
Limitation	High energy cost/safety

(Source: [46,47]).

EOD: End-of-discharge (EOD) is the reading of the lower battery capacity that is occasioned by the energy loss that occurs inside the battery, and a drop in the voltage that causes the battery to reach the low-end voltage cut-off sooner. EOD means the battery is empty under discharge, and that the SOC should be 0%. It can be an absolute voltage level or a variable which is compensated by loading. The prediction of EOD times for a battery has been investigated recently, to predict the time when a predefined cut-off threshold voltage is reached and the power source is no longer available [40], thus indicating that the battery pack has run out of charge [29].

2.2. Architecture of the BMS

The architecture of the BMS comprises both the hardware and software parts. This system is built to control the operational conditions of the battery to prolong its life, guarantee its safety, and provide an accurate estimation of different states of the battery for the energy management modules [6] at all battery states, whether it is in use or not, during charging/discharging. Consequently, the BMS improves battery security, reliability, and prevents overshooting and overcharge, while optimizing the performance of EVs. To meet this, the hardware architecture of the BMS for LIBs comprises six components, namely: cell monitoring (e.g., temperature, charge/discharge monitoring); passive/active cell balancing; current measurement; contactor and interlock control and monitoring; isolation monitoring; and communication interfaces to peripherals and the environment [13]. The software structure of the BMS consists of four functionalities to aid in state determination, power capability prediction, load balancing, and safety monitoring [2]. All of these tasks are run by the BMS controller, which also extracts high-level battery pack information from the individual cell within the pack, and serves as an interface between the battery pack and the vehicle system controller [48].

2.3. Stages of Performing BMS

In relation to battery management, the functional tree of a state-of-the-art BMS for large lithium-ion battery packs, as shown in Figure 1, consists of five main stages. Each phase performs a unique task from each other, but in a coordinated way to guarantee the overall objective of performing an efficient management of battery health. There are five main stages of performing BMS: condition monitoring, hazard protection, charge/discharge management, diagnosis, and data management and assessment, as explained below. Freischer et al. [49] captured all of these main stages, as shown in Figure 1, in their functional tree of a state-of-the-art BMS for large LIB packs. As a tree with one stem, but many branches and leaves, battery management contains five components, as shown in the functional state-of-the-art BMS for large LIB packs, but these components are broken down into smaller units. It starts with monitoring the battery conditions and extends to data management, which aids the decision-making process about the batteries. The batteries' overall state can reflect the kind of action to be taken, which includes regular maintenance checks. Any BMS product chosen must provide most of the functions identified in the functional tree in Figure 1.

2.3.1. Condition Monitoring

The BMS's first main function is to monitor measurable states of the battery, like pack voltage and current, cell voltage, temperature, isolation, and interlocks [50]. Data from battery states are collected at regular intervals through a procedure of monitoring carefully-selected physical parameters, which indicate the health condition or state of the equipment under given profiles [14]. The battery conditions can be influenced by both internal and external parameters, like temperature and vibrations. The literature shows that vibration load and temperature influences performance of LIBs, leading to a significant decrease in cell capacity, and deterioration inconsistencies [51]. Whenever any abnormal conditions, such as over-voltage or overheating, are detected, the BMS should notify the user and execute the preset correction procedure. In addition to these functions, the BMS also monitors the system temperature to provide a better power consumption scheme, and communicates with individual components and operators [15].

2.3.2. Hazard Protection

The battery is protected against hazards by first defining the system settings which can predict events that are hazardous. The hazards can be as a result of undercharge, overcharge [9], rise in temperature, and other unforeseen factors. The BMS monitors the battery state and the obtainable measurement data can be used to detect or predict these events, which are either internal or external. Timely response to hazardous events can be achieved through running of fault tolerance routines, cooling/heating management, interlock loop management, and external communication. The BMS should contain accurate algorithms to measure and estimate the functional status of the battery and, at the same time, be equipped with state-of-the-art mechanisms to protect the battery from hazardous and inefficient operating conditions [50]. Therefore, it is of importance to identify and quantify substances being released from the battery during tests representing misuse and abuse events, and to ensure that the amounts released are not hazardous to vehicle occupants and first aid responders [52].

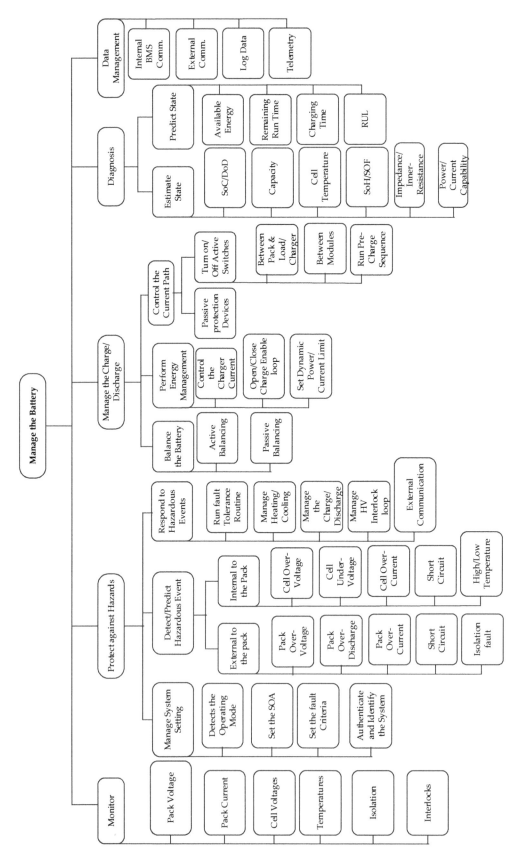

Figure 1. Functional tree of a state-of-the-art BMS for large LIB packs (Reproduced with permission from [49].

2.3.3. Charge/Discharge Management

The requirements of electrified vehicles brought up the challenge in the charge/discharge rate, making battery degradation during charge/discharge optimization extremely important. The charge/discharge starts when the demanded power crosses the threshold, and is evaluated per unit time, and according to [53] cell balancing equalizes the voltage on each cell of the battery pack. This constitutes three major sub-functions for current control, energy management, and battery balancing. The current path is controlled by employing passive protection devices, or by turning on/off active switches between the pack and load/charger, between modules, and running of a pre-charge sequence. On performing energy management the charger current is controlled, as well as opening/closing the charge-enable loop, and sets the dynamic power of current limits. At the same time battery balancing is achieved through active/passive balancing to equalize the states of cell charge. Since the main energy source of EVs is the battery, it is very crucial to have proper battery protection during charge and discharge. More so, when the EVs are travelling long distances, it is proper to predict the remaining driving distance, as this usually involves discharge of up to 80% or more. Consequently, if batteries are discharged in brine, their initial voltage will be above the electrolysis voltage of water, and hydrogen gas will be produced, thus requiring ventilation to avoid an explosion. If discharged by a resistor the current must be kept low enough so that the batteries do not overheat [54].

2.3.4. Diagnosis

Battery health diagnosis is the process of monitoring the underlying degradation to be able to track the actual performance and take countermeasures if developing faults occur [55]. In this phase the battery states are mainly estimated or predicted based on capacity, cell temperature, charge/discharge time, impedance, and power. This diagnostic mode helps to determine or predict the RUL or to observe safe and reliable battery operation while aging. The diagnosis contains functions to estimate and predict battery states. Therefore, information is used, on the one hand, to observe the safe operation of the battery while aging and, on the other hand, to perform complex algorithms, e.g., for a range estimation in EVs [2]. When a component suddenly fails and the system cannot perform its functions, maintenance actions are automatically carried out to restore the system to working order [56]. Various techniques, such as electrochemical impedance spectroscopy, slow rate cyclic voltammetry, differential thermal voltammetry, incremental capacity (IC) and differential voltage (DV) could identify and quantify degradation modes in real-time applications and could be suitable for implementation within the BMS [57].

2.3.5. Data Management and Assessment

Data are important sources of information to build prognostics models, but accuracy of prognostics suffers from inherent data uncertainties. This is attributed to factors, like lack of sufficient data, sensor noise, and unknown environmental and operating conditions, together with engineering variations [14]. Thus, the battery system data are managed in order to make decisions or take actions which can deter system failures or which can lead to catastrophes. This is achieved by internal or external communication that involves human and machine. The algorithm that performs all the tasks is shown in Figure 2.

There are various indirect methods that propose connecting the measured battery parameters (voltage, current, and temperature) with the battery SOC employing a battery model. A high-fidelity battery model is required to capture the characteristics of the real-life battery and predict its behavior under a wide variety of conditions. In the BMS state estimation algorithm, using the parameters as model inputs, the model can be used to calculate the SOC, and other states of the battery [6] and, therefore, determines the battery's more critical states [58]. A great deal of research is being done to improve the performance of estimation algorithms, as shown in Figure 2.

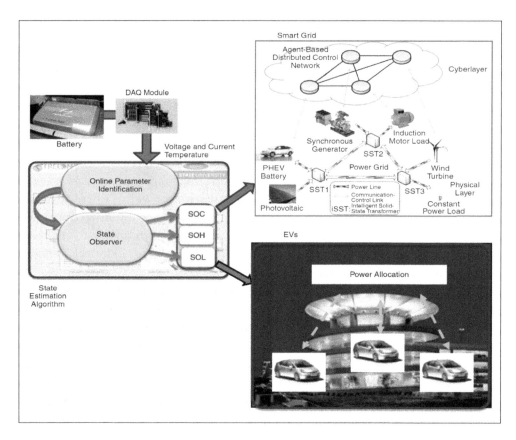

Figure 2. Battery management algorithms function in both the smart grid and EVs (Reproduced with permission from [50].

The battery parameters are captured through online parameter identification. The BMS calculates the percentage of the cell health, by monitoring the cell internal resistance, together with tracking down the weakest cell in the battery pack. The user is, therefore, provided with the information on the overall SOH of the battery pack. This demonstrates the internality of the BMS in the EVs. The EVs receive their power from the centralized power allocation system, which ensures that every vehicle's battery pack is charged to full capacity before it starts its operations. The smart grid (SG) contains various power sources, like the plug-in hybrid electric vehicle (PHEV) battery, synchronous generator, photovoltaic cell, constant power load, and wind turbine.

The emergence of the SG presents the next generation of electrical power systems, and will enable residents to have the opportunities to manage their home energy usage and reduce expenditures on energy [59]. This is due to considering sustainable development and the crisis of energy, and renewable energy production becomes an important factor in the electricity generation system. The power load data is communicated through the cyber layer, which is an agent-based distributed control network. However, if kinetic energy generated when the vehicle is in motion is transformed into charging the battery, then a great deal of savings can be made on the side of power usage, and efficiency can be improved.

2.4. Issues of the BMS

The hardware and software of the BMS have various issues that have to be addressed in order to meet the demand for secure and safe usage of LIBs. For an ideal BMS and its development process, several issues arise that must be taken into consideration seriously during the design of the BMS. These issues are illuminated below.

2.4.1. Diversity of Battery Management Applications

There are a diverse number of battery management applications. These applications include applications for monitoring of battery tests, evaluation of various state diagnosis and state predictions, rapid assembly of battery system demonstrators, tests for new sensor technologies, like sensor-less temperature measurement, etc. [60]. This variant in battery management applications calls for the development of a modular and flexible BMS. The modularized balancing system should have different equalization systems that operate inside and outside of the modules [8].

2.4.2. Handling of Potential, but Unprecedented, Hazards

LIB cells need to be monitored continuously. To maintain the safety and reliability of battery cells and the safety of people, the use of battery systems is of great use. However, some unprecedented hazards may occur which may turn out to be catastrophic. There is a missing literature on the safety analysis of the hazards of opening lithium batteries, and how discharging them mitigates these hazards. The main safety hazard of opening LIB in air is the exothermic reaction of lithium ions or lithium metal (if present) with oxygen. Opening in water results in an exothermic reaction and the generation of hydrogen gas, which is explosive [54]. Since damaged cells cannot be safely opened, this calls for recycling. Another safety hazard is discharging of the batteries themselves. If the batteries are discharged in brine, their initial voltage will be above the electrolysis voltage of water and hydrogen and oxygen gases will be produced. These gases must be ventilated to avoid an explosion. If the batteries are discharged by a resistor the current must be kept low enough so that the batteries do not overheat.

Some spent batteries are classified as hazardous wastes, increasing transportation, treatment, and disposal costs, as well as the effort needed to achieve regulatory compliance [61]. These hazards form the major function of the BMS to protect the battery against hazardous situations, while maintaining each cell of the battery within its safe and reliable operating range. However, despite the hazardous nature of spent LIBs, they also contain valuable metals, such as copper, aluminum, and cobalt with commercial potential, and the increased mining of natural ores for these metals is leading to shortages, creating a market for recycled LIBs [62,63].

2.4.3. Lack of Safe Operating Areas for Specific Battery Cells

Due to the continuous change of both the internal and external environment for the batteries, there is no single existence of a safe operating area for specific battery cells. When cells are connected in series, some discrepancies in cell internal resistance and differences in cell capacity may occur, and might lead to cell overcharging or undercharging. This inconsistency leads to unreliability and unstable efficiency of the cells, and this poses a serious problem. To solve this problem cell balancing can be used to achieve long battery life and to ensure reliability and safety [60,64,65]. Two types of algorithms are used in cell balancing: voltage-based algorithm and state of charge-based balancing algorithms [66]. The BMS is imperative for active or passive balance circuits to overcome any inconsistency problems among the serially-connected cells [67]. Therefore, it is required that the BMS is designed and developed to take control of the unending control of the change of environment that is sometimes unpredictable.

2.4.4. Ensuring an Efficient Operational State of the Peripheral Control Units and the Power Converters

Since the meaningful hardware and electrochemical properties of the battery cells are impacted by many diverse factors, it is difficult to ensure an efficient operational state of the peripheral control units [60,68]. This will, consequently, impact the design and development of the BMS for EVs. Prognostics, themselves, are useful because they supply the decision-maker with an early warning about the expected time to system/subsystem/component failure and let them decide the appropriate actions to deal with the failure. The benefit from prognostics can flourish if its information is used as the main source of system health management. The BMS not only controls the operational conditions of the battery to prolong its life and guarantee its safety, but also provides accurate estimation of the SOC and SOH for the energy management modules in the smart grid and EVs. To fulfill these tasks, a BMS has several features to control and monitor the operational state of the battery at different battery cell, battery module, and battery pack levels [50].

2.5. Prognostic Methods

Prognostics and health management (PHM) is a set of activities that monitor and estimate the system's SOH throughout its entire life and take suitable decisions at favorable times to extend the system's RUL [69]. Prognostics, itself, are useful because they supply the decision-maker with an early warning about the expected time to system/subsystem/component failure and let them decide the appropriate actions to deal with this failure [70]. The benefits from prognostics can flourish if their information are used as the main source for system health management. The least mature element, and chief component of PHM, are prognostics, which attempt to estimate the RUL of a component when a given abnormal condition has been detected [71]. The key factor is to estimate the RUL, as well as assess the confidence estimate. This makes a prognostic failure a relatively recent area of research to which the scientific community is beginning to give increasing importance, contrary to diagnostics [56,72].

Various factors, like storage voltage, internal and external battery temperatures, rate of discharge, depth-of-discharge, vibrations, etc. [73], must be taken into account when performing battery capacity degradation monitoring. According to Wu et al. [74], this LIB degradation is a nonlinear and time-varying dynamic electrochemical process, and in-depth mechanism analysis is clear in physical significance and concepts. It involves a large number of parameters and complex calculations for accurate modelling making it unsuitable for real-time monitoring and accurate modeling. The capacity degradation of LIBs is often used as a health indicator to establish degradation models. LIB failure, however, occurs when the capacity drops below a normal capacity value or failure threshold value [75]. To perform this task for safe and reliable use of LIBs [76], there are basically three methods classified as physical methods, data driven methods, and hybrid methods that are used to realize an accurate BMS. The summary of these methods is illustrated in Table 4.

Table 4. Summary of prognostics methods.

Prognostic Approaches	Categories of Approaches	Pros	Cons
Physical Approaches ([5,6,37])	Electrical Circuit Model-Based Estimation (ECM) and Electro-chemical Model-Based Estimation (EChM)	- Gives accurate predictions of the temperature distribution - Shows better performance - Simplicity	- The test has to be conducted under exact conditions - Some measurements must be conducted via invasive operation - Some instruments cannot be utilized into real application - Hard to identify the parameters in the model - Parameters may change along with the working condition.
Data-based Approaches ([6,72,74])	Machine Learning Approaches	- Does not need a data model - The algorithms are simple and feasible - The algorithms are the best solution for non-linear systems	- The point estimated value of RUL - Does not describe the uncertainty of measurement results
	Filtering Approaches	- Can be used in any form of state-space model - Best solution for non-linear, Gaussian, and non-Gaussian systems	- Needs data mode (state-space model) - The point estimated value of RUL

Table 4. *Cont.*

Prognostic Approaches	Categories of Approaches	Pros	Cons
	Stochastic Approaches	- Considers the time-dependence of the degradation process - Describes the uncertainty of predictable results	- Higher calculation complexity - Considers uncertain factors
Hybrid Approaches [26,77–82]	Series/Parallel Approach	- Achieves higher accuracy than conventional methods - Increases process reliability and robustness	- Reliability is valid only for given conditions and a period of time.

2.5.1. Physical Methods

The physical method is also called the model-based method. It comprises of an electrical circuit model-based estimation (ECM) method, and an electrochemical circuit model-based estimation (EChM) method. In the ECM discrete-time identification methods are less robust due to undesired sensitivity issues in the transformation of discrete domain parameters. This method promises simplicity by way of enabling easy implementation on a low-cost target microcontroller. It shows better performance in a low SOC range compared with one that uses average SOC in the ECM. The battery's nonlinear dynamic behavior identification could increase significantly as this method is quite accurate. When it comes to temperature distribution through the cell surface, and the behavior under various operating conditions, the ECM gives accurate predictions as it could be used in enhanced SOC estimation procedures.

The EChM includes dependence of the battery behavior on SOC and temperature. However excess temperature can greatly accelerate the battery aging process and even cause fire or explosion in the battery pack under severe cases. The tests are to be conducted under exact conditions despite the same measurements being conducted under invasive operations. The parameters are difficult to be identified in this model, as they change along with working conditions. In general the battery degradation increases if it is kept at a high SOC. Currently, countries and vehicle manufacturers are announcing aggressive targets for completely phasing out internal combustion engines, and EVs will get a great boost [83].

2.5.2. Data-Based Methods

The data-driven prognostics typically require sufficient offline training datasets for accurate remaining useful life for engineering products [84]. Data-based approaches of battery modeling use the battery's SOH data, which can be measured through advanced sensor technology to extract effective feature information, and construct the degradation model to predict RUL [74]. The data-based models are based on three methods, namely, machine learning or artificial intelligence (AI), filtering, and stochastic approaches. The machine learning method is a probabilistic method meant to improve the performance of estimation algorithms. There are four approaches under this algorithm, namely, particle swarm optimization (PSO), genetic algorithm-based estimation (GA), fuzzy-based neural networks (ANFIS), and fuzzy logic-based estimation (FL). The AI method does not need a data model, is simple, feasible, and is the best solution for non-linear systems [37], but this method does not describe the uncertainty of measurement results.

The filtering technique is used in any form of space model and is the best solution for non-linear, Gaussian, and non-Gaussian systems. However, this model needs a data model (state-space model). On the other hand, the stochastic technique is desired because it describes the uncertainty of the predictable results and considers the time dependence of the degradation process. However, this approach involves calculation complexity, and considers uncertain factors. The main advantage of this approach is its precision, since the predictions are achieved based on a mathematical model of the degradation. However, the derived degradation model is specific to a particular kind of component or material and, thus, cannot be generalized to all the system's components. In addition to this, obtaining a mathematical model of degradation is not an easy task and needs well-instrumented test benches, which can be expensive [85]. The advantage of using a data-driven prognostic approach is its applicability, cost, and implementation.

2.5.3. Hybrid Methods

This method constitutes the series and parallel approaches. This method is usually based on combining various physical and data-based approaches to leverage the strengths from both categories. This method proves to be exhibiting more strength than its predecessors since it narrows down their weaknesses. Thus, this method increases process reliability and robustness by combining the complementary information from different prognostic methods in intelligent ways compared to a single model-based method [77,78,81,82]. However, this method has a limitation of offering reliable validity only for certain given conditions and periods of time. These benefits of hybrid methods explain why it is gaining popularity compared to its counterparts.

2.6. Battery Management System Framework

Figure 3 is an elaborate basic framework with descriptions of the software and hardware of the BMS of EVs. The hardware component is embedded into the EV equipment, and is coupled with instructions on how to perform certain basic operations to ensure smooth running of EVs. This system will send signal warnings to the driver and risk responders whenever they sense some element of danger so that the necessary action can be taken. Close monitoring and frequent maintenance operations for this kind of system is crucial, so that any eventuality can be countered well in advance. The BMS achieves this by rigorously opening the contactors in the case of harsh limits violations to prevent the battery operating beyond its limits [12].

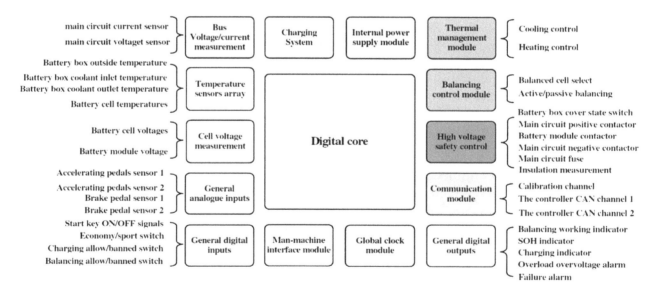

Figure 3. Basic framework of software and hardware of the BMS in vehicles (Reproduced with permission from [39].

The inputs that the BMS should have are a main circuit current sensor and voltage sensor to measure the main current and voltage; temperature sensors to measure the cell's temperature, the temperature outside the battery box, and maybe also the temperature at the battery coolant inlet and outlet; general analog inputs, like an accelerator pedal sensor and brake pedal sensor; and general inputs, like start key (ON/OFF) signals, charging allow/banned switch, etc. Consequently, the BMS outputs are to the thermal management modules, like fans and electric heaters, to provide cooling and heating control; balancing modules, like capacitors plus switch arrays and dissipation resistance to provide battery equalization; voltage safety management, like a main circuit contactor and battery module contactor; general digital outputs, like a charging indicator and failure alarm;

and a communication module [39]. The global clock module and the internal power supply module, together with the charging system and man-machine interface module also exist in the BMS.

In Figure 3, the software of the BMS covers various functions, as shown. First the battery parameters are detected, which includes total and individual cell current and voltage, temperature, smoke, insulation, collision, impedance, and so on. This to prevent overcharge or over-discharge. Once all parameters have been detected, the internal battery states (SOH, DOD, SOC, SOF) are estimated and the temperature is controlled according to such conditions as working current, voltage, and temperature. This is done to prevent the battery from operating in hazardous conditions and maintains its performance for a long time [86]. However, the variations of internal parameters can be very high and, thus, affect the battery performance. This can be due to production processes as each cell in a battery pack differs and does not reach the same full charge at the same time during the charge period, and also reaches different SOC during the discharge time. This calls upon the BMS to perform cell balancing.

The BMS also performs an onboard diagnosis to identify any fault that may arise from sensors, actuation, battery, network, overcharge, over-discharge, overload, insulation, extreme temperature rise or fall, loose connections, and so on. After the faults have been diagnosed the EV's system control unit is informed through the network. If a certain threshold value is exceeded, or is likely to be, the BMS can also cut off the power supply to prevent any possible damage from taking place.

The BMS also controls the charger when charging the batteries and adopts reliable battery equalization methods based on the information of each cell that is available. This is done so that each cell's SOC is made as consistent as possible during the charging and discharging period. This ensures that the circuit should deliver a current high enough to perform the required charge redistribution during battery runtime. In order to evaluate the functionality of the overcharge/over-discharge protection system, charging or discharging of the battery is performed beyond the recommended limits by the manufacturer [52].

The heating and cooling process is determined by the BMS based on the overall temperature distribution within the battery pack and the requirements of charge/discharge. However, due to variability in manufacturing, cooling, heating, and other operating conditions, some of the crucial thermal and electrical parameters can vary from cell to cell [87]. Excessively high, low, or uneven temperature will do harm to battery performance, thus, a reasonable battery thermal management system with local thermal control management must be designed, with which it cools down under high temperature conditions, or heats up under low temperature conditions [88]. A safe battery operation is ensured for both surface temperature and internal temperature, which proves to be crucial since the battery's internal temperature can reach a critical condition much quicker than the surface temperature [6]. The EV's LIB system is monitored online, and it is usually real-time. The overall data from the EVs, such as SOC, SOH, charge/discharge, faults, and so on, is stored by the BMS. In the battery arrangement for EVs, one battery within the pack must be assigned as the master controller, while the rest are assigned as the slaves. In summary, the real-time battery terminal voltage, cell working temperatures, and load current are measured by the BMS, and the online parameters of the battery pack model can be identified by the reduced labeled samples (RLS) based on the real-time data provided by the BMS [89].

3. Opportunities and Challenges on Prognosis of LIB Health

Since prognostics are still considered relatively immature (as compared to diagnostics), more focus, so far, has been on developing prognostic methods rather than evaluating and comparing their performances. Consequently, there is a need for dedicated attention towards developing standard methods to evaluate prognostic performance from a viewpoint of how post-prognostic reasoning will be integrated into the health management decision-making process [90]. There are many opportunities for prognostics and health management of LIBs, but they have been met with various challenges in equal measure. In this research we will look at these opportunities and challenges from four perspectives, namely, technological, financial/cost, security, and environmental. This paper evaluates each aspect from the opportunities and challenges perspective that LIB research and development is facing. Table 5 shows a summary of opportunities and challenges of LIBs.

Table 5. Summary of opportunities and challenges of LIBs.

Aspect	Opportunities	Challenges	References
Technological	- Existence of recycling technologies - Growth in demand for LIB	• Performance of higher power, larger energy capacity • RUL Prediction, SOH diagnosis, Aging • Enduring adverse circumstances • Development of low weight battery • Reliability • Degradation uncertainties	[1,9,23,60–70]
Financial/Cost	- Intelligent cost management - Cost savings	• Lower cost • Battery degradation costs	[14,24,38,61,62,71]
Security	- Battery state monitoring system	• More safety assurance	[48,52,74,88,91]
Environmental	- 3R-principle: Recycle, reuse, reduce - Mitigated emissions - Integration of renewable energies	• Disposing waste batteries • How to reduce production of new batteries	[64,68,72–85]

3.1. Technological Aspects

Currently, the vehicle manufacturers use the high-power LIB technology to supply electric and hybrid vehicles [92]. These EVs are supplied with power from a battery pack made up of modules connected in series and/or parallel, depending on the desired voltage supply and storage capacity. Out of this, there are many technological opportunities in the use of LIBs. This includes the existence of recycling technologies, and extensive demand of LIBs. The LIB technology is considered as one of the most promising for the near future by a majority of literary sources [11]. However, a great deal has to be done to ensure that batteries of higher power performance and larger energy capacity are realized, as well as improving the batteries to endure adverse circumstances. Consequently, the prediction of RUL, monitoring of battery aging, and SOH diagnosis requires technologies that are accurate in order to avoid catastrophic failures. LIBs provide a lower range of kilometers for EVs as compared to gasoline vehicles, thus, technologies require improvements towards ensuring that higher power energy density is achieved [93].

LIB technologies are very promising for the development of future-generation EVs. These sets of batteries exhibit several advantages as demonstrated, together with various appealing features making them a darling use in many applications. This is explained by the way it has caught the market share in commercialization in consumer electronics, such as cell phones, laptops, video cameras, digital cameras, power tools, and other portable electronic devices [46]. LIBs have many advantages which include [1,34,46,91,94,95]:

• Light weight: applications that make use of LIB go farther and faster due to their lightweight.
• High energy density: EV operates longer between charges while still consuming the same amount of power. LIBs are highly efficient and can be charged with electricity or renewable energies [11].
• Low self-density: the rate of self-discharge is far lower than that of lead acid batteries [96].
• No maintenance: LIBs require little to no maintenance to maintain high-performing products.
• Faster recharge: LIBs have little to no resistance, which allows you to charge at a much higher rate.
• Customizable: not only are LIBs more powerful, lighter, and hold charge longer than lead acid batteries, but they are customizable to fit your needs.

The distance travelled by EVs after full charging is determined by the size of the battery pack (Wh) against the energy consumption (Wh/km):

$$\text{Distance travelled (km)} = \frac{\text{Battery pack size(Wh)}}{\text{Energy consumption (Wh/km)}} \tag{5}$$

According to the study conducted by Martinez et al. it is revealed through the literature that the total distance an EV can cover after a full recharge process is 200 km due to their weight and number of batteries [97]. This signifies that more research is required on how the power efficiency can be increased to offer higher distances travelled by EVs.

3.2. Cost Aspects

The cost and performance of the LIBs are the most expensive component in a vehicle, and is directly linked with the adoption of EVs [18,98], however, in the recent past this cost have been reduced significantly by almost 65% from 2010 [65]. Several countries are working around the clock to see how best they can substitute fossil-based powered energies for technological options of greener/renewable energy to further cut the cost, as well as to conserve the environment. Manufacturers are working on how to employ intelligent cost management methods in order to produce low-cost LIBs. This is partly achieved by recycling used LIBs, which could have gone into waste dumped in a landfill. However, the development of LIBs, and its certification of safety-critical applications, are very expensive.

These costs can be reduced by encapsulating safety-critical components, and safety measures can be restricted to the respective parts [35]. Accounting for battery aging is crucial as the cost of LIBs has a crucial significant impact on overall system cost. The modeled battery degradation cost includes the impacts of the battery temperature, the average SOC, and the DOD on the fading LIB capacity [99]. However, the general cost of batteries reduced tremendously from 1000 $/kWh in 2008, to 268 $/kWh in 2015, which is a 73% reduction in seven years [100]. If this trend continues, it looks promising to consumers of EVs in the future, but if alternative measures to sources of energy are not sought, then the meager resources for the manufacture of LIBs will be depleted.

3.3. Security Aspects

Safety of LIBs is paramount to ensure confidence and widespread adoption of electro-mobility in our society, as they are a proven technology for automotive applications and their continuous use in the future is undeniable. For enhanced security and more accurate SOC estimation, the parameter values of the equivalent circuit models (ECMs) should be continually updated since surface temperature measurements alone might not be sufficient to ensure safe battery operation [58]. During cycling, cells within a pack exhibit non-uniform properties, which may lead to some imbalances (e.g., voltage variations between cells) that may trigger a safety hazard [52]. When EVs are used outdoors, poor pavement conditions, changes in temperature and load can cause performance degradation in LIBs. Battery degradation may lead to leakage, insulation damage, and partial short-circuit. If there is no online detection of degradation, further battery usage will cause serious situations, such as spontaneous combustion and explosions, especially if the current state of health has not been assessed in a timely fashion, or the future battery health state has not been estimated [74]. The main thermal safety issues of LIBs to be addressed are overheating, combustion, explosion, and cycle life. To avoid any catastrophic incidences caused by degradation of LIBs, and to predictively maintain the safety of vehicles, carrying out research on RUL prognostics of LIBs is of great importance [48,52,74,88].

3.4. Environmental Aspects

The use of LIBs will be the next big thing as many governments are fighting against production and sale of vehicles powered only by fossil fuels in favor of cleaner vehicles. This is in a bid to clean up the country's air, or in fighting against global warming, thus ensuring zero-emission vehicles (ZEV).

EVs are seen as an alternative to the conventional transportation based on combustion engines, looking to contribute to solving environmental issues related with zero emissions policies. For this case the EVs are projected as the most sustainable solutions for future transportation [101]. EVs have many advantages over conventional hydrocarbon internal combustion engines, including energy efficiency, environmental friendliness, noiselessness, and less dependence on fossil fuels [102].

According to the International Energy Agency (IEA) report, there is an alarming statistic that shows just how far many other countries have to go in expunging the use of fossil-powered vehicles from their roads. Globally, 95% of EVs are sold in only 10 countries: China, the U.S., Japan, Canada, Norway, the U.K., France, Germany, the Netherlands, and Sweden [103]. This global statistic still poses a challenge to researchers and environmental enthusiasts in regard to the uptake of zero-emission vehicles or green vehicles globally. There are several opportunities in relation to environmental aspects in the use, production, and sale of green vehicles, or at least hybrid ones, which is a boost to the fight against global warming and puts pressure on the extensive use, production, and sale of EVs. Several environmental opportunities that exist include, but are not limited to, the 3R principle: recycle, reuse, reduce, mitigated emissions, and integration of renewable energies.

3.4.1. 3R Principle: Recycle, Reuse, and Reduce

The sharp growing volume of spent LIBs [63,104,105], requires a well-functioning collection and recycling infrastructure to minimize associated environmental impacts and maximize the batteries' reuse potential [62]. The recycling of LIBs reduces energy consumption, reduces greenhouse gas emissions, and results in considerable natural resource savings when compared to landfill [106]. However, it is unclear which recycling processes have the least impact on the environment. There is need for incentives from government and non-governmental agencies to LIB recyclers as a motivating factor to improve recycling, thus mitigating the pressure on the scarce raw material. If the raw materials come from ores, significant negative environmental factors can occur from ore mining and processing, and these can be avoided if the material can be recycled.

Spent $LiFePO_4$ battery packs will retain approximately 80% of their performance, allowing the pack to be applied in a second application, such as a stationary energy storage system [98]. In these spent $LiFePO_4$ batteries, whenever they are recycled at the end of their useful life, a great deal of valuable metals can be recovered, such as copper, aluminum, magnesium, nickel, cobalt, and lithium, thus reducing the pressure on mining the ore, environmental contamination problems [107], and the costs of production. Consequently the LIB consumers require awareness on taking part on the 3R-principle. Since many consumers prefer new batteries, resulting in spent batteries having little potential for reuse, end up being dumped along with other urban solid waste.

Taking into account the importance of key parameters for the environmental performance of LIBs, research efforts should not only focus on energy density, but also on maximizing cycle life and charge/discharge efficiency [42]. The application of the 3R principle to LIBs will bring savings quantified in terms of energy and cumulative energy extracted from the natural environment [108]. It will be seen how material or cell recovery from existing cells will be another source of future materials for LIBs [83]. Therefore, the 3R rule for reuse, recycle, and reduce should be employed purposely to reduce the extinction of the rare iron ores, and this will go along with the conservation of the environment. Current research is aimed towards these principles to improve on the technologies around it.

3.4.2. Mitigated Emissions

The benefits of mitigated emissions include reduced air pollution and climate change, and increased integration and penetration of renewable sources of energy [109]. There are many government policies which are being set out to mitigate emissions from vehicles, as shown in Table 5. One of the policies is the encouragement of EVs, green energy-powered vehicles, and/or hybrid vehicle production, sale, and usage. Some states worldwide have set timelines and others are planning on the shift to EVs. Table 6 shows a summary of the progress so far.

Table 6. Countries that want to ban gas and diesel.

Country	Year	Expectations
Norway	2025	• All new passenger cars and vans sold by 2025 should be zero emission vehicles. As per 2016, 40% of cars sold were electric and hybrid. it is leading the way
India	2030	• Projection of having every vehicle sold to be powered by electricity
France	2040 After-2040	• To end the sale of gas and diesel powered vehicles as it fights global warming • Automakers will only be allowed to sell cars that run on electricity/other cleaner power. Hybrid cars will also be permitted
Britain	2040 2050	• To ban sales of new gasoline and diesel cars in a bid to clean up the country's air. • All cars on the road will need to be have zero emissions
China	-	• Working on a plan to ban production and sale of vehicles powered only by fossil fuels.

(Source: [103,110]).

Other countries in the league are Austria, Denmark, Ireland, Netherlands, Japan, Portugal, Korea, and Spain, who have set official targets for EV sales. The USA does not have a federal policy, but at least eight states have set out goals. According to the IEA, India will join China and the USA to account for 2/3 of the world's expansion in renewable power sources from solar and wind [110]. This is a clear indicator that the demand for LIBs will soar high when new players join the league, as well as the current ones up their game.

3.4.3. Integration of Renewable Energies

The demand for electricity storage capability for EVs is on the rise, and it will increase even more in the future. To support EVs' electric storage there have recently been increases in the contribution of renewable energies to the electrical supply mainly from the installation of photovoltaic modules and wind turbines [111]. Therefore, the internal combustion engine can be replaced with small-sized photovoltaic (PV) modules located on the roof of the EVs, and a micro-wind turbine located in front of the EVS, behind the condenser of the air conditioning system [112]. This technique will improve the power efficiency, regulate the DC-link accurately, and produce suitable stator currents for the traction motor. This is followed by the fact that, by 2030, the demand for energy consumption for EVs will increase to 50% and 40% in USA and Europe, and double for India and China, respectively. Therefore, renewable energy remains the only important resource to consider [59].

4. Future Research Agenda

Currently, there is a scarcity of real prognostics to meet industrial challenges. This may be due to inherent uncertainties associated with deterioration processes, lack of sufficient quantities of data, sensor noise, unknown environmental and operating conditions, and engineering variations, etc., which prevents building prognostic models that can accurately capture the evolution of degradation [14]. This makes research in battery health management a worthy area of future research. This is attributed to the over-emphasis on electrification of our vehicles on our roads today in many countries so as to reduce emissions, and the environmental impact on the depletion of fossil fuels [113]. However, this vehicle electrification has to be regulated to ensure that desirable benefits are achieved in the long run. The solution is, perhaps, the investment on EV powered by alternative solutions.

Cost Effective Production: this is mainly an industrial topic, but a very important future research direction to reduce the price of EVs on the market and thereby widen the customer base [113]. Investment on better technologies which can cause less degradation, but lead to higher energy efficiency on a large scale, and lower long-term costs, is required to be researched. The cost and performance of the battery, the most expensive component in a vehicle, is directly linked with the adaption of electric vehicles. The adoption towards battery electric vehicles mainly depends on the willingness to pay for the extra cost of the traction battery [93]. However the cost can significantly be reduced by economies of scale, and implementing an accurate SOC estimation strategy [114]. The technological breakthroughs in battery life, abuse tolerance and drive range will eventually result in the development of cost effective, long-lasting LIBs.

Disposal of replaced battery: Since spent LiBs retain some of their electrical power, their improper disposal can cause explosions, posing massive environmental, human health or safety hazards and necessitating expensive clean-up and mitigation measures [62]. A lot has to be done to see how this challenge is over done in future. How will the replaced batteries be disposed without posing an environmental impact is another research area in future? Research has to be conducted on how to further extract their useful energy capacity to make an extra profit, while saving out the rare battery resource. Also since pent batteries are defined as hazardous waste, further improvements on basic requirements for packaging, collection, storage, transport, and disposal should be addressed adequately, mainly from scientific research [115] and industrial practices.

Life cycle assessment: research on the LIBs used in EVs must be conducted so that the life cycle study on system boundary [115] can be assessed. Materials used in LIBs production that are rare, toxic, and difficult to recycle should be avoided for improvement of the environment. According to [43], if the model simulations show that cell change-out extends pack life indefinitely while maintaining pack performance at a steady-state, the concept would be of interest to EV and battery manufacturers for its economic benefits, and will hopefully lead to a reduced load of batteries on the recycling and disposal infrastructure.

Identifiability: identifiability in battery SOH estimation is required to establish how to estimate the progress of different battery aging effects. This is perhaps one of the largest challenges in battery SOH estimation [116]. Specifically, the fact that different battery aging dynamics are complex, intertwined, and similar in their time constants means that it is fundamentally very difficult to estimate the progress of different battery aging effects online using voltage, current, and temperature measurements. This, in turn, makes it very challenging to estimate the health of LIBs online, predict their death, and control them in a manner that postpones such death. Battery identifiability remains a very open research area whose exploration can shed light on the extremely important question of what additional sensors, beyond terminal current, temperature, and voltage, can provide with respect to the best means for onboard battery health prognostics, diagnostics, and control [117], using various model estimation algorithms. There are a number of SOH estimation algorithms in battery management systems, but one of the important classes of estimation algorithms is the equivalent circuit model [51].

Developing second-use technology of retired EVs' LIBs: the already retired EVs' LIBs require some research to establish how they can be best put into meaningful second use. According to Wang et al. [94] recycling of spent LiFePO$_4$ batteries is important not only for the treatment of waste, but also for the recovery of useful resources. However they further observed that the treatment of spent LiFePO$_4$ batteries is challenging because LiFePO$_4$ batteries do not contain any precious metals, treatment is complex using traditional recycling processes, and the number of spent batteries recovered from the public has been very small recently, and is faced with challenges. The major challenge that must to be addressed by the recycling industry is developing economical ways to extract and process these metals [118]. Furthermore, there is still a lack of adequate policy and feasible technology for addressing retired LIBs [104], thus, there needs to be recycling processes developed that have economic advantages in terms of chemical costs and added value [34].

Lowering capacity degradation: the LIB capacity is influenced by many factors; among them are temperature, vibrations, and other unforeseen environmental factors. Thus, the efficiency of energy required keeps varying over time [119]. This energy efficiency in LIBs reflects charging and discharging energy powers of the same cycle, and so it is closely related to the battery capacity. Energy efficiency is defined as the percentage of energy use which actually achieves the energy service required [16]. It is given by:

$$\eta = \frac{W_D}{W_C} \times 100\% \tag{6}$$

where W_D is the energy efficiency during discharge, and W_C is the energy efficiency during charging. Due to high energy density per weight, LIBs are a better option than lead acid batteries in EV application [34,36,120,121]. However, much degradation cannot be prevented, and the real lifetime of the LIBs can be extended by using various other types of approaches.

5. Summary and Closing Remarks

EVs are considered one of the novel solutions to the transport system since they reduce over-dependence on fossil energy. This, in return, will reduce carbon emissions, as EVs act as greener solutions in the transport industry. However, research has shown that the major challenge of the LIBs is battery life. This, however, can be extended by cell balancing to ensure safety of the systems, as well as reliability. The manner in which BMS offers this safety is two-fold: safety of persons and safety of cells. The current growth of EVs is anticipated to lead to enormous penetration into the electric power grid, thus calling for innovative ways of supplementing the EV's power. This is feared because the over-dependence on electric power may add to extensive loads on the power grid, which will have extensive effects on existing distribution networks.

This paper presented a comprehensive review in terms of battery health management for EVs. First the health management systems for batteries are introduced by battery terminologies, BMS architecture, stages of performing BMS, and the prognostic methods used in performing battery health. Furthermore, the opportunities and challenges of BMS from three perspectives, namely technological, cost, security, and environmental aspects was reviewed. Finally, the future research agendas are discussed. In the future the production of cost-effective LIBs, the disposal of replaced LIBs, life-cycle assessment, development of second-use technologies of retired EV batteries, and how battery degradation can be reduced should be the focus of research. Therefore, in regard to the issues raised for research in this paper, further exploration is essential.

Author Contributions: All authors contributed to the paper. Z.B.O. wrote the manuscript with the supervision from L.Z. and D.S., L.Z. acted as a corresponding author.

References

1. Raszmann, E.; Baker, K.; Shi, Y.; Christensen, D. Modeling stationary lithium-ion batteries for optimization and predictive control. In Proceedings of the 2017 IEEE Power and Energy Conference at Illinois (PECI), Champaign, IL, USA, 23–24 February 2017. [CrossRef]
2. Liu-Henke, X.; Scherler, S.; Jacobitz, S. Verification oriented development of a scalable battery management system for lithium-ion batteries. In Proceedings of the 2017 Twelfth International Conference on Ecological Vehicles and Renewable Energies (EVER), Monte Carlo, Monaco, 11–13 April 2017. [CrossRef]
3. Ordoñez, J.; Gago, E.J.; Girard, A. Processes and technologies for the recycling and recovery of spent lithium-ion batteries. *Renew. Sustain. Energy Rev.* **2016**, *60*, 195–205. [CrossRef]
4. Dubarry, M.; Devie, A.; Liaw, B.Y. The value of battery diagnostics and prognostics. *J. Energy Power Sources* **2014**, *1*, 242–249.
5. Song, Y.; Liu, D.; Yang, C.; Peng, Y. Data-driven hybrid remaining useful life estimation approach for spacecraft lithium-ion battery. *Microelectron. Reliab.* **2017**, *75*, 142–153. [CrossRef]

6. Muñoz-Galeano, R.-B.N.; Pablo, J.; Sarmiento-Maldonado, H.O. SoC estimation for lithium-ion batteries: review and future challenges. *Electronics* **2017**, *6*, 102. [CrossRef]

7. Zhang, L.; Mu, Z.; Sun, C. Remaining useful life prediction for lithium-ion batteries based on exponential model and particle filter. *IEEE Access* **2018**, *6*, 17729–17740. [CrossRef]

8. Daowd, M.; Antoine, M.; Omar, N.; Lataire, P.; van den Bossche, P.; van Mierlo, J. Battery management system-balancing modularization based on a single switched capacitor and bi-directional DC/DC converter with the auxiliary battery. *Energies* **2014**, *7*, 2897–2937. [CrossRef]

9. Fujita, Y.; Hirose, Y.; Kato, Y.; Watanabe, T. Development of battery management system. *FUJITSU TEN Tech. J.* **2016**, *42*, 68–80.

10. Allam, A.; Onori, S.; Marelli, S.; Taborelli, C. Battery health management system for automotive applications: A retroactivity-based aging propagation study. In Proceedings of the American Control Conference (ACC), Chicago, IL, USA, 1–3 July 2015; pp. 703–716. [CrossRef]

11. Andwari, A.M.; Pesiridis, A.; Rajoo, S.; Martinez-Botas, R.; Esfahanian, V. A review of battery electric vehicle technology and readiness levels. *Renew. Sustain. Energy Rev.* **2017**, *78*, 414–430. [CrossRef]

12. Akdere, M.; Giegerich, M.; Wenger, M.; Schwarz, R.; Koffel, S.; Fühner, T.; Waldhör, S.; Wachtler, J.; Lorentz, V.R.H.; März, M. Hardware and software framework for an open battery management system in safety-critical applications. In Proceedings of the IECON 2016—42nd Annual Conference of the IEEE Industrial Electronics Society, Florence, Italy, 23–26 October 2016; pp. 5507–5512. [CrossRef]

13. Brandl, M.; Gall, H.; Wenger, M.; Lorentz, V.; Giegerich, M.; Baronti, F.; Fantechi, G.; Fanucci, L.; Roncella, R.; Saletti, R.; et al. Batteries and battery management systems for electric vehicles. In Proceedings of the 2012 Design, Automation & Test in Europe Conference & Exhibition (DATE), Dresden, Germany, 12–16 March 2012; pp. 971–976. [CrossRef]

14. Javed, K.; Gouriveau, R.; Zerhouni, N. State of the art and taxonomy of prognostics approaches, trends of prognostics applications and open issues towards maturity at different technology readiness levels. *Mech. Syst. Signal Process.* **2017**, *94*, 214–236. [CrossRef]

15. Xing, Y.; Ma, E.W.M.; Tsui, K.L.; Pecht, M. Battery management systems in electric and hybrid vehicles. *Energies* **2011**, *4*, 1840–1857. [CrossRef]

16. Wang, S.; Zhao, L.; Su, X.; Ma, P. Prognostics of lithium-ion batteries based on battery performance analysis and flexible support vector regression. *Energies* **2014**, *7*, 6492–6508. [CrossRef]

17. Bruen, T.; Hooper, J.M.; Marco, J.; Gama, M.; Chouchelamane, G.H. Analysis of a battery management system (BMS) control strategy for vibration aged Nickel Manganese Cobalt Oxide (NMC) Lithium-Ion 18650 battery cells. *Energies* **2016**, *9*, 255. [CrossRef]

18. Shareef, H.; Islam, M.M.; Mohamed, A. A review of the stage-of-the-art charging technologies, placement methodologies, and impacts of electric vehicles. *Renew. Sustain. Energy Rev.* **2016**, *64*, 403–420. [CrossRef]

19. Castro, T.S.; de Souza, T.M.; Silveira, J.L. Feasibility of electric vehicle: electricity by grid × photovoltaic energy. *Renew. Sustain. Energy Rev.* **2017**, *69*, 1077–1084. [CrossRef]

20. Coffman, M.; Bernstein, P.; Wee, S. Integrating electric vehicles and residential solar PV. *Transp. Policy* **2017**, *53*, 30–38. [CrossRef]

21. Gough, R.; Dickerson, C.; Rowley, P.; Walsh, C. Vehicle-to-grid feasibility: A techno-economic analysis of EV-based energy storage. *Appl. Energy* **2017**, *192*, 12–23. [CrossRef]

22. Hannan, M.A.; Azidin, F.A.; Mohamed, A. Hybrid electric vehicles and their challenges: A review. *Renew. Sustain. Energy Rev.* **2014**, *29*, 135–150. [CrossRef]

23. Palinski, M. A Comparison of electric vehicles and conventional automobiles: Costs and quality perspective. Bachelor's Thesis, Novia University of Applied Sciences, Vaasa, Finland, 2017.

24. Salisbury, M. Economic and Air Quality Benefits of Electric Vehicles in Nevada Executive Summary. Southwestern Energy Efficiency Project (SWEEP). 2014. Available online: http://www.swenergy.org/data/sites/1/media/documents/publications/documents/Economic_and_AQ_Benefits_of_EVs_in_NV-Sept_2014.pdf (accessed on 10 April 2018).

25. Helmers, E.; Patrick, M. Electric cars: Technical characteristics and environmental impacts. *Environ. Sci. Eur.* **2012**, *24*, 14. [CrossRef]

26. Pappas, J.C.K. A New Prescription for electric cars. *Energy Law J.* **2014**, *35*, 151–198.

27. Hall, D.; Moultak, M.; Lutsey, N. *Electric Vehicle Capitals of the World: Demonstrating the Path to Electric Drive*; International Council on Clean Transportatio: Washington, DC, USA, 2017.

28. Varun, M.; Kumar, C. Problems in electric vehicles. *Int. J. Appl. Res. Mech. Eng.* **2012**, *2*, 63–73.
29. Saha, B.; Quach, C.C.; Hogge, E.F.; Strom, T.H.; Hill, B.L.; Goebel, K. Battery health management system for electric UAVs. In Proceedings of the 2011 Aerospace Conference, Big Sky, MT, USA, 5–12 March 2011. [CrossRef]
30. Cazzola, P.; Gorner, M.; Munuera, L. Global EV Outlook 2017: Two Million and Counting. International Energy Agency (IEA). 2017. Available online: https://webstore.iea.org/global-ev-outlook-2017 (accessed on 22 February 2018).
31. Zhang, L.; Peng, H.; Ning, Z.; Mu, Z.; Sun, C. Comparative research on RC equivalent circuit models for lithium-ion batteries of electric vehicles. *Appl. Sci.* **2017**, *7*, 1002. [CrossRef]
32. Rezvani, M.; Abuali, M.; Lee, S.; Lee, J.; Ni, J. A comparative analysis of techniques for electric vehicle battery Prognostics and Health Management (PHM). *SAE Int.* **2011**, *11*. [CrossRef]
33. Xi, Z.; Jing, R.; Yang, X.; Decker, E. State of charge estimation of lithium-ion batteries considering model Bias and parameter uncertainties. In Proceedings of the ASME 2014 International Design Engineering Technical Conferences & Computers and Information in Engineering Conference, Buffalo, NY, USA, 17–20 August 2014; pp. 1–7. [CrossRef]
34. Li, L.; Bian, Y.; Zhang, X.; Xue, Q.; Fan, E.; Wu, F.; Renjie Chen, R. Economical recycling process for spent lithium-ion batteries and macro- and micro-scale mechanistic study. *J. Power Sources* **2018**, *377*, 70–79. [CrossRef]
35. Li, J.; Lai, Q.; Wang, L.; Lyu, C.; Wang, H. A method for SOC estimation based on simplified mechanistic model for LiFePO$_4$ battery. *Energy* **2016**, *114*, 1266–1276. [CrossRef]
36. Podder, S.; Khan, M.Z.R. Comparison of lead acid and Li-ion battery in solar home system of Bangladesh. In Proceedings of the 2016 5th International Conference on Informatics, Electronics and Vision, ICIEV, Dhaka, Bangladesh, 13–14 May 2016; pp. 434–438. [CrossRef]
37. Leone, G.; Cristaldi, L.; Turrin, S. A data-driven prognostic approach based on statistical similarity: An application to industrial circuit breakers. *Measurement* **2017**, *108*, 163–170. [CrossRef]
38. Wei, J.; Dong, G.; Chen, Z. Remaining useful life prediction and state of health diagnosis for lithium-ion batteries using particle filter and support vector regression. *IEEE Trans. Ind. Electron.* **2017**, *65*, 5634–5643. [CrossRef]
39. Lu, L.; Han, X.; Li, J.; Hua, J.; Ouyang, M. A review on the key issues for lithium-ion battery management in electric vehicles. *J. Power Sources* **2013**, *226*, 272–288. [CrossRef]
40. Saxena, A.; Celaya, J.R. Roychoudh loading profiles: Some lessons learned. In Proceedings of the European Conference on Prognostics and Health Management Society, Dresden, Germany, 3–5 July 2012.
41. Chen, A.; Sen, P.K. Advancement in battery technology: a state-of-the-art review. In Proceedings of the 2016 IEEE Industry Applications Society Annual Meeting, Portland, OR, USA, 2–6 October 2016. [CrossRef]
42. Peters, J.F.; Baumann, M.; Zimmermann, B.; Braun, J.; Weil, M. The environmental impact of Li-Ion batteries and the role of key parameters—A review. *Renew. Sustain. Energy Rev.* **2017**, *67*, 491–506. [CrossRef]
43. Mathew, M.; Kong, Q.H.; McGrory, J.; Fowler, M. Simulation of lithium ion battery replacement in a battery pack for application in electric vehicles. *J. Power Sources* **2017**, *349*, 94–104. [CrossRef]
44. Su, X.; Wang, S.; Pecht, M.; Zhao, L.; Ye, Z. Interacting multiple model particle filter for prognostics of lithium-ion batteries. *Microelectron. Reliab.* **2017**, *70*, 59–69. [CrossRef]
45. Khorasgani, H.; Biswas, G.; Sankararaman, S. Methodologies for system-level remaining useful life prediction. *Reliab. Eng. Syst. Saf.* **2016**, *154*, 8–18. [CrossRef]
46. Chen, X.; Shen, W.; Vo, T.T.; Cao, Z.; Kapoor, A. An overview of lithium-ion batteries for electric vehicles. In Proceedings of the 2012 10th International Power & Energy Conference, Ho Chi Minh City, Vietnam, 12–14 December 2012; pp. 230–235. [CrossRef]
47. Piromjit, P.; Tayjasanant, T. Peak-demand management for improving undervoltages in distribution systems with electric vehicle connection by stationary battery. In Proceedings of the 2017 IEEE Transportation Electrification Conference and Expo, Asia-Pacific (ITEC Asia-Pacific), Harbin, China, 7–10 August 2017. [CrossRef]
48. Zhang, F.; Rehman, M.M.U.; Zane, R.; Maksimovic, D. Hybrid balancing in a modular battery management system for electric-drive vehicles. In Proceedings of the 2017 IEEE Energy Conversion Congress and Exposition (ECCE), Cincinnati, OH, USA, 1–5 October 2017; pp. 578–583. [CrossRef]

49. Fleischer, C.; Sauer, D.U.; Barreras, J.V.; Schaltz, E.; Christensen, A.E. Development of software and strategies for battery management system testing on HIL simulator. In Proceedings of the 2016 Eleventh International Conference on Ecological Vehicles and Renewable Energies (EVER), Monte Carlo, Monaco, 6–8 April 2016. [CrossRef]

50. Rahimi-Eichi, H.; Ojha, U.; Baronti, F.; Chow, M.-Y. Battery management system: An overview of its application in the smart grid and electric vehicles. *IEEE Ind. Electron. Mag.* **2013**, 4–16. [CrossRef]

51. Zhang, L.; Ning, Z.; Peng, H.; Mu, Z.; Sun, C. Effects of vibration on the electrical performance of lithium-ion cells based on mathematical statistics. *Appl. Sci.* **2017**, *7*, 802. [CrossRef]

52. Ruiz, V.; Pfrang, A.; Kriston, A.; Omar, N.; van den Bossche, P.; Boon-Brett, L. A review of international abuse testing standards and regulations for lithium ion batteries in electric and hybrid electric vehicles. *Renew. Sustain. Energy Rev.* **2018**, *81*, 1427–1452. [CrossRef]

53. Artakusuma, D.D.; Afrisal, H.; Cahyadi, A.I.; Wahyunggoro, O. Battery management system via bus network for multi battery electric vehicle. In Proceedings of the 2014 IEEE International Conference on Electrical Engineering and Computer Science, Kuta, Indonesia, 24–25 November 2014; pp. 179–181. [CrossRef]

54. Sonoc, A.; Jeswiet, J.; Soo, V.K. Opportunities to improve recycling of automotive lithium ion batteries. *Procedia CIRP* **2015**, *29*, 752–757. [CrossRef]

55. Nuhic, A.; Terzimehic, T.; Soczka-guth, T.; Buchholz, M.; Dietmayer, K. Health diagnosis and remaining useful life prognostics of lithium-ion batteries using data-driven methods. *J. Power Sources* **2013**, *239*, 680–688. [CrossRef]

56. Belkacem, L.; Simeu-abazi, Z.; Dhouibi, H.; Gascard, E.; Messaoud, H. Diagnostic and prognostic of hybrid dynamic systems: Modeling and RUL evaluation for two maintenance policies. *Reliab. Eng. Syst. Saf.* **2017**, *164*, 98–109. [CrossRef]

57. Pastor-Fernández, C.; Widanage, W.D.; Chouchelamane, G.H.; Marco, J. A SoH diagnosis and prognosis method to identify and quantify degradation modes in li-ion batteries using the IC/DV technique. In Proceedings of the 6th Hybrid and Electric Vehicles Conference (HEVC 2016), London, UK, 2–3 November 2016. [CrossRef]

58. Hannan, M.A.; Lipu, M.S.H.; Hussain, A.; Mohamed, A. A review of lithium-ion battery state of charge estimation and management system in electric vehicle applications: Challenges and recommendations. *Renew. Sustain. Energy Rev.* **2017**, *78*, 834–854. [CrossRef]

59. Melhem, F.Y.; Grunder, O.; Hammoudan, Z.; Moubayed, N. Optimization and energy management in smart battery storage system with integration of electric vehicles. *Can. J. Electr. Comput. Eng.* **2017**, *40*, 128–138. [CrossRef]

60. Giegerich, M.; Akdere, M.; Freund, C.; Fuhner, T.; Grosch, J.L.; Koffel, S.; Schwarz, R.; Waldhor, S.; Wenger, M.; Lorentz, V.R.H.; et al. Open, flexible and extensible battery management system for lithium-ion batteries in mobile and stationary applications. In Proceedings of the 2016 IEEE 25th International Symposium on Industrial Electronics (ISIE), Santa Clara, CA, USA, 8–10 June 2016; pp. 991–996. [CrossRef]

61. Gaines, L. The future of automotive lithium-ion battery recycling: Charting a sustainable course. *Sustain. Mater. Technol.* **2014**, *1–2*, 2–7. [CrossRef]

62. Zeng, X.; Li, J.; Liu, L. Solving spent lithium-ion battery problems in China: Opportunities and challenges. *Renew. Sustain. Energy Rev.* **2015**, *52*, 1759–1767. [CrossRef]

63. Wei, J.; Zhao, S.; Ji, L.; Zhou, T.; Miao, Y.; Scott, K.; Li, D.; Yang, J.; Wu, X. Reuse of Ni-Co-Mn oxides from spent Li-ion batteries to prepare bifunctional air electrodes. *Resour. Conserv. Recycl.* **2018**, *129*, 135–142. [CrossRef]

64. Yusof, M.S.; Toha, S.F.; Kamisan, N.; Hashim, N.N.W.N.; Abdullah, M. Battery cell balancing pptimisation for battery management system. *IOP Conf. Ser. Mater. Sci. Eng.* **2017**, *184*, 012021. [CrossRef]

65. Rahman, A.; Rahman, M.; Rashid, M. Wireless battery management system of electric transport. *IOP Conf. Ser. Mater. Sci. Eng.* **2017**, *260*, 012029. [CrossRef]

66. Piao, C.; Wang, Z.; Cao, J.; Zhang, W.; Lu, S. Lithium-ion battery cell-balancing algorithm for battery management system based on real-time outlier detection. *Math. Probl. Eng.* **2015**, *2015*. [CrossRef]

67. Lin, J.-C.M. Development of a new battery management system with an independent balance module for electrical motorcycles. *Energies* **2017**, *10*, 1289. [CrossRef]

68. Liang, Y.; Wang, Y.; Han, D. Design of energy storage management system based on FPGA in design of energy storage management system based on FPGA in Micro-Grid. *IOP Conf. Ser. Earth Environ. Sci.* **2017**, *108*, 052040. [CrossRef]

69. Sutharssan, T.; Montalvao, D.; Chen, Y.K.; Wang, W.-C.; Pisac, C.; Elemara, H. A review on prognostics and health monitoring of proton exchange membrane fuel cell. *Renew. Sustain. Energy Rev.* **2017**, *75*, 440–450. [CrossRef]

70. Elattar, H.M.; Elminir, H.K.; Riad, A.M. Prognostics: A literature review. *Complex Intell. Syst.* **2016**, *2*, 125–154. [CrossRef]

71. Goebel, K.; Saha, B.; Saxena, A.; Celaya, J.R.; Christophersen, J.P. Prognostics in battery health management. *IEEE Instrum. Meas. Mag.* **2008**, *11*, 33–40. [CrossRef]

72. Tsui, K.L.; Chen, N.; Zhou, Q.; Hai, Y.; Wang, W. Prognostics and health management: A review on data driven approaches. *Math. Probl. Eng.* **2014**, *2015*. [CrossRef]

73. Wang, D.; Yang, F.; Zhao, Y.; Tsui, K.L. Battery remaining useful life prediction at different discharge rates. *Microelectron. Reliab.* **2017**, *78*, 212–219. [CrossRef]

74. Wu, L.; Fu, X.; Guan, Y. Review of the remaining useful life prognostics of vehicle lithium-ion batteries using data-driven methodologies. *Appl. Sci.* **2016**, *6*, 166. [CrossRef]

75. Zhou, D.; Xue, L.; Song, Y.; Chen, J. On-Line Remaining useful life prediction of lithium-ion batteries based on the optimized Gray Model GM(1,1). *Batteries* **2017**, *3*, 21. [CrossRef]

76. Arachchige, B.; Perinpanayagam, S.; Jaras, R. Enhanced prognostic model for lithium ion batteries based on particle filter state transition model modification. *Appl. Sci.* **2017**, *7*, 1172. [CrossRef]

77. Zhou, D.; Gao, F.; Breaz, E.; Ravey, A.; Miraoui, A. Degradation prediction of PEM fuel cell using a moving window based hybrid prognostic approach. *Energy* **2017**, *138*, 1175–1186. [CrossRef]

78. Chang, Y.; Fang, H.; Zhang, Y. A new hybrid method for the prediction of the remaining useful life of a lithium-ion battery. *Appl. Energy* **2017**, *206*, 1564–1578. [CrossRef]

79. Skima, H.; Medjaher, K.; Varnier, C.; Dedu, E.; Bourgeois, J. Microelectronics Reliability A hybrid prognostics approach for MEMS: From real measurements to remaining useful life estimation. *Microelectron. Reliab.* **2016**, *65*, 79–88. [CrossRef]

80. Wu, X.; Ye, Q.; Wang, J. A hybrid prognostic model applied to SOFC prognostics. *Int. J. Hydrog. Energy* **2017**, *42*, 25008–25020. [CrossRef]

81. Li, Z.; Wu, D.; Hu, C.; Terpenny, J. An ensemble learning-based prognostic approach with degradation-dependent weights for remaining useful life prediction. *Reliab. Eng. Syst. Saf.* **2018**. [CrossRef]

82. Khalastchi, E.; Kalech, M.; Rokach, L. A hybrid approach for improving unsupervised fault detection for robotic systems. *Expert Syst. Appl.* **2017**, *81*, 372–383. [CrossRef]

83. Olivetti, E.A.; Ceder, G.; Gaustad, G.G.; Fu, X. Lithium-ion battery supply chain considerations: Analysis of potential bottlenecks in critical metals. *Joule* **2017**, *1*, 229–243. [CrossRef]

84. Xi, Z.; Zhao, X. Data driven prognostics with lack of training data sets. In Proceedings of the ASME 2015 International Design Engineering Technical Conferences & Computers and Information in Engineering Conference, Boston, MA, USA, 2–5 August 2015. [CrossRef]

85. Medjaher, K.; Zerhouni, N.; Baklouti, J. Data-driven prognostics based on health indicator construction: Application to PRONOSTIA's data. In Proceedings of the 2013 European Control Conference (ECC), Zurich, Switzerland, 17–19 July 2013; pp. 1451–1456.

86. Orcioni, S.; Ricci, A.; Buccolini, L.; Scavongelli, C.; Conti, M. Effects of variability of the characteristics of single cell on the performance of a lithium-ion battery pack. In Proceedings of the 2017 13th Workshop on Intelligent Solutions in Embedded Sysems (WISES), Hamburg, Germany, 12–13 June 2017; pp. 15–21. [CrossRef]

87. Dey, S.; Perez, H.E.; Moura, S.J. Model-based battery thermal fault diagnostics: Algorithms, analysis, and experiments. *IEEE Trans. Control Syst. Technol.* **2017**, 1–12. [CrossRef]

88. Ye, X.; Zhao, Y.; Quan, Z. Thermal management system of lithium-ion battery module based on micro heat pipe array. *Int. J. Energy Res.* **2017**, *42*, 648–655. [CrossRef]

89. Zhang, X.; Wang, Y.; Liu, C.; Chen, Z. A novel approach of remaining discharge energy prediction for large format lithium-ion battery pack. *J. Power Sources* **2017**, *343*, 216–225. [CrossRef]

90. Saxena, A.; Celaya, J.; Saha, B.; Saha, S.; Goebel, K. Metrics for offline evaluation of prognostic performance. *Int. J. Progn. Health Manag.* **2010**, *1*, 1–20.

91.   Sreejith, R.; Rajagopal, K.R. An insight into motor and battery selections for three-wheeler electric vehicle. In Proceedings of the 2016 IEEE 1st International Conference on Power Electronics, Intelligent Control and Energy Systems (ICPEICES), Delhi, India, 4–6 July 2016; pp. 9–14. [CrossRef]

92.   Rizoug, N.; Sadoun, R.; Mesbahi, T.; Bartholumeus, P.; LeMoigne, P. Aging of high power Li-ion cells during real use of electric vehicles. *IET Electr. Syst. Transp.* **2017**, *7*, 14–22. [CrossRef]

93.   Fotouhi, A.; Auger, D.J.; Propp, K.; Longo, S.; Purkayastha, R.; O'Neill, L.; Walus, S. Lithium-sulfur cell equivalent circuit network model parameterization and sensitivity Analysis. *IEEE Trans. Veh. Technol.* **2017**, *66*, 7711–7721. [CrossRef]

94.   Wang, W.; Wu, Y. An overview of recycling and treatment of spent LiFePO$_4$ batteries in China. *Resour. Conserv. Recycl.* **2017**, *127*, 233–243. [CrossRef]

95.   Gong, Y.; Yu, Y.; Huang, K.; Hu, J.; Li, C. Evaluation of lithium-ion batteries through the simultaneous consideration of environmental, economic and electrochemical performance indicators. *J. Clean. Prod.* **2018**, *170*, 915–923. [CrossRef]

96.   Amjad, S.; Neelakrishnan, S.; Rudramoorthy, R. Review of design considerations and technological challenges for successful development and deployment of plug-in hybrid electric vehicles. *Renew. Sustain. Energy Rev.* **2010**, *14*, 1104–1110. [CrossRef]

97.   Martínez, D.A.; Poveda, J.D.; Montenegro, D. Li-Ion battery management system based in fuzzy logic for improving electric vehicle autonomy. In Proceedings of the 2017 IEEE Workshop on Power Electronics and Power Quality Applications (PEPQA), Bogota, Colombia, 31 May–2 June 2017. [CrossRef]

98.   Berckmans, G.; Messagie, M.; Smekens, J.; Omar, N.; Vanhaverbeke, L.; van Mierlo, J. Cost projection of state of the art lithium-ion batteries for electric vehicles up to 2030. *Energies* **2017**, *10*, 1314. [CrossRef]

99.   Badawy, M.O.; Sozer, Y. Power flow management of a grid tied PV-battery system for electric vehicles charging. *IEEE Trans. Ind. Appl.* **2017**, *53*, 1347–1357. [CrossRef]

100. Cazzola, P.; Gorner, M.; Yi, J.T.; Yi, W. Global EV Outlook 2016 Electric Vehicles Initiative. International Energy Agency. 2016. Available online: https://www.iea.org/publications/freepublications/.../Global_EV_Outlook_2016.pdf (accessed on 10 February 2018).

101. Purwadi, A.; Shani, N.; Heryana, N.; Hardimasyar, T.; Firmansyah, M.; Sr, A. Modelling and analysis of electric vehicle DC fast charging infrastructure based on PSIM. In Proceedings of the 2013 1st International Conference on Artificial Intelligence, Modelling and Simulation, Kota Kinabalu, Malaysia, 3–5 December 2013; pp. 359–364. [CrossRef]

102. Saw, L.H.; Tay, A.A.O.; Zhang, L.W. Thermal management of lithium-ion battery pack with liquid cooling. In Proceedings of the 2015 31st Thermal Measurement, Modeling & Management Symposium (SEMI-THERM), San Jose, CA, USA, 15–19 March 2015; pp. 298–302. [CrossRef]

103. Petroff, A. These Countries Want to Ditch Gas and Diesel Cars–26 July 2017. *CNNMoney*, 11 September 2017. Available online: http://money.cnn.com/2017/07/26/autos/countries-that-are-banning-gas-cars-for-electric/index.html (accessed on 4 February 2018).

104. Wang, M.M.; Zhang, C.C.; Zhang, F.S. Recycling of spent lithium-ion battery with polyvinyl chloride by mechanochemical process. *Waste Manag.* **2017**, *67*, 232–239. [CrossRef] [PubMed]

105. Chen, X.; Ma, H.; Luo, C.; Zhou, T. Recovery of valuable metals from waste cathode materials of spent lithium-ion batteries using mild phosphoric acid. *J. Hazard. Mater.* **2017**, *326*, 77–86. [CrossRef] [PubMed]

106. Boyden, A.; Soo, V.K.; Doolan, M. The environmental impacts of recycling portable lithium-ion batteries. *Procedia CIRP* **2016**, *48*, 188–193. [CrossRef]

107. Fan, B.; Chen, X.; Zhou, T.; Zhang, J.; Xu, B. A sustainable process for the recovery of valuable metals from spent lithium-ion batteries. *Waste Manag. Res.* **2016**, *34*, 474–481. [CrossRef] [PubMed]

108. Dewulf, J.; Van der Vorst, G.; Denturck, K.; Van Langenhove, H.; Ghyoot, W.; Tytgat, J.; Vandeputte, K. Recycling rechargeable lithium ion batteries: Critical analysis of natural resource savings. *Resour. Conserv. Recycl.* **2010**, *54*, 229–234. [CrossRef]

109. Sovacool, B.K.; Axsen, J.; Kempton, W. The future promise of Vehicle-to-Grid (V2G) integration: A sociotechnical review and research agenda. *Annu. Rev. Environ. Resour.* **2017**, *42*, 377–406. [CrossRef]

110. OECD/IEA. World Energy Outlook 2017 Executive Summary. International Energy Agency, 2017. Available online: http://www.iea.org/publications/freepublications/publication/world-energy-outlook-2017---executive-summary---english-version.html (accessed on 7 February 2018).

111. Helbig, C.; Bradshaw, A.M.; Wietschel, L.; Thorenz, A.; Tuma, A. Supply risks associated with lithium-ion battery materials. *J. Clean. Prod.* **2018**, *172*, 274–286. [CrossRef]

112. Fathabadi, H. Plug-in Hybrid Electric Vehicles (PHEVs): Replacing Internal Combustion Engine with Clean and Renewable Energy Based Auxiliary Power Sources. *IEEE Trans. Power Electron.* **2018**. [CrossRef]

113. Grunditz, E.A.; Thiringer, T. Performance analysis of current BEVs based on a comprehensive review of specifications. *IEEE Trans. Transp. Electrification* **2016**, *2*, 270–289. [CrossRef]

114. Bashash, S.; Moura, S.J.; Fathy, H.K. Charge trajectory optimization of plug-in hybrid electric vehicles for energy cost reduction and battery health enhancement. In Proceedings of the 2010 American Control Conference, Baltimore, MD, USA, 30 June–2 July 2010. [CrossRef]

115. Wang, Q.; Liu, W.; Yuan, X.; Tang, H.; Tang, Y.; Wang, M.; Zuo, J.; Song, Z.; Sun, J. Environmental impact analysis and process optimization of batteries based on life cycle assessment. *J. Clean. Prod.* **2018**, *174*, 1262–1273. [CrossRef]

116. Hatzell, K.B.; Sharma, A.; Fathy, H.K. A survey of long-term health modeling, estimation, and control of Lithium-ion batteries: Challenges and opportunities. In Proceedings of the 2012 American Control Conference (ACC), Montreal, QC, Canada, 27–29 June 2012; pp. 584–591.

117. Berecibar, M.; Gandiaga, I.; Villarreal, I.; Omar, N.; van Mierlo, J.; van den Bossche, P. Critical review of state of health estimation methods of Li-ion batteries for real applications. *Renew. Sustain. Energy Rev.* **2016**, *56*, 572–587. [CrossRef]

118. Winslow, K.M.; Laux, S.J.; Townsend, T.G. A review on the growing concern and potential management strategies of waste lithium-ion batteries. *Resour. Conserv. Recycl.* **2018**, *129*, 263–277. [CrossRef]

119. Kong, Q.; Ruan, M.; Zi, Y. A health management system for marine cell group. *IOP Conf. Ser. Earth Environ. Sci.* **2017**, *69*, 012081. [CrossRef]

120. Hu, J.; Zhang, J.; Li, H.; Chen, Y.; Wang, C. A promising approach for the recovery of high value-added metals from spent lithium-ion batteries. *J. Power Sources* **2017**, *351*, 192–199. [CrossRef]

121. Peng, X.; Garg, A.; Zhang, J.; Shui, L. Thermal management system design for batteries packs of electric vehicles: A Survey. In Proceedings of the 2017 Asian Conference on Energy, Power and Transportation Electrification (ACEPT), Singapore, 24–26 October 2017; pp. 1–5. [CrossRef]

10

FPGA Implementation of a Three-Level Boost Converter-fed Seven-Level DC-Link Cascade H-Bridge inverter for Photovoltaic Applications

Nagaraja Rao Sulake [1,*], Ashok Kumar Devarasetty Venkata [2] and Sai Babu Choppavarapu [1]

[1]   Department of Electrical and Electronics Engineering, Jawaharlal Nehru Technological University, Kakinada 533003, Andhra Pradesh, India; chs_eee@yahoo.co.in

[2]   Department of Electrical and Electronics Engineering, Rajeev Gandhi Memorial College of Engineering and Technology, Nandyal 518501, Andhra Pradesh, India; rgmdad09@gmail.com

*   Correspondence: nagarajraomtech@gmail.com or nagarajarao.ee.et@msruas.ac.in

Abstract: This paper presents an optimized single-phase three-level boost DC-link cascade H-bridge multilevel inverter (TLBDCLCHB MLI) system to generate a seven-level stepped output voltage waveform for photovoltaic (PV) applications. The proposed TLBDCLCHB MLI system is obtained by integrating a three-level boost converter (TLBC) with a seven-level DC-link cascade H-bridge (DCLCHB) inverter. It consists of a TLBC, level generation unit (LGU) and phase sequence generation unit (PSGU). When compared with traditional boost converter-fed multilevel inverter systems, the proposed TLBDCLCHB MLI system requires a single DC source, fewer power switches and gate drivers. Reduction in the switch count and number of DC sources makes the system cost effective and requires a smaller installation area. Pulse generation for the power switches of an LGU in a DCLCHB inverter is accomplished by providing proper conducting angles that are generated by optimized conducting angle determination (CAD) techniques. In this paper two CAD techniques i.e., equal-phase CAD (EPCAD) and step pulse wave CAD (SPWCAD) techniques are proposed to evaluate the performance of the proposed system in terms of the total harmonic distortion (THD) and the quality of the stepped output voltage waveform. The proposed system has been modeled and simulated using MATLAB/SIMULINK software. Results are presented and discussed. Also, a prototype model of a single-phase TLBDCLCHB MLI system is developed using a field-programmable gate array (FPGA)-based pulse generation with a resistive load and its performance is analyzed for various operating conditions.

Keywords: three-level boost converter (TLBC); DC-link cascade H-bridge (DCLCHB) inverter; conducting angle determination (CAD) techniques; total harmonic distortion (THD)

1. Introduction

Development in power electronics lay down a widespread scope for the resourceful operation of power converters. A few setups of power converters are produced to do the sun-powered photovoltaic (PV) applications with enhanced adequacy [1,2]. PV power generation is an encouraging elective source of energy and has numerous focal points compared to the other elective energy sources like wind, ocean, biomass, fuel, geothermal, and so on. In PV power generation, boost converters and multilevel inverters (MLIs) are playing a major role in power conversion. These power converters are broadly being utilized as a connection between load and supply. As most of the renewable power source generation is DC in nature, the DC-DC boost converters are utilized to increase the voltage level, and the DC must be changed over to AC for grid connection. Therefore, MLIs are used for DC to AC conversion [3,4]. The power generation using a traditional boost converter and inverter consists of a greater number of components, requires a larger installation area, is bulky in size, and costly. Also, the traditional boost converters are unable to produce a high boost ratio [5,6]. This paper proposes a three-level boost converter (TLBC) with a high boost ratio, based on one switch, one inductor, (2N-1) capacitors, and (2N-1) diodes for 'N' levels. It is a pulse width modulation (PWM)-controlled boost converter capable of maintaining an equal voltage in all 'N' output levels and controlling the input current.

In this paper, the structure of single-phase three-level boost DC-link cascade H-bridge multilevel inverter (TLBDCLCHB MLI) system is proposed to generate a seven-level stepped output voltage waveform for PV applications. The proposed system is obtained by integrating a three-level boost converter (TLBC) with a seven-level DC-link cascade H-bridge (DCLCHB) [7,8]. Also, the objective of the proposed work is to investigate the performance of a single-phase seven-level DCLCHB inverter [9,10] using conducting angle determination (CAD) techniques [11] in terms of the total harmonic distortion (THD) and the quality of the stepped output voltage waveform. Here, equal-phase CAD (EPCAD) based on equal-area distribution and step pulse wave CAD (SPWCAD) based on volt-second area equal to step pulse wave techniques are proposed to evaluate the performance of the DCLCHB inverter [12]. The proposed TLBDCLCHB MLI system is modeled, simulated and validated through experimental setup using field-programmable gate array (FPGA)-based pulse generation.

2. Structure of TLBDCLCHB MLI System

The block diagram of the proposed TLBDCLCHB MLI system is shown in Figure 1. The proposed TLBDCLCHB MLI system consists of single DC voltage source, TLBC, and DCLCHB inverter to generate a seven-level stepped output waveform. The DCLCHB inverter is composed of level generation unit (LGU) and phase sequence generation unit (PSGU). LGU is used to generate the required number of levels and PSGU is used to generate positive and negative sequence voltage levels [13]. The equivalent structure of TLBDCLCHB MLI system is shown in Figure 2.

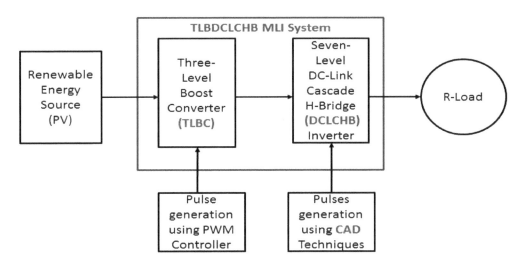

Figure 1. Block diagram of the proposed (TLBDCLCHB MLI) system.

The number of power switches '$N_{Switches}$' and the number of levels 'm' for a single-phase TLBDCLCHB MLI system are calculated using Equations (1) and (2), respectively.

$$m = (2C + 1)^H \tag{1}$$

$$N_{Switches} = (m - 1) + 5 \tag{2}$$

where 'C' is the number of DC link capacitors integrated to a DCLCHB inverter and 'H' is the number of H-Bridge circuits.

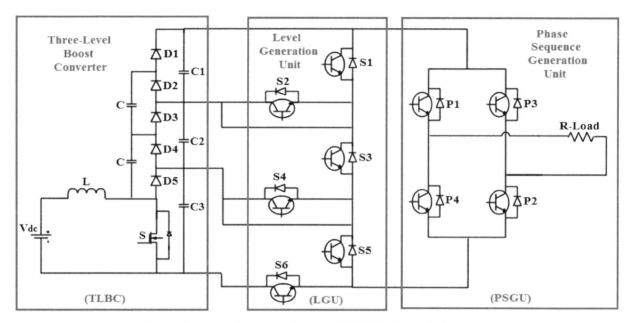

Figure 2. Equivalent structure of TLBDCLCHB MLI system.

Table 1 demonstrates the required number of power switches and DC sources for the traditional and proposed boost DC-link-based inverter to generate a stepped output waveform from seven levels to fifteen levels. From the investigation, it is gathered that the proposed TLBDCLCHB MLI system has reduced the switch count and requires just a single DC source compared to traditional boost multilevel inverter (MLI) systems [14,15].

Table 1. Component requirements for existing and proposed boost based MLI systems.

Number of Levels	Boost Cascade MLI (Conventional)		TLBDCLCHB MLI (Proposed)	
	Number of Power Switches	Number of DC Sources	Number of Power Switches	Number of DC Sources
7	15	3	11	1
9	20	4	13	1
11	25	5	15	1
13	30	6	17	1
15	35	7	19	1

3. Three-Level Boost Converter (TLBC)

The circuit configuration of DC-DC TLBC is represented in Figure 2. It consists of a traditional boost converter, (2N-1) capacitors, and (2N-1) diodes. The main advantages of using the TLBC topology are; it can be extended to any number of levels by adding only diodes and capacitors without changing the main circuit; no need of additional voltage balance circuit; and voltage gain can be increased without the use of a transformer by operating at minimum duty ratio. The TLBC circuit consists of three stages which are operated by varying duty cycles of 0.4, 0.5, and 0.6. The operation of the TLBC is explained in [16].

3.1. Analysis of DC-DC TLBC

Considering the presence and absence of inductor power loss for both traditional and proposed TLBCs gives important information to designers. From basic principles, the voltage gain or the boost factor of the traditional boost converter [14] is given by Equation (3):

$$\text{Voltage gain,} \ \frac{V_o}{V_{dc}} = \frac{1}{1-D} \tag{3}$$

where 'V_o' is the output voltage, 'V_{dc}' is the input voltage and 'D' is the duty cycle.

By considering the lossless system, the voltage gain for the N-level boost converter can be expressed as:

$$\frac{V_o}{V_{dc}} = \frac{N}{1-D} \tag{4}$$

For a lossless system, the input current, I_{dc}, can be expressed as:

$$I_{dc} = \frac{N\,I_o}{1-D} \tag{5}$$

From Equation (5), the input current, I_{dc}, can be controlled using duty cycle 'D' in the PWM. Now the voltage gain or boost factor expression for the N-level boost converter can be derived as follows:

Based on the condition that the average voltage across the inductor 'L' is equal to zero. The total inductor voltage during the ON—OFF condition can be expressed as,

$$V_L = D(V_{dc} - I_L R_L) + (1-D)(V_{dc} - V_C - I_L R_L) = 0 \tag{6}$$

where, 'I_L' is the inductor current which is equivalent to 'I_{dc}' and 'R_L' is the inductor resistance or parasitic resistance.

Here, the first term of Equation (6) is valid when the switch 'S' is turned ON, and the second term can write when the switch 'S' is turned OFF. From Equation (6) it can be written as,

$$V_{dc}(D + 1 - D) + I_L R_L(-D - 1 + D) = (1-D)V_C \tag{7}$$

From Equation (7),

$$V_{dc} = (1-D)V_C + I_L R_L \tag{8}$$

Therefore, from Equations (3)–(7), the input voltage, V_{dc} can be expressed as,

$$V_{dc} = (1-D)\frac{V_o}{N} + \frac{N\,V_o}{(1-D)R_O}R_L \tag{9}$$

Therefore, Equation (9) can be expressed as follows:

$$\frac{V_{dc}}{V_O} = \frac{1}{\frac{(1-D)}{N} + \frac{NR_L}{(1-D)R_O}} \tag{10}$$

Equation (10) is equal to Equation (3) if $N = 1$ and $R_L = 0$. From Equation (10), it can be observed that the voltage gain reaches a maximum before $D = 1$, and then becomes 0. The effect of parasitic resistance 'R_L' is responsible for the limitation in the boost factor. The actual boost factor or voltage gain against the duty cycle is analyzed by varying 'R_L' in Equation (10). Figures 3 and 4 describe the graph between voltage gain versus duty cycle for the traditional boost converter, i.e., $N = 1$ and for the proposed TLBC, i.e., $N = 3$ using different values of R_L/Ro.

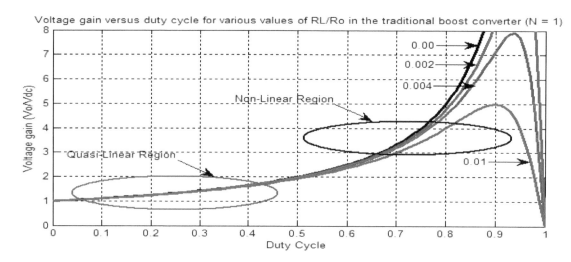

Figure 3. Duty cycle versus voltage gain of a traditional boost converter for various values of R_L/Ro ($N = 1$).

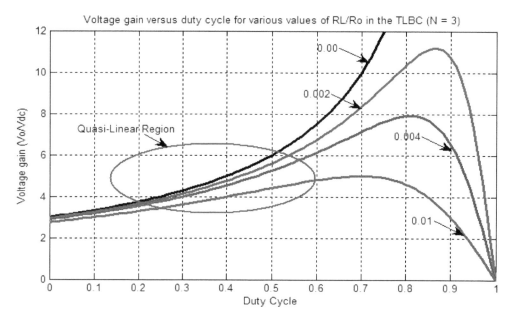

Figure 4. Duty cycle versus voltage gain of TLBC for various values of R_L/Ro ($N = 3$).

From the Figures 3 and 4, it can be noticed that the voltage gain of the traditional boost converter is quasi-linear when the duty cycle varies from 0 to 0.5, but beyond that, the boost factor of a traditional boost converter becomes non-linear; therefore, the control of a traditional boost converter is complicated.

Similarly, from Figure 4, i.e., when $N = 3$, it can be observed that the quasi-linear region is extended with a high voltage gain for the TLBC. Therefore, the TLBC achieves a high voltage gain compared to the traditional boost converter, and, also a better operating point of the duty cycle for the TLBC, which is from 0.4 to 0.6.

In the next section, the effect of the voltage drop across the switch and diodes is studied.

3.2. TLBC Voltage Drop Across Switch and Diodes

In actual operation of TLBC, the voltage drop across the switch and diodes must be considered since it avoids full charge across the capacitors [17,18]. Therefore, it reduces the conversion efficiency of the TLBC topology. In general, the voltage drop in the power switches and diodes can be around 2 V, and it can be neglected in medium- and high-voltage applications but must be considered in low-voltage applications. Here, the voltage drop across the switch and diodes is assumed to be equal to 'V_d'.

From Figure 5, it can be noticed that the voltage across the C_5 becomes,

$$V_{C5} = V_{C3} - V_{switch} - V_{diode} = V_{C3} - 2V_d \qquad (11)$$

where, '$2V_d$' is the voltage drop across the switch 'S' and diode 'D_5'.

Similarly, V_{C2} and V_{C1} can be written as,

$$V_{C2} = 2V_{C3} - 4V_d \qquad (12)$$

$$V_{C1} = 3V_{C3} - 8V_d \qquad (13)$$

where, V_{C2} and V_{C1} are the expressions for the voltage output of two-level and three-level boost converters, respectively. The generalized output voltage expression for the N-level boost converter can be expressed as follows:

$$V_o = NV_C - (N - 1)4V_d \qquad (14)$$

where, 'V_C' is the lower capacitor output voltage, and follows Equation (3).

Equation (14) gives the output voltage for the multiple stages of the boost converter. The efficiency of the proposed converter for the N-level is given by the Equation (15).

$$Efficiency, \eta = \frac{V_o}{NV_C} = 1 - \frac{(N - 1)4V_d}{NV_C} \qquad (15)$$

Form Equation (15), the efficiency of the TLBC circuit can be reduced by considering the voltage drop across the switch and diode.

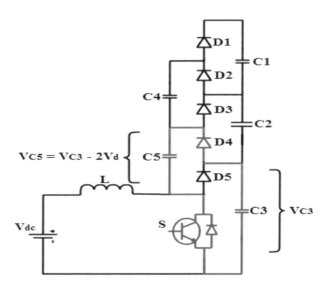

Figure 5. Charging C5 of a TLBC with switch and diode's voltage drop.

3.3. Closed-Loop Control of TLBC

From Figure 5 and Equation (10), it can be observed that the output voltage gain depends on the ratio of load resistance (R_O) and source resistance (R_L), i.e., if there is any variation between the load and source resistances, the output voltage of TLBC is not kept constant, and, from Equation (13), there will be a variation in the duty cycle to get the required amount of output voltage by considering the voltage drop across the switch and diodes. Therefore, the proposed TLBC circuit is modeled in closed-loop mode using an integral controller to maintain the constant output voltage.

Figure 6 represents the TLBC circuit in closed-loop mode. In the case of any variation in the load side or source side, the output changes, so a suitable controller is designed to change the duty cycle by comparing 'V_{ref}' with 'V_{out}' in order to maintain the required output voltage.

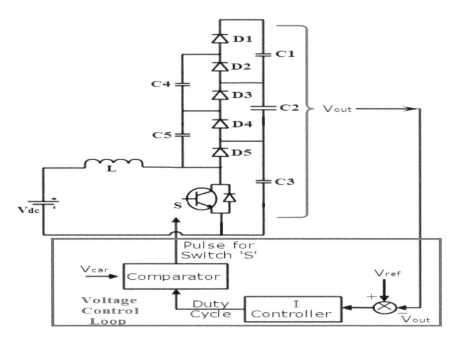

Figure 6. Model of TLBC in closed-loop mode.

4. DC-Link Cascade H-Bridge (DCLCHB) Inverter

Figure 7 depicts the DCLCHB inverter topology for the generation of a single-phase seven-level output voltage waveform. It is composed of LGU and PSGU [12].

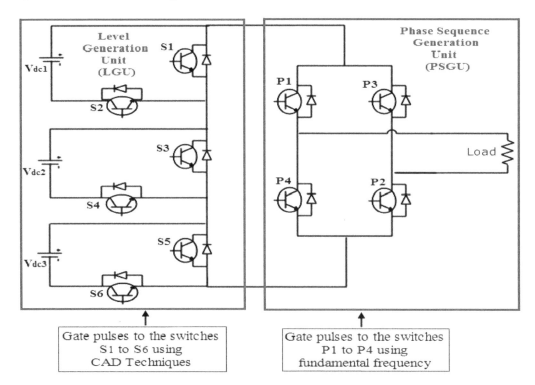

Figure 7. Single-phase seven-level DCLCHB inverter.

Switches in the LGU are used to generate the required number of levels and it is formed by connecting half-bridge cells in series. Each half-bridge cell consists of a DC source controlled by two switches. PSGU consists of an H-bridge circuit, which is used to generate the positive and negative sequence voltage levels. Table 2 gives a component requirement to generate a seven-level output voltage for the proposed and existing MLIs. It clearly shows a substantial component reduction when using a DCLCHB structure [7,8].

Table 2. Component requirements for existing and proposed cascade MLI systems.

Components	Traditional MLI	Proposed MLI
Switches	12	10
Clamping diodes	0	0
DC sources	3	3
Capacitors	0	0

In this DCLCHB inverter topology, all the magnitudes of DC voltage sources are equal ($V_{dc1} = V_{dc2} = V_{dc3}$).

$$i.e., V_{dci} = V_{dc}, \text{ where } i = 1, 2, \text{ and } 3 \tag{16}$$

The maximum value of the output phase voltage is obtained by using Equation (17).

$$V_{max} = \sum_{i=1}^{S} V_{dci} \tag{17}$$

The number of output phase voltage levels can be obtained from Equation (1). The number of power switches for the DCLCHB inverter can be calculated using Equation (18).

$$N_{Switches} = (m - 1) + 4 \tag{18}$$

Equation (16) gives the output level of the LGU. By using PSGU, the positive and negative levels are obtained at the load (V_o), the synthesized stepped AC output phase voltage will be obtained by using the Equations (19) and (20).

$$V_{o, max} = \sum_{i=1}^{3} + V_{dci}, \text{ If } P_1, P_2 = 1 \tag{19}$$

$$V_{o, max} = \sum_{i=1}^{3} - V_{dci}, \text{ If } P_3, P_4 = 1 \tag{20}$$

For a single-phase seven-level DCLCHB inverter, the switching sequences to generate the required levels are given in Table 3.

Table 3. Switching sequence to generate seven-level output for a DCLCHB inverter.

S.No	LGU Switches						PSGU Switches				Voltage Levels (Volts)
	S_1	S_3	S_5	S_2	S_4	S_6	P_1	P_2	P_3	P_4	
1	1	1	1	0	0	0	0	0	0	0	0
2	0	1	1	1	0	0	1	1	0	0	100
3	0	0	1	1	1	0	1	1	0	0	200
4	0	0	0	1	1	1	1	1	0	0	300
5	0	1	1	1	0	0	0	0	1	1	−100
6	0	0	1	1	1	0	0	0	1	1	−200
7	0	0	0	1	1	1	0	0	1	1	−300

5. Conducting Angle Determination (CAD) Techniques

CAD techniques are a vital part of any inverter since they are directly related to the efficiency of the entire system [6,12]. It is used to control the proposed DCLCHB inverter output phase voltage and also for the calculation of the two main parameters such as %THD and V_{rms}. In this paper, a step pulse wave CAD (SPWCAD) technique has been employed to trigger the switches of LGU in the DCLCHB inverter and is compared with a conventional CAD technique, i.e., equal-phase CAD (EPCAD) technique. The generation of the seven-level stepped voltage waveform using CAD techniques is shown in Figure 8.

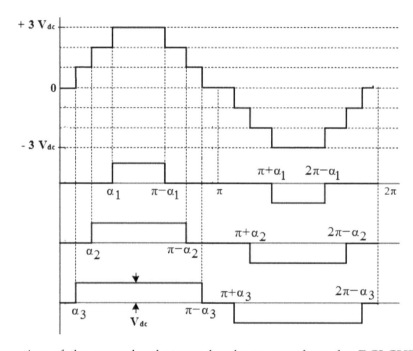

Figure 8. Generation of the seven-level stepped voltage waveform for DCLCHB inverter using CAD techniques.

In the presented EPCAD and SPWCAD techniques, for an m-level stepped waveform in the period of the first quadrant, i.e., 0 to 90°, $2(m - 1)/2$ conducting angles need to be determined. From Figure 8, to generate a seven-level stepped waveform in the first quadrant, i.e., 0 to 90°, three conducting angles need to be determined. They are defined as the main conducting angles, i.e., α_1, α_2 and α_3 using the time-sequence. From Figure 8, it can be noticed that only the main conducting angles need to be determined; the rest of the conducting angles can be derived from the main conducting angles. The solution of the main conducting angles, i.e., α_1, α_2 and α_3 must satisfy the following condition:

$$0 \leq \alpha_1 \leq \alpha_2 \leq \alpha_3 \leq \frac{\pi}{2} \tag{21}$$

5.1. EPCAD Technique

In the EPCAD technique the main conducting angles are derived by taking an average distribution of the conducting angles from 0 to180°. In this technique, the main conducting angles are obtained by using Equation (22).

$$\alpha_i = i\left(\frac{180°}{m}\right) \tag{22}$$

where $i = 1, 2, \ldots \left(\frac{m-1}{2}\right)$, m = number of levels.

For a seven-level stepped waveform using the EPCAD technique, three main conducting angles need to be determined using Equation (22), i.e., α_1, α_2 and α_3, the values of which are 25.71°, 51.43°, and 77.14°, respectively.

5.2. SPWCAD Technique

In the proposed SPWCAD technique, conducting angles are acquired by computing the volt-second areas of the sine reference voltage waveform that is equivalent to the stepped output phase voltage waveform of the DCLCHB inverter. In the seven-level DCLCHB inverter, since three half-bridge cells are connected in series, the reference voltage 'V_{ref}' and output-phase voltage '$V_{out\text{-}phase}$' can be obtained by the Equations (23) and (24) respectively:

$$V_{ref} = 3 \left(\frac{4V_{dc}}{\pi} \right) (M_i \sin \omega t) \tag{23}$$

$$V_{out-phase} = i.V_{dc} \ (1 \leq i \leq 3) \tag{24}$$

where, 'M_i' is the modulation index, and 'i' is the integer number.

Figure 9 demonstrates the dummy conducting angles (α_i^1) in the case of $M_i = \pi/4$. The areas of A_1^1, A_2^1, and A_3^1 are encompassed by the sine reference voltage wave and the stepped output-phase voltage levels in the positive half-cycle of the seven-level DCLCHB inverter.

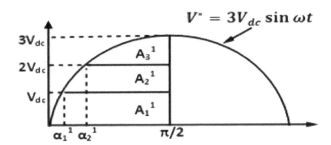

Figure 9. Reference voltage waveform with dummy conducting angles.

The generation of a step pulse wave in the DCLCHB inverter to meet the equivalent areas as A_1^1, A_2^1, and A_3^1. Here, main conducting angles (α_i) are defined as the switching timing angles of step pulse waves in the DCLCHB inverter. Figure 10 represents the main conducting angles and the stepped output phase voltage of the seven-level DCLCHB inverter during the positive half cycle.

Assuming that the areas A_1, A_2, and A_3 made by the main conducting angles in Figure 10 are equivalent to A_1^1, A_2^1, and A_3^1 made by dummy conducting angles which are obtained as follows:

Area, A_1^1 can be obtained from Equation (25):

$$A_1^1 = (A_1^1 + A_2^1 + A_3^1) - (A_2^1 + A_3^1) \tag{25}$$

Therefore, from Equation (25), A_1^1 can be written as:

$$A_1^1 = \int_0^{\frac{\pi}{2}} 3\,V_{dc} \sin \omega t d(\omega t) - \left\{ \int_{\alpha_1^1}^{\frac{\pi}{2}} 3V_{dc} \sin \omega t d(\omega t) - \left(\frac{\pi}{2} - \alpha_1^1 \right) V_{dc} \right\} \tag{26}$$

From Figure 10, Area, A_1 can be expressed as,

$$A_1 = \left(\frac{\pi}{2} - \alpha_1 \right) V_{dc} \tag{27}$$

Since $A_1{}^1$ is equal to A_1, the real conducting angle α_1 of the proposed SPWCAD technique is obtained as,

$$\alpha_1 = \frac{\pi}{2} - \left\{ \int_0^{\frac{\pi}{2}} 3\, V_{dc} \sin \omega t\, d(\omega t) - \left(\int_{\alpha_1^1}^{\frac{\pi}{2}} 3\, V_{dc} \sin \omega t\, d(\omega t) - \left(\frac{\pi}{2} - \alpha_1^1 \right) \right) \right\} \tag{28}$$

Similarly, by equating $A_2{}^1$ to A_2, the angle α_2 can be expressed as,

$$\alpha_2 = \frac{\pi}{2} - \left\{ \int_0^{\frac{\pi}{2}} 3\, V_{dc} \sin \omega t\, d(\omega t) - \left(\int_{\alpha_2^1}^{\frac{\pi}{2}} 3\, V_{dc} \sin \omega t\, d(\omega t) - \left(\frac{\pi}{2} - \alpha_2^1 \right) \right) \right\} \tag{29}$$

α_3 can be obtained by equating $A_3{}^1$ to A_3 or by using Equation (30):

$$\int_0^{\frac{\pi}{2}} 3\, V_{dc} \sin \omega t\, d(\omega t) - \left(A_1^1 + A_2^1 \right) = \left(\frac{\pi}{2} - \alpha_3 \right) V_{dc} \tag{30}$$

For a seven-level DCLCHB inverter, the conducting angles are calculated by using Equations (31)–(33).

$$\alpha_1 = \frac{12 M_i}{\pi} \left\{ \cos\left(\sin^{-1}\left(\frac{\pi}{12 M_i} \right) \right) - 1 \right\} + \sin^{-1}\left(\frac{\pi}{12 M_i} \right) \tag{31}$$

$$\alpha_2 = \frac{12 M_i}{\pi} \left\{ \cos\left(\sin^{-1}\left(\frac{2\pi}{12 M_i} \right) \right) - \cos\left(\sin^{-1}\left(\frac{\pi}{12 M_i} \right) \right) \right\} + 2\sin^{-1}\left(\frac{2\pi}{12 M_i} \right) - \sin^{-1}\left(\frac{\pi}{12 M_i} \right) \tag{32}$$

$$\alpha_3 = \frac{3\pi}{2} - \frac{12 M_i}{\pi} \cos\left(\sin^{-1}\left(\frac{2\pi}{12 M_i} \right) \right) - 2\sin^{-1}\left(\frac{2\pi}{12 M_i} \right) \tag{33}$$

Therefore, for a seven-level stepped waveform using the SPWCAD technique, three main conducting angles need to determined using Equations (31)–(33), i.e., α_1, α_2 and α_3 and their values are 9.43°, 29.59°, and 55.88°, respectively, for the $M_i = 0.8$.

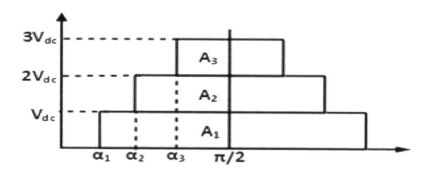

Figure 10. Output phase-voltage of a seven-level DCLCHB inverter in the positive half-cycle voltage.

5.3. Comparison of the SPWCAD and EPCAD Technique

The conducting angles acquired by the SPWCAD technique are different from those acquired by the EPCAD technique. In the SPWCAD technique, conducting angles are acquired based on the modulation index (M_i) whereas the EPCAD technique gives the same conducting angles irrespective of the M_i. The proposed SPWCAD technique method can be applied to different modulation indices. The range of M_i and the number of conducting angles are listed in Table 4. Also, the values of conducting angles for the proposed SPWCAD technique using various modulation indices are listed in Table 5 for the DCLCHB inverter.

Table 4. Number of conducting angles and steps in output waveform SPWCAD technique for various M_i.

Range of M_i	Number of Conducting Angles	Number of Steps in Output Waveform
$0 < M_i < 0.33$	1	3
$0.33 \leq M_i < 0.66$	2	5
$0.66 \leq M_i < 1$	3	7

Table 5. Conducting angles of SPWCAD technique for various M_i.

Conducting Angles (Degrees)	Modulation Indices (M_i)		
	0.3	0.6	0.8
α_1	27.17	12.7	9.439
α_2	–	41.65	29.59
α_3	–	–	55.88

5.4. THD Calculation of the Seven-Level Stepped Output Phase Voltage using CAD Techniques

The general THD expression for a periodic output phase voltage for a proposed DCLCHB inverter can be expressed as,

$$THD = \sqrt{\left(\frac{V_{rms}}{V_1}\right)^2 - 1} \tag{34}$$

where V_1 is the RMS (root mean square) value of the fundamental component and V_{rms} is the RMS value of the output phase voltage. For the proposed seven-level RV MLI, V_{rms} and V_1 can be obtained by using Equations (35) and (36).

$$V_{rms} = V_{dc}\sqrt{\left[\frac{2}{\pi} \cdot \left((\alpha_2 - \alpha_1) + 4(\alpha_3 - \alpha_2) + 9(\alpha_3 - \alpha_2) \right)\right]} \tag{35}$$

$$V_1 = \frac{4V_{dc}}{\pi\sqrt{2}}[(\cos \alpha_1 + \cos \alpha_2 + \cos \alpha_3)] \tag{36}$$

The output phase voltage THD expression for the proposed seven-level DCLCHB inverter can be obtained by substituting Equations (35) and (36) into Equation (34) and is given by:

$$V_1 = \frac{4V_{dc}}{\pi\sqrt{2}}[(\cos \alpha_1 + \cos \alpha_2 + \cos \alpha_3)] \tag{37}$$

Theoretical values of the output phase voltage THD for the seven-level DCLCHB inverter using EPCAD and the proposed SPWCAD techniques with the corresponding main conducting angles are given in Table 6.

Table 6. Conducting angles, theoretical output phase voltage THD, and V_{rms} values for a seven-level DCLCHB inverter (m = 7).

CAD Technique	Conducting Angles (in Degrees)			% THD (Theoretical)	V_{rms} (V)
	α_1	α_2	α_3		
EPCAD	25.71	51.43	77.14	31.05	165.8
SPWCAD	9.43	29.59	55.88	11.95	219.1

6. Simulation and Experimental Validation of the Proposed TLBDCLCHB Inverter System

The simulation of the proposed single-phase seven-level TLBDCLCHB inverter system is analyzed using MATLAB Simulink and validated experimentally through FPGA-based pulse generation.

6.1. TLBC Simulation Results in Open-Loop Mode

The simulation of the TLBC is carried out and analyzed in open-loop mode by considering the DC input voltage V_{dc} of 50 V, which should be boosted to a total DC-link voltage of 250 V, 300 V, and 375 V for the duty cycles of 0.4, 0.5, and 0.6, respectively, and the voltage across each of the capacitors at the output should be boosted to 83.33 V, 100 V, and 125 V for the duty cycles of 0.4, 0.5, and 0.6, respectively, as shown in Figures 11–13.

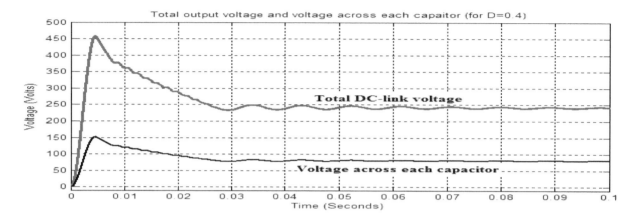

Figure 11. Output of TLBC in open-loop for $D = 0.4$.

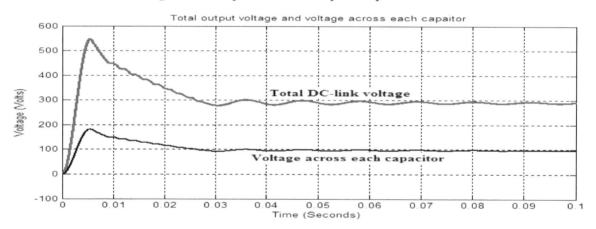

Figure 12. Output of TLBC in open-loop for $D = 0.5$.

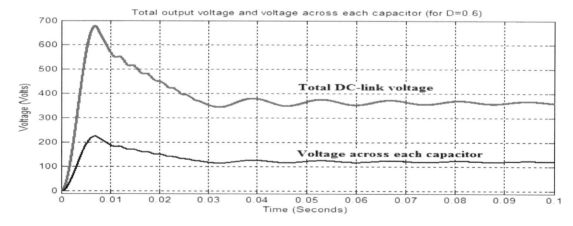

Figure 13. Output of TLBC in open-loop for $D = 0.6$.

From Figures 11–13, it is observed that the total DC-link output voltage for the duty cycles of 0.4, 0.5, and 0.6 has been boosted to 242.6, 291.4 V, and 361.4 V as opposed to 250 V, 300 V, and 375 V, respectively, due to the voltage drop across the switch and diodes. Similarly, the voltage drops across each of the capacitors are boosted to 82.17 V, 98.5 V, and 121.8 V as opposed to 83.33 V, 100 V, and 125 V. Also, it is noticed that the ripple in the total DC-link output voltage and voltage across each capacitor are increased by increasing the duty cycle from 0.4 to 0.6 to achieve the maximum DC-link output voltage with minimum ripple. Further, the open-loop mode of TLBC is extended to operate in closed-loop mode to maintain the required amount of DC-link output voltage by compensating the voltage drop across the switch and diodes, and, also, to reduce the ripple.

6.2. TLBC Simulation Results in Closed-Loop Mode

The proposed TLBC is implemented in closed-loop to maintain the constant output voltage using the voltage control loop. Here, the output of TLBC is measured and fed to an integral controller by comparing with the required reference output voltage to vary the duty cycle. Figure 14 shows the total DC-link output voltage and the voltage across each capacitor of the TLBC in closed-loop mode for $D = 0.5$. The corresponding change in duty cycle due to the reference output voltage is shown in Figure 15.

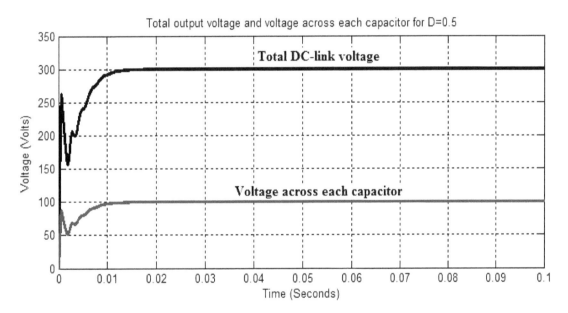

Figure 14. Output of TLBC in closed-loop for $D = 0.5$.

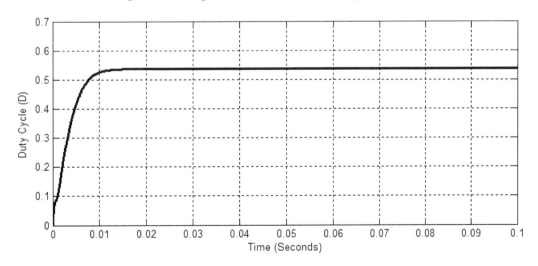

Figure 15. Change in duty cycle of TLBC in closed-loop.

From the Figures 14 and 15, it is observed that the total DC-link output voltage and the voltage across each capacitor of the TLBC is boosted to 300 V and 100 V, respectively, as per the reference output voltage by changing the duty cycle using an integral controller.

6.3. TLBDCLCHB Inverter Output using CAD Techniques

Simulation results of TLBC-fed DCLCHB inverter to generate a seven-level stepped output waveform using EPCAD and SPWCAD techniques and its THD analysis are shown in Figures 16–21 for $D = 0.5$.

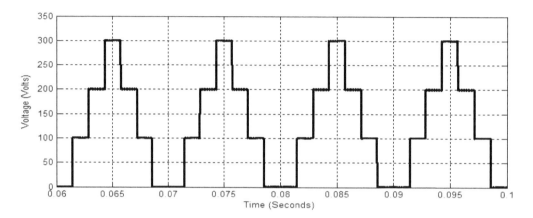

Figure 16. Inverter output across LGU using the EPCAD technique.

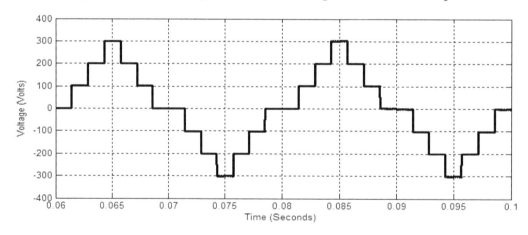

Figure 17. TLBDCLCHB inverter output using EPCAD technique.

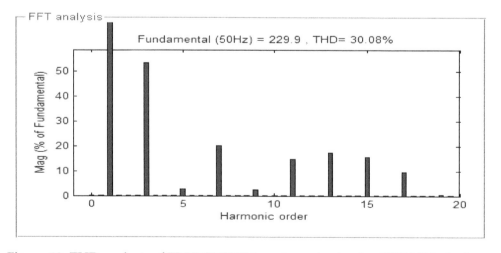

Figure 18. THD analysis of TLBDCLCHB inverter output using EPCAD technique.

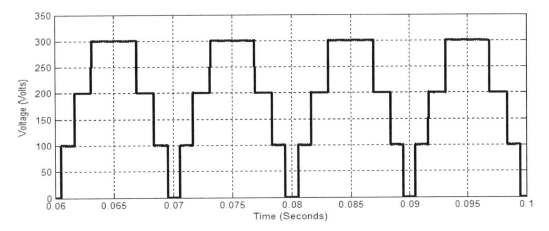

Figure 19. Inverter output across LGU using the SPWCAD technique.

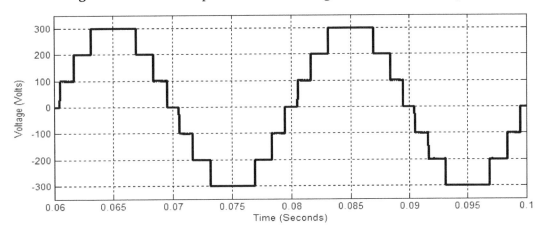

Figure 20. TLBDCLCHB inverter output using the SPWCAD technique.

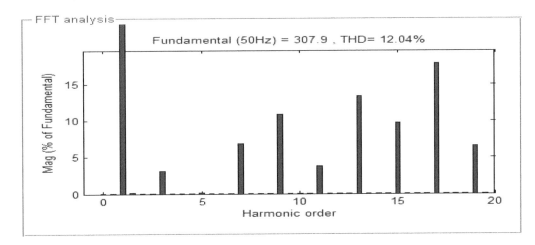

Figure 21. THD analysis of the TLBDCLCHB inverter output using the SWPCAD technique.

6.3.1. Using the EPCAD Technique

Figure 16 shows the output voltage across the LGU in the DCLCHB inverter. Referring to Figure 16, the LGU generates a unipolar stepped waveform, and it can be converted to a bipolar stepped wave using PSGU. Figures 17 and 18 show the seven-level stepped output phase voltage and its THD analysis of the TLBDCLCHB inverter using the EPCAD technique for the duty cycle of 0.5. It is observed that the magnitude of the fundamental output phase voltage and its RMS value is 229.9 V and 162.5 V, respectively. Also, the THD of the proposed TLBDCLCHB inverter output using the EPCAD technique is 30.08%.

6.3.2. Using the SPWCAD Technique

The unipolar output phase voltage across the LGU in the DCLCHB inverter using the SPWCAD technique is shown in Figure 19. Figures 20 and 21 show the seven-level stepped output phase voltage and its THD analysis of the TLBDCLCHB inverter using the SPWCAD technique for the duty cycle of 0.5 with $M_i = 0.8$. It is observed that the magnitude of the fundamental output phase voltage and its RMS value is 307.9 V and 217.7 V, respectively. Also, the THD of the proposed TLBDCLCHB inverter output using the SPWCAD technique is 12.04%.

6.4. Experimental Validation of the TLBDCLCHB Inverter System Using an FPGA-Based Pulse Generation

The model of the proposed TLBDCLCHB inverter system-fed R-load is implemented employing Xilinx Spartan FPGA-based pulse generation [19,20] to validate the Simulink results. The block diagram, Xilinx Spartan6 development board, and the prototype model of the TLBDCLCHB inverter system using an FPGA is shown in Figures 22–24, respectively. It consists of TLBC, the DCLCHB inverter, a personal computer (PC), an FPGA controller, R-load, buffer circuit, optocoupler, and driver circuit. The output of TLBC-fed DCLCHB inverter using an FPGA controller is shown in Figure 25.

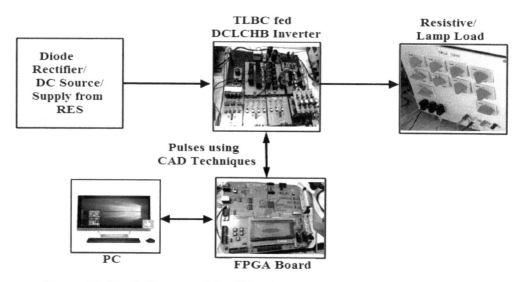

Figure 22. Block diagram of the TLBDCLCHB hardware implementation.

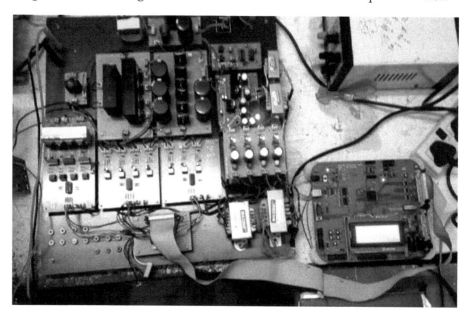

Figure 23. Hardware implementation of the TLBDCLCHB inverter system with an FPGA controller.

Figure 24. Xilinx Spartan6 development board.

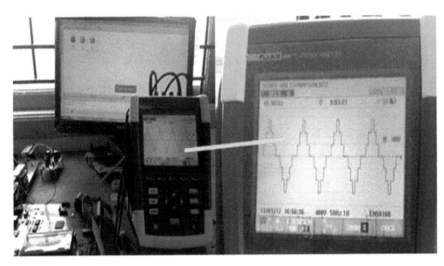

Figure 25. Seven-level stepped output voltage of TLBC-fed DCLCHB inverter system.

Referring to Figure 26, the TLBC output voltage across each capacitor and the total DC-link voltage are shown in Channel 1 (CH1) and Channel (CH2), respectively, for $D = 0.5$. It is observed that CH1 and CH2 voltages are boosted to 100 V and 300 V, respectively.

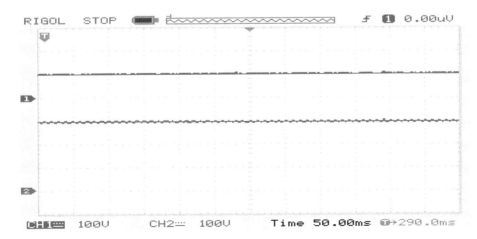

Figure 26. Voltage across the capacitor CH1 and total DC-link voltage of TLBC CH2 for $D = 0.5$.

Figures 27 and 28 show the generation of the pulse for the LGU switches (S_1 to S_6) in the DCLCHB inverter to generate the stepped output waveform using EPCAD and SPWCAD techniques, respectively, for the modulation index of 0.8. Figure 29 shows the generation of the pulse for the PSGU switches (P_1 to P_4) in the DCLCHB inverter to generate the positive and negative levels using an H-bridge inverter.

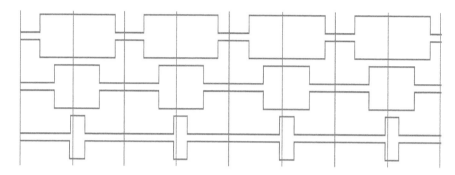

Figure 27. Generation of pulses for the LGU switches (S_1 to S_6) in the DCLCHB inverter using the EPCAD technique through Xilinx ISE.

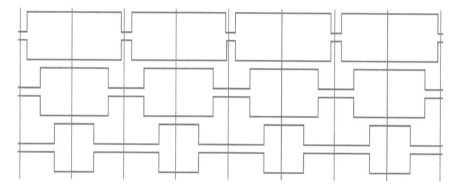

Figure 28. Generation of pulses for the LGU switches (S_1 to S_6) in the DCLCHB inverter using the SPWCAD technique through Xilinx ISE.

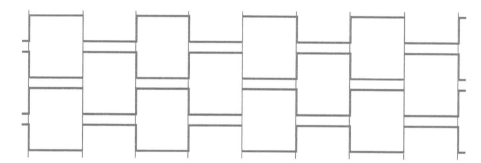

Figure 29. Generation of pulses for the PSGU switches (P_1 to P_4) in the DCLCHB inverter using a pulse generator through Xilinx ISE.

Figures 30 and 31 represent the TLBDCLCHB inverter system experimental output phase voltage and its harmonic spectrum using the EPCAD technique for the generation of a seven-level stepped output voltage. Referring to Figures 30 and 31, it is observed that the RMS value of the output phase voltage is 161.7 V and its THD is 31.5%.

Figures 32 and 33 represent the TLBDCLCHB inverter system experimental output phase voltage and its harmonic spectrum using the SPWCAD technique for the generation of a seven-level stepped output voltage. Referring to Figures 32 and 33, it is observed that the RMS value of the output phase voltage is 216 V and its THD is 11.5%.

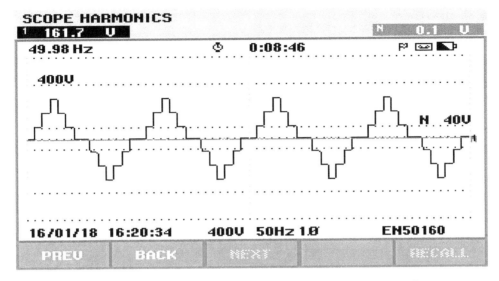

Figure 30. TLBDCLCHB inverter output using the EPCAD technique.

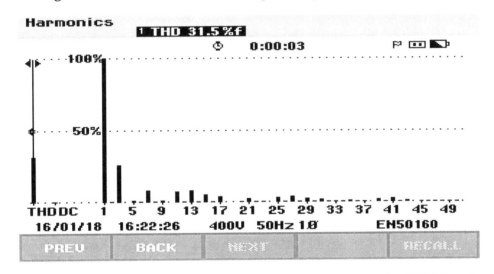

Figure 31. THD analysis of the TLBDCLCHB inverter output using the EPCAD technique.

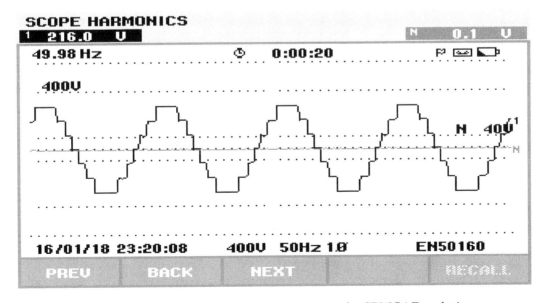

Figure 32. TLBDCLCHB inverter output using the SPWCAD technique.

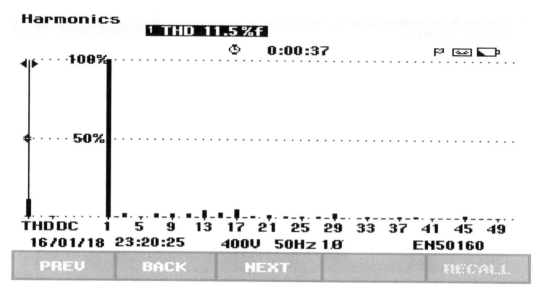

Figure 33. THD analysis of the TLBDCLCHB inverter output using the SPWCAD technique.

6.5. TLBDCLCHB Inverter System Analysis and Comparison Using CAD Techniques

In this study, theoretical and simulation results of the proposed TLBDCLCHB inverter system for the generation of a seven-level stepped output phase voltage using the EPCAD and SPWCAD techniques have been validated experimentally through an FPGA-based pulse generation. Tables 7 and 8 analyze the output phase voltage (V_{rms}) and THD of the TLBDCLCHB inverter system for different duty cycles, i.e., $D = 0.4$, 0.5, and 0.6. The THD of the prototype model for the proposed CAD techniques are conceded using a Fluke 435 power quality analyzer, and the results are presented in Figures 16–29 for the EPCAD and SPWCAD techniques.

Table 7. Simulation comparison of (V_{rms}) and % THD.

Duty Cycle (D)	EPCAD Technique		SPWCAD Technique	
	V_{rms} (V)	THD (%)	V_{rms} (V)	THD (%)
0.4	135.7	29.71	185.5	12.02
0.5	162.5	30.08	217.7	12.04
0.6	203.6	29.90	272	12.07

Table 8. Experimental comparison of (V_{rms}) and %THD.

Duty Cycle (D)	EPCAD Technique		SPWCAD Technique	
	V_{rms} (V)	THD (%)	V_{rms} (V)	THD (%)
0.4	134.6	31.8	184.7	11.8
0.5	161.7	31.5	216	11.5
0.6	202.1	31.5	270.9	11.7

Referring to Tables 6–8, it is inferred that the most extreme output phase voltage and lower THD are accomplished by utilizing the SPWCAD technique rather than the EPCAD technique. From the simulation, it is obtained that V_{rms} and %THD content in the EPCAD technique are 161.7 V and 31% for $D = 0.5$. Whereas, in the case of the proposed SPWCAD technique, V_{rms} is 216 V and %THD is only 11.5% for $D = 0.5$ by considering $M_i = 0.8$. From the experimental results, it is observed that the V_{rms} and %THD content in the EPCAD technique are 161.7 V and 31% for $D = 0.5$. For the proposed SPWCAD technique, V_{rms} is 216 V and %THD is only 11.5% for $D = 0.5$ by considering $M_i = 0.8$. Therefore, theoretical values V_{rms} and %THD shown in Table 6 are validated with the simulation and experimental results with a tolerable error of $\pm 2\%$. From the analysis, it is noticed that the magnitude of V_{rms} varies with respect to the duty cycle, but there is only a slight deviation in THD from 0.4 to 0.6.

7. Conclusions

In this paper, a TLBC-fed DCLCHB inverter system has been suggested to generate a seven-level stepped output phase voltage using a single DC source for better performance, efficiency, and reduced cost and size of the inverter. It also presented two control techniques for the DCLCHB inverter based on conducting angle determination, namely, EPCAD and SPWCAD techniques. Here, the SPWCAD technique gives the most extreme output phase voltage and lower THD compared to the EPCAD technique but the SPWCAD technique involves several trigonometric functions. However, same trigonometric functions are repeated; therefore, it is easy to acquire the conducting angles once the equations are derived based on the volt-second balance. In addition, TLBC has been suggested to achieve auto voltage balance and high voltage gain. Therefore, upon considering all the advantages, the proposed TLBDCLCHB inverter system is a good alternative for PV applications compared to the conventional boost-based MLI systems.

Author Contributions: A.K.D.V. and S.B.Ch. contributed to the main idea of this article. N.R.S. developed the Simulink model and performed the experiments, N.R.S., A.K.D.V. and S.B.Ch. all contributed to manuscript writing and revisions. All authors approved the final version to be published.

Abbreviations

The following abbreviations are used in this manuscript:

CAD	conducting angle determination
DCLCHB	DC-link cascade H-bridge
EPCAD	equal-phase CAD
LGU	level generation unit
THD	total harmonic distortion
TLBC	three-level boost converter
TLBDCLCHB	three-level boost DC-link cascade H-bridge
MLI	multilevel inverter
PSGU	phase sequence generation unit
PWM	pulse width modulation
PV	photovoltaic
SPWCAD	step pulse wave CAD

References

1. Rodriguez, J.; Lai, J.S.; Peng, F.Z. Multilevel inverters: A survey of topologies, controls, and applications. *IEEE Trans. Ind. Electron.* **2002**, *49*, 724–738. [CrossRef]
2. Kjaer, S.B.; Pedersen, J.K.; Blaabjerg, F. A review of single-phase grid-connected inverters for photovoltaic modules. *IEEE Trans. Ind. Appl.* **2005**, *41*, 1292–1306. [CrossRef]
3. Rodríguez, J.; Bernet, S.; Wu, B.; Pontt, J.O.; Kouro, S. Multilevel voltage-source-converter topologies for industrial medium-voltage drives. *IEEE Trans. Ind. Electron.* **2007**, *54*, 2930–2945. [CrossRef]
4. Najafi, E.; Yatim, A.H. Design and implementation of a new multilevel inverter topology. *IEEE Trans. Ind. Electron.* **2012**, *59*, 4148–4154. [CrossRef]
5. Davari, P.; Yang, Y.; Zare, F.; Blaabjerg, F. A review of electronic inductor technique for power factor correction in three-phase adjustable speed drives. In Proceedings of the 2016 IEEE Energy Conversion Congress and Exposition (ECCE), Milwaukee, WI, USA, 18–22 September 2016; pp. 1–8.
6. Klumper, C.; Blaabjerg, F.; Thøgersen, P. Alternate ASDs: Evaluation of the converter topologies suited for integrated motor drives. *IEEE Ind. Appl. Mag.* **2006**, *12*, 71–83. [CrossRef]
7. Su, G.J. Multilevel DC-link inverter. *IEEE Trans. Ind. Appl.* **2005**, *41*, 848–854. [CrossRef]
8. Rao, S.N.; Kumar, D.A.; Babu, C.S. New multilevel inverter topology with reduced number of switches using advanced modulation strategies. In Proceedings of the 2013 International Conference on Power, Energy and Control (ICPEC), Sri Rangalatchum Dindigul, India, 6–8 February 2013; pp. 693–699.

9.   Uthirasamy, R.; Chinnaiyan, V.K.; Ragupathy, U.S.; Karpagam, J. Investigation on three-phase seven-level cascaded DC-link converter using carrier level shifted modulation schemes for solar PV system applications. *IET Renew. Power Gener.* **2017**, *12*, 439–449. [CrossRef]

10. Prabaharan, N.; Palanisamy, K. A comprehensive review on reduced switch multilevel inverter topologies, modulation techniques and applications. *Renew. Sustain. Energy Rev.* **2017**, *76*, 1248–1482. [CrossRef]

11. Luo, F.L. Investigation on best switching angles to obtain lowest THD for multilevel DC/AC inverters. In Proceedings of the 2013 IEEE 8th Conference on Industrial Electronics and Applications (ICIEA), Melbourne, Australia, 19–21 June 2013; pp. 1814–1818.

12. Kang, D.W.; Kim, H.C.; Kim, T.J.; Hyun, D.S. A simple method for acquiring the conducting angle in a multilevel cascaded inverter using step pulse waves. *IEE Proc.-Electr. Power Appl.* **2005**, *152*, 103–111. [CrossRef]

13. Rao, S.N.; Kumar, D.A.; Babu, C.S. Implementation of Multilevel Boost DC-Link Cascade based Reversing Voltage Inverter for Low THD Operation. *J. Electr. Eng. Technol.* **2018**, *13*, 1527–1537.

14. Uthirasamy, R.; Ragupathy, U.S.; Chinnaiyan, V.K. Structure of boost DC-link cascaded multilevel inverter for uninterrupted power supply applications. *IET Power Electron.* **2015**, *8*, 2085–2096. [CrossRef]

15. Prasad, G.D.; Jegathesan, V.; Moorthy, V. Minimization of power loss in newfangled cascaded H-bridge multilevel inverter using in-phase disposition PWM and wavelet transform based fault diagnosis. *Ain Shams Eng. J.* **2016**. [CrossRef]

16. Rosas-Caro, J.C.; Ramírez, J.M.; García-Vite, P.M. Novel DC-DC multilevel boost converter. In Proceedings of the 2008 IEEE Power Electronics Specialists Conference, Rhodes, Greece, 15–19 June 2008; pp. 2146–2151.

17. Rosas-Caro, J.C.; Ramirez, J.M.; Valderrabano, A. Voltage balancing in DC/DC multilevel boost converters. In Proceedings of the 2008 40th North American Power Symposium, Calgary, AB, Canada, 28–30 September 2008; pp. 1–7.

18. Rosas-Caro, J.C.; Ramirez, J.M.; Peng, F.Z.; Valderrabano, A. A DC–DC multilevel boost converter. *IET Power Electron.* **2010**, *3*, 129–137. [CrossRef]

19. Maruthupandi, P.; Devarajan, N.; Sebasthirani, K.; Jose, J.K. Optimum control of total harmonic distortion in field programmable gate array-based cascaded multilevel inverter. *J. Vib. Control* **2015**, *21*, 1999–2005. [CrossRef]

20. Cong, J.; Liu, B.; Neuendorffer, S.; Noguera, J.; Vissers, K.; Zhang, Z. High-level synthesis for FPGAs: From prototyping to deployment. *IEEE Trans. Comput.-Aided Des. Integr. Circuits Syst.* **2011**, *30*, 473–491. [CrossRef]

11

Digital Control Techniques Based on Voltage Source Inverters in Renewable Energy Applications

Sohaib Tahir [1,2], **Jie Wang** [1,*], **Mazhar Hussain Baloch** [1,3] **and Ghulam Sarwar Kaloi** [1,4]

[1] School of Electronic, Information and Electrical Engineering, Shanghai Jiao Tong University,
    Shanghai 200000, China; sohaibchauhdary@sjtu.edu.cn (S.T.); mazhar.hussain08ele@gmail.com (M.H.B.);
    Sarwar.kaloi59@gmail.com (G.S.K.)

[2] Department of Electrical Engineering, COMSATS Institute of Information Technology, Sahiwal 58801, Pakistan

[3] Department of Electrical Engineering, Mehran University of Engineering & Technology,
    Khairpur Mirs 67480, Pakistan

[4] Department of Electrical Engineering, Quaid e Awam University of Engineering & Technology,
    Larkana 77150, Pakistan

* Correspondence: jiewangxh@sjtu.edu.cn

Abstract: In the modern era, distributed generation is considered as an alternative source for power generation. Especially, need of the time is to provide the three-phase loads with smooth sinusoidal voltages having fixed frequency and amplitude. A common solution is the integration of power electronics converters in the systems for connecting distributed generation systems to the stand-alone loads. Thus, the presence of suitable control techniques, in the power electronic converters, for robust stability, abrupt response, optimal tracking ability and error eradication are inevitable. A comprehensive review based on design, analysis, validation of the most suitable digital control techniques and the options available for the researchers for improving the power quality is presented in this paper with their pros and cons. Comparisons based on the cost, schemes, performance, modulation techniques and coordinates system are also presented. Finally, the paper describes the performance evaluation of the control schemes on a voltage source inverter (VSI) and proposes the different aspects to be considered for selecting a power electronics inverter topology, reference frames, filters, as well as control strategy.

Keywords: voltage source inverters (VSI); voltage control; current control; digital control; predictive controllers; advanced controllers; stability; response time

1. Introduction

Nowadays, energy demand is getting increased with the passage of time and distributed generation (DG) power systems especially through wind, solar and fuel cells as well as their related power conversion systems are conferred immensely. Many problems like grid instability, low power factor and power outage etc. for power distribution have also been increased with increase in energy demand [1]. However, DG power systems are found to be a sensible solution for such problems as they have relatively robust stability and causes additional flexibility balance. Moreover, their utilization can also improve the distribution networks management and carbon release is also reduced. VSIs are extensively necessitated for the commercial purpose as well as for the industrial applications as they play a key role in converting the DC voltage and current, usually produced by various DG applications, into AC before being discharged into the grid or consumed by the load. Several control systems are introduced, various schemes are proposed and numerous techniques are updated in order to facilitate the control of three-phase VSI. The objectives of these control schemes are to constrain the high and low-frequency electromagnetic pollution and to inject the active power with zero power factor into the grid [2]. The smooth and steady sinusoidal waveform can be a good input to a load for getting the most suitable response, therefore, the output of the inverter, which normally enjoys special standards and characteristics, should be controlled for providing an aforementioned waveform to load and grid.

Generally, it is observed that several problems are caused in linking the DG power system to a grid or grid to load in bidirectional inverters, i.e., grid instability, distortion in the waveform, attenuation as well as major and minor disturbances. Hence, in order to overcome these problems and to provide high-quality power, appropriate controllers with rapid response, compatible algorithm, ability to remove stable errors, less transit time, high tracking ability, less total harmonic distortion, THD value and smooth sinusoidal output should be designed. Various controllers are designed for achieving these qualities. The cascade technologies are introduced in the literature comprises of an inner current loop and outer voltage loop [3–12]. As the inner-loop current controller plays a fundamental role in closed-loop performance, various control approaches like PI [3–6], $H\infty$ [7,8], deadbeat [9–11,13] and μ-synthesis [11] are extensively applied. Outer voltage loop in the aforementioned cases refines the tracking ability and decreases the tracking error. In case of no input limitations, aforesaid PI controllers are the best choice for stabilizing the inner loop performance. However, input constraints restrict their performance and no optimization is usually observed by using PI controllers. The deadbeat control method is proposed in [9] to enhance the closed-loop performance but unfortunately, it was found highly sensitive to the disturbances, parameters mismatches and measurement noise. Later on, some observed based deadbeat controllers are introduced in order to provide compensation for these discrepancies, however, a trade-off was observed between phase margin and closed-loop performance [9,10]. Afterwards, $H\infty$ controllers in [7,8] are offering robust output response instead of input constraints, however, guaranteeing only the local stability like the μ-synthesis controller in [12].

Several other manuscripts are also amalgamated with literature for fulfilling the demand of electric power regarding fulfilling the environmental principles concerning green-house effects [14–18]. Various structures and topologies for interconnecting DGs are presented in [19–21] for parallel operation and in [22–24] for independent operation. For this reason, various control strategies are anticipated for stabilizing the system to control the voltage and frequency in case of unbalanced load and nonlinear loads. Many researchers have proposed several schemes for designing the controller in order to refine the quality of output voltage of DC to AC inverter. In [25], a control scheme is presented for a DG unit in islanded mode, this control technique is suitable for balanced load conditions for a DG unit when it is electronically coupled. However, this technique is constrained to small load variations and remain unable to stabilize the system in large load variations. A robust controller is proposed in [26] for balanced as well as unbalanced systems. However, it fails to address non-linear load properly. In [27], a repetitive control is implemented for controlling the inverters but the relatively slow response and absence of a systematical technique for stabilizing the error dynamics are the core problems. In [28], the uncomplicatedly designed controller is used to mitigate the load disturbances up to a significant extent through a feedforward compensation element, however, it is only restricted to balanced load conditions. In [29], a spatial repetitive control technique is implemented for controlling the current in a single-phase inverter. The results are satisfactory under non-linear load conditions; however, it is not guaranteeing the optimal tracking ability for a three-phase inverter. In [30], a discrete-time sliding mode current controller is proposed, it is optimally operating to control the system at a sudden load change, an unbalanced load and a nonlinear load, however, the system is quite intricate. In [31], the voltage and frequency controller is presented through a discrete-time mathematical model in order to operate the distributed resource units. This technique is achieving good voltage regulations under different load conditions but the results are not verified through the experimental setup. In [32], a controller is proposed having an adaptive feedforward compensation method applied through a Kalman filter for estimating the variation in parameters, the response was robust; however, tuning of covariance matrices are not appropriately described in the paper. In [33], a corresponding controller is recommended for distributed generation systems in grid applications, the anticipated controller is good in handling the grid disturbances and handling the nonlinearities, however, it is not suitable in stand-alone mode due to the nonexistence of voltage loop. In [34,35], the adaptive controller is used and voltage tracking is achieved precisely. The system is guaranteed under systems parameter variations, however, complexity in computation exists and a certain pre-defined value is needed for parameters. In [36], an output voltage controller based on the resonant harmonic filters is presented. It measures the capacitor current and load current in the same sensor. Unbalanced voltage condition and harmonic distortion are compensated in this controller. However, THD value is not defined appropriately, therefore, it is complicated to assess the quality of the controller. An adaptive control technique based proportional derivative controller is presented in [37], for a pulse width modulated inverter operation in islanded distributed generation system, voltage regulation under numerous load conditions is evaluated, though it is not easy to achieve the suitable control gains as par the designing procedure specified in the paper. Moreover, voltage and frequency are optimally controlled, active and reactive power unbalancing is aptly compensated through small signal modeling of inverters in [38].

The key purpose of this study is to provide a comprehensive review of the digital control strategies for different types of three-phase inverters in stand-alone as well as grid-connected modes. Correspondingly, explanation, discussion and comparison of the various control strategies are described in this manuscript in detail.

The manuscript is organized as: classification of voltage source inverters is described in Section 2. Section 3 discusses the characteristics of control systems, followed by a depiction of reference frames in Section 4. The control strategy in decoupled dq frame and time-delay sampling scheme for VSI are depicted in Sections 5 and 6 respectively. An overview of the most commonly used filters and damping techniques is illustrated in Sections 7 and 8 respectively. The grid synchronization techniques followed by modulation techniques are described in Sections 9 and 10 of the manuscript, respectively. Moreover, control Techniques along with their pros. & cons. are described in Section 11. In Section 12, comparative analysis and future goals for the researchers are elaborated. Whereas, conclusions are drawn in Section 13.

2. Classification of VSIs

There are various types, in which the inverters are categorized. Figure 1 shows the complete detail of categories in which voltage source inverters are classified.

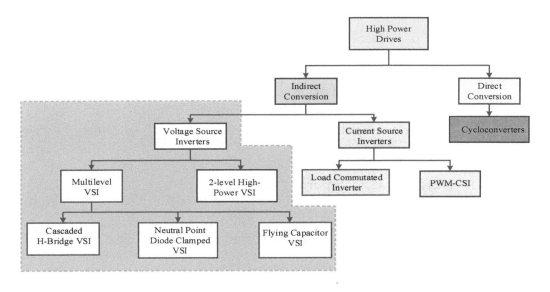

Figure 1. Classification of voltage source inverters (VSIs) in high power drives.

2.1. Multilevel Diode Neutral-Point Clamped Inverter

Multilevel inverter (MLI) was proposed in 1975, its design was like a cascade inverter with diodes facing the source. This inverter was later transformed into a Diode Clamped Multilevel Inverter, which is also named as a Neutral-Point Clamped Inverter (NPC) [39]. In this type of multilevel inverters, the integration of voltage clamping diodes is indispensable. An ordinary DC-bus is separated by an even number of bulk capacitors connected in series with a neutral point in the middle of the line that is dependent on the voltage levels of the inverter. In Figure 2, a five-level NPC-MLI is shown, here the clamping diodes are interlinked to M-1 regulatory pairs if M is considered as voltage levels of the inverter.

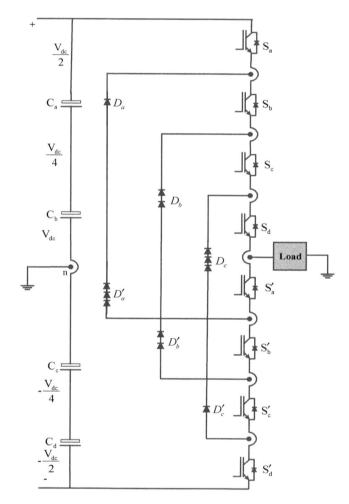

Figure 2. Five-level diode neutral-point clamped inverter.

The neutral point converter was designed by Nabae, Takahashi and Akagi in 1981, this was basically a three-level diode-clamped inverter [40]. A three-phase Three-level diode-clamped inverter is shown in Figure 3.

The NPC-MLI is considered as an important device in conventional high-power ac motor drive applications like mills, fans, pumps and conveyors, moreover, it also offers solutions for industries including chemicals, gas, power, metals, oil, marine, water and mining. The back-to-back configuration of inverters for reformative applications is also considered as a major plus point of this topology, used, for example, in regenerative conveyors, mining industry and grid interfacing of renewable energy sources like wind power [41].

There are several benefits as well as drawbacks of multilevel diode-clamped [39,42]. A common dc bus is shared by all the phases, this results in the reduction of capacitance requirements of the inverter.

Due to this reason, implementation of a back-to-back topology is not only credible but can also be applied practically for performing different operations in an adjustable speed drive and a high-voltage back-to-back inter-connection. The capacitors can be recharged as a group. On fundamental frequency, switching efficacy is relatively higher. However, real power flow is problematic in case of a single inverter as the intermediate dc levels will tend to overcharge or discharge due to inappropriate monitoring and control. The number of clamping diodes are quadratically associated with the number of levels, which can be unwieldy for units with a high number of levels.

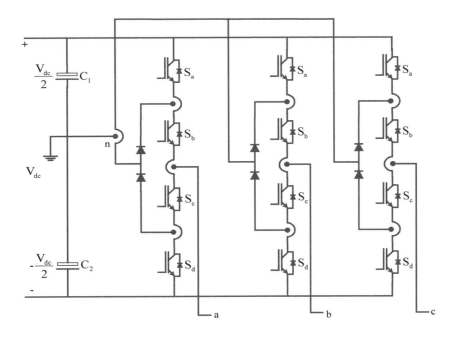

Figure 3. Three-level diode neutral-point clamped inverter.

2.2. Multilevel Capacitor Clamped/Flying Capacitor Inverter

A corresponding topology for the NPC-MLI topology is the Flying Capacitor (FC), or Capacitor Clamped, MLI topology, it is depicted in Figure 4. As an alternative to clamping diodes, capacitors are used for holding the voltages to the referred values. In the NPC-MLI, $M - 1$ number of capacitors are integrated on a shared DC-bus, where M is the level number of the inverter and $2(M - 1)$ switch-diode regulatory pairs are used. Though, for the FC-MLI, instead of clamping diodes, one or more capacitors are used to produce the output voltages depends upon the position and the level of the inverter. They are coupled to the midpoints of two regulatory pairs on the same position on each side of a midpoint [42], see capacitors C_a, C_b and C_c in Figure 5.

The basic difference is the usage of clamping capacitors in place of clamping diodes, as using them increases the number of switching combinations as capacitors do not block reverse voltages [42]. Numerous switching states would be able to produce the same voltage level and the redundant switching states would also be available.

DC side capacitors in this topology have a ladder-like structure and the voltage on each capacitor deviates from that of the other capacitor. The voltage increment between two adjacent legs of the capacitors provides the size of the voltage steps in the output waveform. One advantage of the flying-capacitor-based inverter is the redundancies for inner voltage levels; i.e., two or more effective switching amalgamations can produce an output voltage.

Unlike the diode-clamped inverter, the flying-capacitor inverter never requires all of the switches to be on (conducting state) in a consecutive series. Moreover, the flying-capacitor inverter has phase redundancies, while the diode-clamped inverters have only the line-line redundancies [40]. These redundancies provide selective charging and discharging of specific capacitors and it can be incorporated in the control system for the voltage balancing across the various levels.

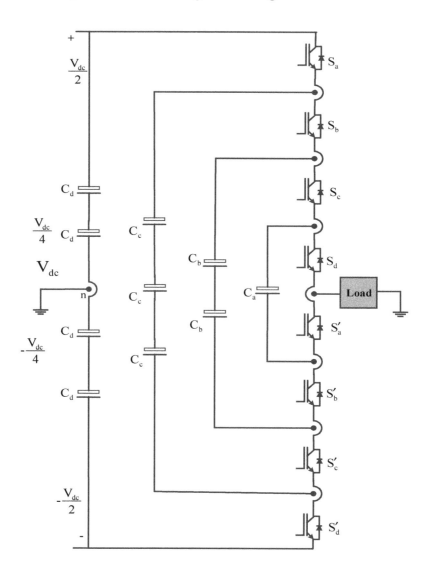

Figure 4. Multilevel (Five-level) capacitor clamped/flying capacitor inverter.

There are several advantages and disadvantages of multilevel flying capacitor inverters [41,43]. Phase redundancies are offered for balancing the voltage levels between the capacitors. Active and reactive power flow can be regulated. The presence of various capacitors allows the inverter to ride through outages for short duration and deep voltage sags. However, the control system is complex for tracking the voltage levels for all of the capacitors. Correspondingly, recharging all the capacitors to the same voltage level and startup are complex. Switching operation and efficacy are poor for real power transmission. The installation of large numbers of capacitors is not much economical and it also makes the system bulky as compared to the clamping diodes in multilevel diode-clamped converters. Likewise, packing is also tougher in the inverters with a higher number of levels. The five-level and three-level FC-MLIs are represented in Figures 4 and 5 respectively.

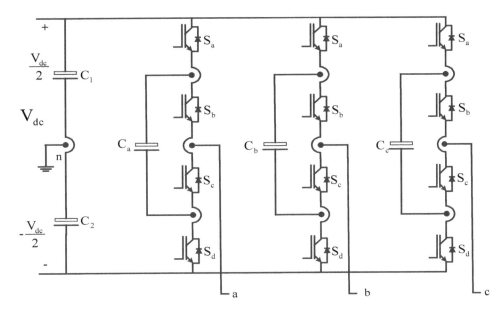

Figure 5. Three-level capacitor clamped/flying capacitor inverter.

2.3. Cascaded H-Bridge Inverter

There are minimum three voltage levels for a multilevel inverter using cascaded topologies. In order to attain a three-level waveform, a single full-bridge or H-bridge inverter is considered. Each inverter is provided with a separate DC source. A three-level cascaded inverter is shown in Figure 6.

By using different combinations of the four switches, S_a, S_b, S_c and S_d, each inverter level can produce three different outputs of voltage, i.e., V_{dc}, 0 and $-V_{dc}$ by connecting the dc source to the ac output. $-V_{dc}$ can be obtained by turning on switches S_b and S_c whereas for obtaining V_{dc}, switches S_a and S_d can be turned on. However, for achieving the output voltage on 0 level either S_a and S_b or S_c and S_d can be turned on. The different full-bridge inverters must be connected in series in the way that the finally produced voltage waveform should be the sum of the inverter outputs. Multilevel cascaded inverters are proposed for the applications such as static VAR generation (reactive power control), an interface with renewable energy sources and for battery-based applications. The main reasons for preferring a cascaded multilevel H-bridge inverter are the availability of possible output levels more than twice the number of dc sources [42–44]. The series of H-bridges enables the manufacturing and packaging process more easy, quick and economical. However, the requirement of a separate dc source for each H-bridge constrains the applications of these inverters to the products having multiple separate DC sources already or readily available.

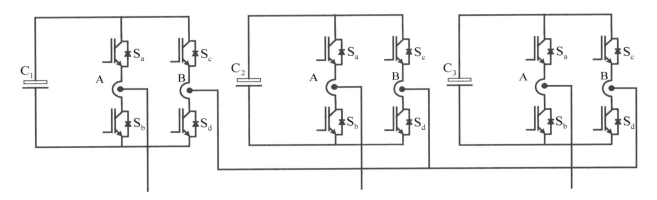

Figure 6. Cascaded H-bridge multilevel inverter.

2.4. Two-Level Three Phase VSI with an Output Filter

A simple two-level inverter is used to convert dc to ac output. It consists of six switches, IGBTs and MOSFETs are the two most suitable switching components for these inverters. Due to simplicity in their structure and ability to handle the voltage by keeping the system stable, they are preferred utmost in the industry and for commercial purpose due to their support in uninterruptible power supply applications. These are usually connected to the load or the grid by using LC or LCL filter. Various types of control systems are implemented by the researchers to improve their performance, robustness and stabilization, compensating the power losses and lowering the THD value. SPWM or SVPWM are mostly applied to these types of inverters for getting appropriate values. Two level three phase VSI is shown in Figure 7. In Figure 7, the S_1 to S_3 and S_1' to S_3' shows the switches of the inverter. Whereas, u_c represents the voltage across the capacitors, C.

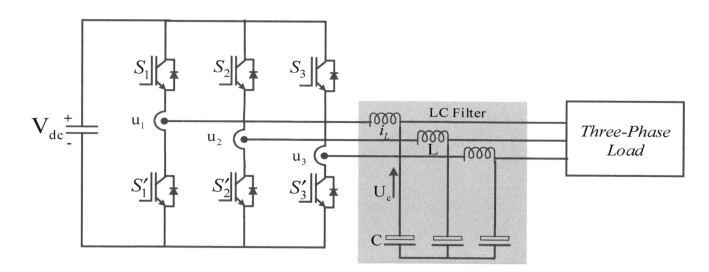

Figure 7. Two-level three phase VSI with an output LC filter.

2.5. Three Phase Four-Leg VSI with an Output Filter

Nowadays, a growing interest in using the three-phase four-leg inverters is observed from the researchers' due to their ability to handle the unbalanced loads efficaciously in four-wire systems [45,46]. In this topology, the neutral point is proposed by connecting the neutral path to the mid-point of the additional fourth leg, as shown in Figure 8. In Figure 8, u_o represents the output voltage of the LC filter, whereas M represents the point neutral point between two switches, S_M and S_M'. Even though the configuration in this topology does not need expensive and large capacitors and produces lower ripple on the DC link voltage, however, using two extra switches lead to a complex control system [47]. Additionally, the split DC-link voltage is about 15% less as compared to the AC voltage in this configuration [48].

Another topology can be using split DC link, which is the most common way of providing a neutral point to three-phase VSIs. This configuration can be provided by using two capacitors i.e., splitting the DC-bus into two parts by using a pair of capacitors and by connecting a neutral path to the mid-point of these capacitors, as shown in Figure 9. Both these configurations have several advantages and disadvantages, however, the split dc-link is found unsuitable for handling the unbalanced loads, whereas, three-phase four leg inverter is found most appropriate for handling the non-linear and unbalanced load conditions. A comparison of different types of VSIs with respect to their characteristics, control contents and complexity is described in Table 1.

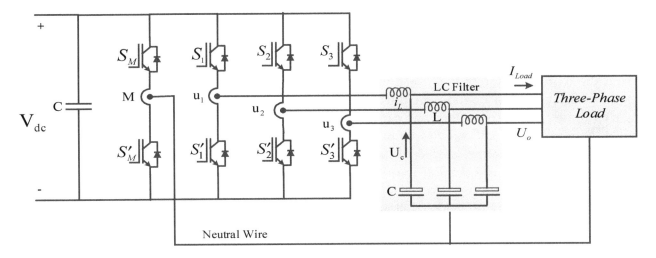

Figure 8. Three-phase four-leg inverter with an output LC filter.

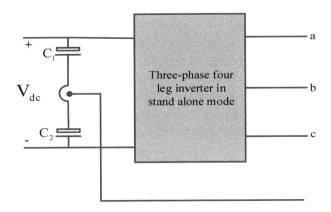

Figure 9. Schematic of a Three-phase four-leg inverter with split DC-link.

Table 1. Comparison of different types of VSI in terms of design, implementation & complexity.

Characteristic	Cascaded H-Bridge VSI	NP-Diode Clamped VSI	Flying Capacitor VSI	2L-VSI
Design & implementation complexity	High	Low	Medium	Low
Specific Requirements	Separate DC sources	Clamping diodes	Additional capacitors	IGBTs/MOSFETs
Control Concerns	Power Sharing	Voltage balancing	Voltage Setup	Voltage/current regulation
Modularity	High	Low	High	Low
Fault tolerance ability	Easy	Difficult	Easy	Easy
Reliability	Medium	Medium	Medium-High	High
Converter Complexity	Medium	Medium	Low-Medium	High
Controller Complexity	Medium-High	Medium-High	Medium-High	Medium
Power Quality	Good	Good	Good	Medium
Operational Power (MW)	3–6	3–7	3–6	3
Switching devices	MV-IGBT, IGCT	MV-IGBT, IGCT	MV-IGBT, IGCT	LV-IGBT

3. Characteristics of Control Systems

There are several parameters and characteristics through which a particular control system is identified. Mainly, there are two characteristics of a control system are found i.e., analog or digital control systems. Both are having some advantages and disadvantages, described as follows:

3.1. Analog Control System

The control systems in which the input and output are designed and analyzed by continuous time analysis or Laplace transform (in s-domain) using state-space formulations. In analog control systems, the representation of the time domain variable is assumed to have infinite precision. Hence, the equations of state space model are differential equations. These systems can be designed without using a computer, microcontrollers or a programmable logic control (PLC). Implementation of analog signals is generally done by using Op-amps, capacitors etc. Robustness against crash or breakdown, having a wide dynamic range, analytical composition accessibility and continuous processing indicate numerous advantages of the analog control systems. However, slow processing speed, interference, complicated implementation in comparative logic, intelligent control systems, neural networks and MIMO are several disadvantages of analog control systems.

3.2. Digital Control System

In digital control systems, modeling, designing, implementation and analysis is carried out in discrete-time or z-transformation domain. In digital control systems, as the name depicts that digital signals are analyzed. Therefore, time is sampled and quantized for state space equations. Additionally, as a digital computer has finite precision, extra attention is needed to ensure that error in coefficients, i.e., A/D conversion, D/A conversion etc. are not producing any disturbances or inadequate effects. In a digital controller, the output is a weighted sum of current as well as previous input and output samples, therefore, its implementation requires the storage of relevant values in a digital controller.

Mostly, a digital controller is implemented via a computer, so, found most economical to control the plants. Moreover, it is relatively easier to constitute and reconstitute through software. Likewise, programs can be leveled to the confines of storage without any additional cost. Correspondingly, digital controllers are compliant with constraints of the program can be changed. Furthermore, the digital controllers are less responsive to the changes in environmental conditions, unlike the analog controllers. Flexibility, swift expansion, uncomplicated implementation in comparative logic, intelligent systems and MIMO, high accuracy as well as robustness against interference are several advantages of these systems. Though, low processing speed, low dynamic range and non-user-friendly interface are the several drawbacks of the digital control systems. The digital controllers are implemented with various technologies which are classified into three categories expressed as follows:

1.   Microcontroller Based implementation (MC) [49–51]
2.   Digital Signal processing-based implementation (DSP) [52–54]
3.   Field programmable gate array-based implementation (FPGA) [55–57]

In reliable scientific research, generally, DSP is used. Fixed point arithmetic and floating-point algorithms are mostly used in implementing the digital control technique by DSP. A traditional slow microprocessor is used normally in slow applications. However, an FPGA is found adequate in fast controllers, due to its abilities of bug fixing and to be reprogrammed in complex structures.

A general structure of a closed loop grid connected digital control system, with an inner current loop and an outer voltage loop, is depicted in Figure 10. In this figure, a voltage source inverter with an output filter is considered. An AC bus is connected to point of common coupling, PCC. Moreover, coordinates transformation from abc to dq is achieved by a phase angle, PH. However, PLL represents the phase locked loop. The symbols S_1, S_2, S_3, S_1', S_2' and S_3' represents the switches, responsible for positive and negative sequences of the inverter output.

The $v_{dref.}$ and $v_{qref.}$ represents the reference voltages in dq frame. SVPWM shows the space vector pulse width modulation technique for generating drive signals for a voltage source inverter.

The voltage across capacitors, u_c and current across inductors, i_L are measured and transformed into a synchronized dq reference frame. The input voltage is computed in the dq frame on the basis of $v_{ref.}$ in the three-phase reference frame. The computed data is then transformed from rotating dq to abc reference frame. Afterward, the PWM technique would be selected accordingly.

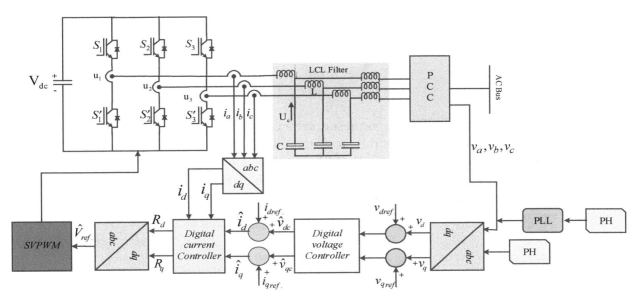

Figure 10. Schematic diagram of a controlled three-phase grid connected VSI with a digital controller.

4. Reference Frames

Control systems are implemented in either a single phase or a three-phase synchronous reference frame. These frames are synchronized with each other through special formulation in order to be compatible for facilitating the modeling, design, analysis and transformation of one phase and three phase systems into other systems. Complex structures, especially for multi-level converters, can be simplified by using these reference frames describes as follows [1,58].

4.1. abc Reference Frame

A general three-phase system is said to be applied to *abc* frame without any transformation. An individual controller is to be used for each phase current in *abc* frame but Delta and star connection has to be considered for designing a control system. Non-linear controllers are used in this system due to their rapid dynamic response.

4.2. dq Reference Frame

This frame is used in three-phase systems. Park's Transformation is used for transforming the *abc* frame into *dq* frame. This transformation causes the current and voltage waveforms to be converted into a frame that rotates synchronously with the grid voltage. As a result, the variables are converted into DC variables and they can easily be controlled and filtered if required.

4.3. αβ Reference Frame

This frame is used in three-phase systems and sometimes sensationally in single phase systems too. Grid current is transformed into a stationary reference frame from *abc* frame or single-phase frame by using Clark's transformation. Therefore, by using this transformation control variable can be transformed into sinusoidal quantities.

5. The Control Strategy in Decoupled dq Frame

In a digital control scheme in dq reference frame, decoupling is the most important issue to be discussed. Generally, a balanced and interrupted sinusoidal waveform can be obtained by adopting ac voltage control in an inverter station. Therefore, the fundamental requirement is to simplify the control design [59]. The controller in an inverter station is based on a mathematical steady-state model in the synchronous reference frame. Moreover, during a balanced network state, the direction of the current injected into the loads is assumed as the reference direction. The mathematical representation of a steady-state model is expressed as following:

$$\begin{cases} u_{bd} = \omega L i_{sq} + u_{sd} \\ u_{bq} = -\omega L i_{sd} + u_{sq} \end{cases} \qquad (1)$$

In Equation (1), the terms u_{bd} and u_{bq} represents the voltages in dq frame under balanced network conditions. Likewise, k_p and k_i represents the proportional and integrated controllers and the equation by using aforementioned coefficients represents a PI controller. Correspondingly, u_{sd} and u_{sq} represents the bus voltages in dq axis. However, i_{sd} and i_{sq} represents the active and reactive current respectively. Commonly, the d-axis is fixed to the voltage source space vector, i.e., the amplitude of the desired ac voltage space vector is kept constant and the value of $u_{sq} = 0$. Then Equation (1) can be simplified as:

$$\begin{cases} u_{bd} = \omega L i_{sq} + u_{sd} \\ u_{bq} = -\omega L i_{sd} \end{cases} \qquad (2)$$

According to Equation (2), the control structure of the inverter station is shown in Figure 11, where a PI controller is employed in the ac voltage control [60]. Moreover, $u_{sref.}$ is the reference voltage which can be set accordingly for the desirable amplitude of AC bus voltage.

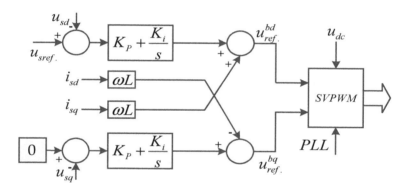

Figure 11. The decoupling control strategy in dq reference frame.

6. Time Delay Sampling Scheme for VSI

Time sampling for digital controllers is done by using a discrete time-domain analysis, i.e., z-domain. Two fundamental advantages of using z-domain analysis over s-domain (continuous time domain) analysis for designing a current controller are: First the control implementation is achieved on a computer-based system, i.e., the control calculation, sampling measurements and PWM signals sequence are updated in discrete time steps. Although, this sample and hold feature is a characteristic of a control system and effects its dynamics as per the referred sampling frequency. Secondly, the multiple time delays can be modeled by using a backshift operator, which affords no simplifications in linear control design, unlike continuous time domain, where the multiple time delays were sampled using an exponential term, which is approximated generally by applying Taylor-series expansion. The sampling effect is a most critical requirement to handle model uncertainties, issues in power supplies and relative disturbances. Therefore, in order to deal with aforementioned issues, zero order hold, ZOH should be incorporated in the control system. In ZOH, a pole or a zero is added into the existed controller through the compensator. The fundamental advantage of this technique is its uncomplicated structure to be implemented on a system, though, it only affects a limited share of the overall delay.

There are two basic sampling routines generally employed in the digital control systems, i.e., single updated sampling and double updated sampling [61]. A single-update sampling method comprises of the measurement samplings, in which calculated modulation indexes are updated once in every switching period. Whereas, a double-update sampling concept conferred to a PWM concept in which the measurement sampling and therefore, the calculated modulation index are updated twice in every switching period [61]. The detailed single-update and double-update sampling are shown in Figure 12, where, T(k) represents the switching period of the present time slot. However, T(k − 1) and T(k − 2) shows the switching period of the former time slots.

A single-update PWM-technique with sampling at the beginning of a switching period is depicted in Figure 12a. In this technique, the modulation index is updated once in beginning of a switching period. A time domain of one sampling time is introduced in the control loops. This effect is modeled with a backshift operator while taking discrete time domain into the account.

Figure 12b shows another scheme of a single-update PWM sampling in which the modulation index is updated in middle of a switching period. Therefore, the time delay due to sampling and updating routine is the mean value of the two converter voltage reference values, i.e., actual and former control cycles. Therefore, the transfer function of a single-update PWM technique with sampling in middle of a switching period is determined.

In the double-update sampling concept, sampling and updating occurs twice in each sampling period. In this technique, the modulation index is updated on the basis of former control cycle's measurements. According to this behavior, the time-delay is one control cycle. The pattern of a double-update sampling is presented in Figure 12c.

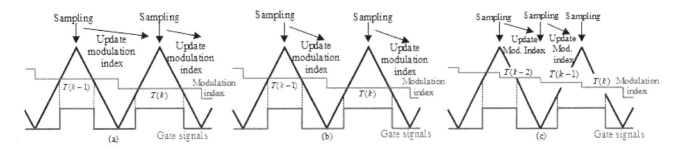

Figure 12. Time delay model of a VSI (**a**) single-update time delay model (sampling at beginning) (**b**) single-update time delay model (sampling at middle) (**c**) double-update time delay model for a VSI.

7. Output Filters for Inverters

The harmonics reduction is the foremost priority of the researchers while designing a power electronics or an electrical system. Therefore, an output filter is used for this purpose. An output filter uses the controlled phenomenon of switching the semiconductor devices for harmonics reduction. There are numerous topologies of such filters introduced in the literature by combining the inductor (L) and capacitor (C) i.e., L, LC and LCL filters unified with the inverters to their output.

7.1. L-Filter

In high switching frequency inverters, the first order L-filter is considered as the most suitable filter. However, inductance decreases the dynamics of the whole system.

7.2. LC-Filter

An LC filter is a second-order filter having substantially sophisticated damping behavior as compared to an L-filter. This filter topology is relatively easier to design and it is a compromise between the values of inductance and capacitance. The cut-off frequency needs the relatively higher value of inductance whereas the voltage quality can be improved through the higher value of capacitance. The value of resonant frequency is dependent on the impedance of the grid when the system is connected to the grid supply. An LC-filter is mostly preferred in standalone mode. The three-phase two-level and three-phase four legs voltage source inverters with an integrated LC filter are shown in Figures 7 and 8 respectively.

7.3. LCL-Filter

An LCL filter is a third order filter, mostly used for the grid-tied inverters. The lower frequency is preferable in presence of aforementioned filters. This filter supports the comparatively healthier decoupling between the filter and the grid impedance. This filter should be precisely designed by taking into consideration the parameters of the inverters. Otherwise even the smaller values of inductance can bring resonance and unstable states into the system. However, the smaller inductance can provide optimized current ripple diminishing values. A three-phase VSI with an LCL filter is shown in Figure 13. Where, V_{th} and Z_{th} represents the Thevenin voltage and Thevenin impedance respectively. However, the complexity of the control system inflated significantly and the dynamic performance of the inverter can perhaps be affected when relatively complex filter structures are employed. Thus, these topologies are most suitable for high power applications, which employ low switching frequencies. However, Figures 14 and 15 show the one-leg block diagram of a single-phase and three-phase grid-connected systems, respectively. Where, K_{pwm} represents the pulse width modulation characteristic of the system, whereas, u_g, u_i, u_c, L_g, L_i, i_i and i_g represents the grid side voltage, inverter side voltage, voltage across capacitor, grid side inductance, inverter side inductance, inverter side current and grid side current respectively.

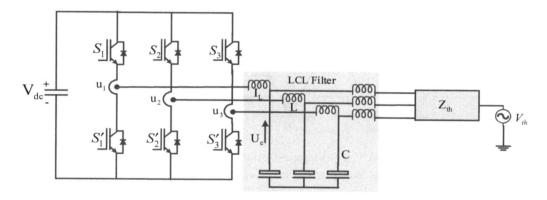

Figure 13. A Three-phase voltage source inverter in grid-connected mode.

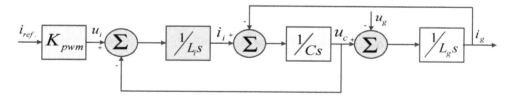

Figure 14. One-leg block diagram of a single-phase grid-connected system.

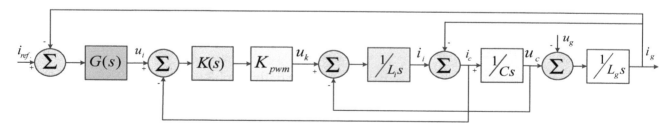

Figure 15. One-leg block diagram of a dual-loop current control strategy for VSI.

8. Damping Techniques for Grid-Connected VSIs

In the grid-connected applications, LCL filter is highly preferred due to its harmonic suppressing capability. In this case, the voltage across the point of common coupling PCC is controlled in synchronism with the current. Therefore, it becomes possible to regulate the active and reactive power injected into the grid according to the requirement. The LCL filter offers a resonance frequency which can be a source of instability in the closed-loop system. This problem is stated by various researchers in the literature and numerous damping strategies are proposed to solve it [62–65]. Damping methods can be classified into two groups. (i) Passive damping and (ii) Active damping.

8.1. Passive Damping

Passive damping is to inserting passive elements in the filter for reduction of the resonant peak in the system [32]. Generally, passive damping schemes never desire any amendments in the control strategy. Though, these approaches change attenuation of the filter, as a result of which losses increases [18,32,34]. The passive damping techniques, presented generally in the literature, results in the addition of a simple resistor in series with the filter capacitor [63]. The major drawback of this technique is a reduction in filter attenuation, increasing power losses and large filter volume [62]. A general schematic of passive damping control strategy for a grid connected VSI is shown in Figure 16.

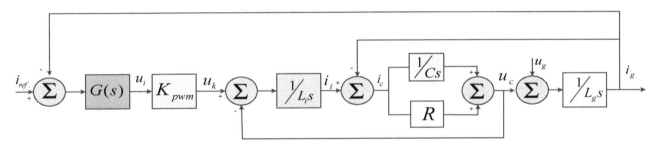

Figure 16. One-leg block diagram of a passive damping control strategy for a grid-connected VSI.

8.2. Active Damping

The active damping methods are proposed to overcome the drawbacks associated with the passive damping techniques. Active damping techniques offer modifications in the control policy in order to afford closed loop damping [65,66]. The active damping techniques are classified into 3 groups, i.e., single loop, multi-loop and complex controllers. Single loop methods are incorporated to damp the LCL filter resonance, without supplementary measurement. These methods comprise of low pass filter-based method, virtual flux estimation method, sensor-less method, splitting capacitor method, notch-filter method and grid current feedback method. Generally, single-loop methods are found relatively robust during uncertainty in parameters and variation in grid inductance [62]. Multiloop methods explore additional measurements. This group comprises of capacitor current feedback, capacitor voltage feedback and weighted average current control techniques. However, the third group of active damping methods is based on complex control structures. This outcome of these techniques is usually a suitable and a robust dynamic response [67]. These techniques include predictive control, state-space controllers, adaptive controllers, sliding mode controller and vector control. Additionally, when LCL filter is selected, there are two options for current control: grid current or converter current. Various techniques are proposed but there exists a disagreement in the literature about the suitable solution of these issues and it is agreed that the current control strategy should be carefully selected. An active damping technique with a damping resistance as well as a harmonic compensator are described in [65]. A general schematic of active damping control strategy for a grid connected VSI is shown in Figure 17.

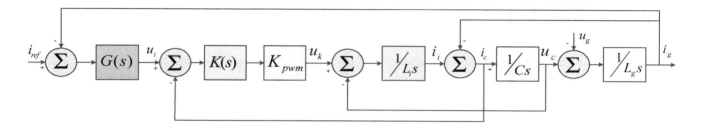

Figure 17. One-leg block diagram of an active damping control strategy for a grid-connected VSI.

9. Grid Synchronization Techniques

The grid voltage must be synchronized with the injected current in a utility network for a significant output. In synchronization algorithm, phase of the grid voltage vector is considered and control variables i.e., grid voltages and grid currents are synchronized by using it. Various methods are introduced in literature for extracting the phase angle [68]. Some commonly used techniques found in credible research articles are discussed as:

9.1. Zero-Crossing Technique

The simplest method to implement is Zero-Crossing method. However, it is not considered on a larger scale due to poor performances reported in the literature. Especially, during voltage variations, ample values of harmonics and notches are observed.

9.2. Filtering of Grid Voltages

The grid voltages can be filtered in the dq frame as well as in the $\alpha\beta$ reference frame. The performance of zero-crossing method is improved by voltage filtering [68]. However, it is a complicated process to extract the phase angle out of utility voltage, especially during a fault condition. This method uses the arctangent function to realize the phase angle. Generally, a delay is observed in processing a signal while using the filtering method. Therefore, designing of the filter must be considered critically.

9.3. Phase Locked Loop Technique

The phase locked loop, PLL technique is considered as the state-of-the-art method to obtain the phase angle of the grid voltages. The PLL is implemented in dq-synchronous reference frame. In this case, the coordinates transformation from abc to dq is preferred and reference voltage, $\hat{u}_d$ would be set to zero for realizing the lock. A general schematic of PLL technique is depicted in Figure 18. A PI regulator is generally used to control the reference variable. Afterward, the grid frequency is integrated in the system and utility voltage angle is acquired after passing through a voltage-controlled oscillator, VCO. This voltage angle is then fed into the $\alpha\beta$ to dq transformation module for transforming into the synchronous reference frame.

This technique is found the most suitable for rejecting notches, grid harmonics and other disturbances. However, additional improvements are needed to handle the unsymmetrical voltage faults. Especially, filtering techniques to filter the negative sequence should be proposed in case of unsymmetrical voltage faults, as second-order harmonics are propagated by the PLL system and reflected in the obtained phase angle. Moreover, it should be assured to estimate the phase angle of the positive sequence of the grid voltages during unbalanced grid voltages [68].

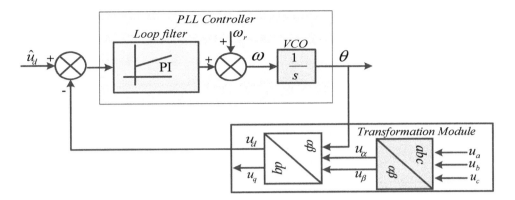

Figure 18. General schematic of a three-phase PLL technique.

10. Modulation Schemes

In power electronics converters, the major problem is the reduction in harmonics. PWM control techniques provide the most suitable solution for harmonics reduction. A sinusoidal output having controlled values of frequency and magnitude is the core purpose for using these PWM techniques. Primarily, PWM techniques are classified into three major categories i.e., Triangular Comparison-based PWM (TC-PWM), Space Vector-based PWM (SV-PWM) and Voltage look-up table-based PWM (VLUT-PWM).

10.1. Triangular Comparison Based PWM

In Triangular Comparison based PWM (TC-PWM) techniques, PWM waves are produced by the combination of an ordinary triangular carrier and a fundamental modulating reference signal. The triangular carrier signal has relatively very higher frequency than that of a fundamental modulating reference signal. The magnitude and frequency of the fundamental modulating reference signal control the magnitude and frequency of the central module in the grid side. PWM and Synchronous PWM (SPWM) are the core techniques to be mentioned in TC-PWM [69].

10.2. Space Vector Based PWM

In SVPWM techniques, the revolving reference vectors provide the reference signals. The magnitude and frequency of central module in grid side are controlled by the frequency and magnitude of the revolving reference vectors respectively. This technique was first introduced to generate vector based PWM in the three-phase inverters. However, nowadays it is expanded to various other newly introduced inverters. SV-PWM is considered to be the more advanced technique for PWM generation for getting qualified sinusoidal output with low THD values [69].

10.3. Voltage Look-Up Table-Based PWM

In VLUT-PWM, a new method is introduced to obtain the voltage reference based on the current reference for an inverter. The major advantage of this technique is its compatibility and simplicity with the load conditions. The switching frequency in this technique is usually taken significantly lower as compared to various other presented techniques [52].

11. Control Techniques

Connecting the grid to the distributed generation system plays a key role and if bit negligence is shown in implementing this procedure, a number of problems can arise i.e., the grid uncertainty and disturbance, so in order to overcome this situation, a suitable controller must be designed for it. In this section, the most appropriate control techniques are described according to their applications. Various single loop and multiloop control systems are discussed in the literature for power droop control, voltage and current control. In which inner loop is for current regulation and outer loop is for voltage regulation [13,70,71]. In Figure 19, the categorization of classical an advanced control technique is depicted clearly.

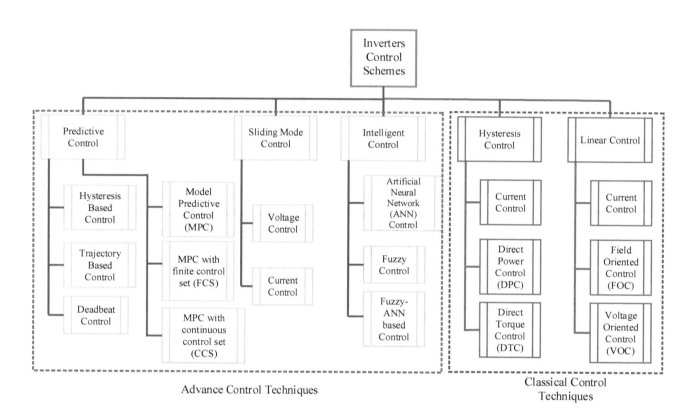

Figure 19. Classification of control techniques for VSIs.

11.1. Classical Control Techniques

The classical controllers include the category of controllers for adding or subtracting a proportion and adjusting the system accordingly. These controllers involve proportional (P), proportional integration (PI), proportional integral derivative (PID) and proportional derivative (PD) controllers. These controllers are considered as the most fundamental controllers in the industry for controlling linear systems and considered as the base of control theory. Lot of work in literature is being done on these controllers [49–52,72–78].

The fundamental benefits of implementing these controllers are their ability to tune themselves according to the requirement of the plant and their simple structure. Moreover, they are the most commonly used controllers on commercial levels, so easily available. However, their tracking ability, response time and ability to handle stable error are relatively lower as compared to modern state-of-the-art controller. The schematic of a digital PI controller for controlling a three-phase VSI with an LC filter in stand-alone mode is shown in Figure 20.

In Figure 20, i_{af}, i_{bf} and i_{cf} represents the filter current across phase a, b and c respectively. Likewise, v_a, v_b and v_c characterizes the voltage across phase a, b and c respectively. Likewise, i_d and i_q represents the current across the d and q axis respectively. Moreover, S_a, S_b and S_c represents the switching commands across phase a, b and c respectively. Correspondingly, V_{ref}^d. and V_{ref}^q. symbolizes the reference voltages along d and q axis respectively.

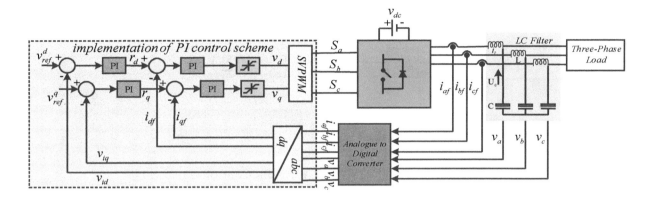

Figure 20. Schematic of a PI control algorithm for a VSI.

11.2. PR Controllers

PR controllers are the combination of proportional and resonant controllers. The frequencies closer to resonant frequency are integrated by the integrator. Therefore, phase shift or stationary error do not occur. This controller can be applied in both *ABC* and αβ frames. Due to high gain near resonant frequencies, this controller has the ability to eliminate the steady-state errors of electrical quantities. The resonant controller maintains the network frequency equal to the resonant frequency. It is capable of adjusting the frequency according to changes in grid frequency. However, an accurate tuning is always needed for optimal results and this technique is found sensitive to the frequency variations [30,31]. These controllers are relatively better than PI controllers in terms of their tracking ability and response time. If used with a harmonic compensator, they can optimally handle THD. Their capability to handle current in grid-connected inverters is also remarkable. However, damping issues still exist. The active and passive damping adjustments and integration in a system with a harmonic compensator are somehow, the complicated issues. Moreover, they do not have outstanding ability to handle stable error and phase shift. The limitation to handle specific frequencies i.e., closer to resonant frequencies is also a drawback of these controllers. A PR controller with a harmonic compensator, HC, in stand-alone mode for a VSI is shown in Figure 21. Structures of a simple PR controller and a discrete PR controller are shown in Figures 22 and 23, respectively. Where, T_s represents the sampling period, ω represents the grid angular frequency. However, K_p and K_r denotes the proportional and resonant coefficients respectively. The PR controller with a harmonic compensator is proposed in [65].

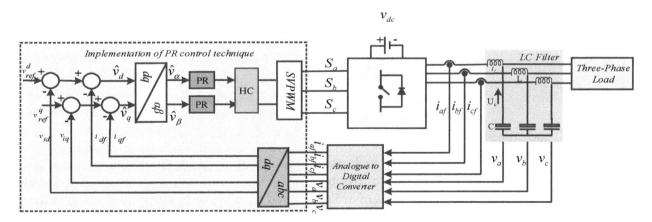

Figure 21. General schematic of a PR Controller and HC for a three-phase VSI.

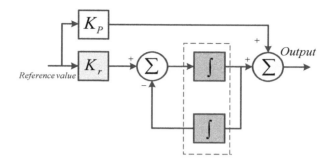

Figure 22. General structure of a PR Controller.

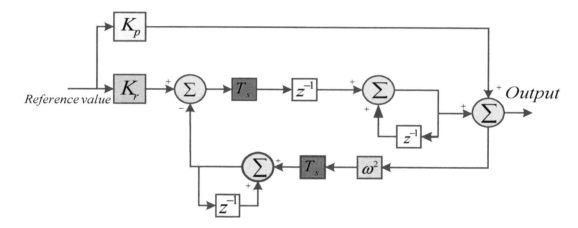

Figure 23. Structure of a discrete PR Controller.

11.3. LQG Control Technique

The integration of Kalman filter with an LQR controller gives rise to an LQG controller. In this technique, Kalman Filter, as well as an LQG controller, can be designed independently of each other. This control scheme is valid for both linear time-invariant systems as well as for linear time-varying systems. LQG control technique facilitates the designing of a linear feedback controller for an uncertain nonlinear control system [79–81]. An LQG control structure with a Kalman estimator is shown in Figure 24. Where, u_e represents the known input and y_c is the estimated noise/disturbance. The Kalman estimator provides the optimal solution to the continuous or discrete estimation problems.

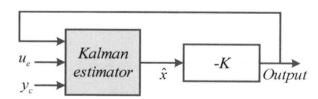

Figure 24. Structure of an LQG controller.

11.3.1. Linear Quadratic Regulator

The linear quadratic regulator (LQR) technique is found optimal for steady as well as transient states [82–84]. As the name depicts, this control technique is a combination of linear and quadratic functions, where the dynamics of the system are described by a set of linear equations and the cost of the system is a quadratic function. The cost function parameters are considered critically while designing the controller. LQR algorithm is an automatic approach for finding a suitable state-feedback controller. Pole placement with state feedback controller provides the system with a high degree of freedom and makes it simpler to implement. This method is characteristically steady and it can be employed even if some of the system parameters are unknown. However, exertion in finding the exact weighting factors limits the applications of LQR control scheme. Moreover, it has a discrepancy of tracking accuracy during load changes [83–85].

11.3.2. Linear Quadratic Integrator

In linear quadratic integrator (LQI) scheme, cost minimization is considered critically. This technique is implemented for nullifying the steady-state error between actual grid voltage and reference grid voltage during load variations [82]. An integral term used with LQ control is for minimizing the tracking error produced by uncertain disturbances in instantaneous reference voltage. Optimal gains for providing adequate tracking with zero steady-state error are relatively simpler to attain by using this technique. The rapid dynamic response, accurate tracking ability and relatively simpler designing procedure provide this technique a benefit over other techniques. However, complications in extracting the model and phase shift in voltage tracking even in normal operative condition are the major drawbacks of this scheme.

11.4. Hysteresis Control Technique

Hysteresis control is considered as a nonlinear method [86–93]. The hysteresis controllers are used to track the error between the referred and measured currents. Therefore, the gating signals are generated on the basis of this reference tracking. Hysteresis bandwidth is adjusted for error removal in reference tracking. This is an uncomplicated concept and has been used since analog control platforms were intensively used. This technique does not require a modulator; therefore, the switching frequency of an inverter is dependent on the hysteresis bandwidth operating conditions and filter parameters [94]. The major drawback of hysteresis controller is its uncontrolled switching frequency; however, researchers are working on improving this controller and several works are presented and several techniques are proposed in the literature. Main advances in this technique are direct torque control (DTC) [87,88,95,96] and direct power control (DPC) [97–99]. In DPC, active and reactive powers are directly controlled, however, in DTC torque and flux of the system are controlled. Error signals are produced by hysteresis controllers and drive signals are generated by the look-up according to the magnitude of the error signals. Hysteresis controllers require very high frequency for constraining the variables in hysteresis band limits, whenever implemented on a digital platform as shown in Figure 25. Moreover, switching losses are very high in this type of controllers. So, Hysteresis controllers are found inappropriate for high power applications.

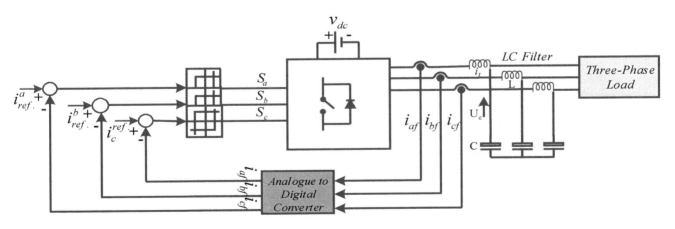

Figure 25. A Hysteresis control technique for VSI.

11.5. Sliding Mode Control

The sliding mode control is considered to be an advanced power control technique for the power converters. It fits into the family of adaptive control and variable structure control [100–104].

Sliding mode control is a non-linear technique, whereas it can be instigated to both non-linear as well as linear systems [100]. In Figure 26, a sliding mode control along with SVM/PWM is presented. Where, β_v represents the gain, λ is a strictly positive constant and ϕ is a trade-off between the tracking error and smoothing of the control discontinuity. The sliding controller produces the voltage references in a converter for generating the drive signals. A predefined trajectory is executed and the control variable is forced to slide along it [102–104]. The robust and stable response is achieved even in the system parameters variation or load disturbances by implementing sliding mode control technique. This controller is more robust and capable of removing the stable error as compared to the classical controllers. However, some drawbacks in implementing a sliding mode control are difficulty in finding a suitable sliding surface and limitation of sampling rate that degrades the performance of SMC will be degraded. Whenever tracking a variable reference, the chattering phenomenon is another drawback of SMC technique. As a result, overall system efficacy is reduced [105,106].

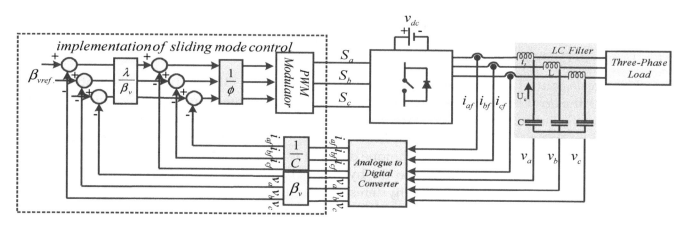

Figure 26. A Sliding mode control technique on a VSI.

11.6. Partial Feedback Controllers

There are several techniques presented for conversion of non-linear systems to linear systems for their uncomplicated computation. Partial feedback controllers are one of the most effective and forthright techniques for transforming the non-linear systems into the linear ones. By this technique, a system can be converted either partially or fully into a linear system, depends on the system constraints. Linearity in a system is attained by the cancellation of the nonlinearities inside the system. So, these systems can be controlled by using the linear controllers whenever a non-linear system is fully transformed into a linear system i.e., exact feedback linearization method. However, if it is partially converted into a linear system then it is known to be partial feedback linearization. PFL controller is implemented in [104,106–109]. In PFL, it is difficult to ensure the stability of complicated renewable energy system applications. However, an independent subsystem can be obtained from PFL for constraining the extensive use of this method. Moreover, in order to deal with these problems, exact feedback linearization (EFL) is a forthright and model-based technique for scheming nonlinear control techniques. EFL receipts the built-in nonlinearity characteristic of the system under deliberation and consents the conversion of a nonlinear structure into a linear one, algebraically. EFL removes nonlinearities of a system through nonlinear feedback, as a result, the transformed system is not reliant on an operating point.

11.7. Repetitive Control

The plug-in scheme (PIS) and internal model (IM) principle are the basic concepts of repetitive control (RC). RC uses an IMP which is in correspondence to the model of a periodic signal. In order to derive this model, trigonometric Fourier series expansion is used. If the model of reference is fed into the closed loop path, optimal reference tracking can be obtained. Moreover, it is found robust against disturbances and has the ability to reject them. RC mostly deals with periodic signals. Closed loop behavior of the system and Magnitude response of the IM are the core factors used for analyzing the performance of the repetitive controller in case of frequency variation or any other uncertainty in the system. Both these factors indicate the performance sagging in case of variation or uncertainty in the reference signal. In presence of a periodic disturbance, RC intends to attain zero tracking error when a periodic or a constant command is referred to it. RC has an ability to locate an error, a time-period before and fine-tunes the next command according to the feedback control system for eliminating the observed error. However, it lacks the ability to handle physical noise. For this purpose, an LPF can be used. Kalman's filtering approach is also noticeable to remove this noise [27,110–113]. The general structure of a repetitive controller is shown in Figure 27.

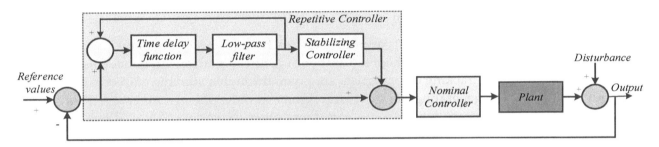

Figure 27. Block diagram of Repetitive control algorithm.

11.7.1. Fuzzy Control

This control technique belongs to the family of intelligent control systems. The PI controller is replaced by a fuzzy logic controller in this technique as shown in Figure 28. Where, v_{fz} is the fuzzified output voltage. However, its block diagram is shown in Figure 29. In a fuzzy controller, the tracking error of load current and its derivative are given as the input. This controller design is dependent on the awareness, knowledge, skills and experience of the converter designer in terms of functions involvement. Due to non-linear nature of the power converters, the system can be stabilized in case of parameters variation even if the exact model of the converter is unknown. Fuzzy logic controllers are also categorized as non-linear controllers and probably the best controllers amongst the repetitive controllers [113–116]. However, strong assumptions and adequate experience are required in fuzzification of this controller. As it is dependent on the system input and draw conclusions according to the set of rules assigned to them during the process of their modeling and designing.

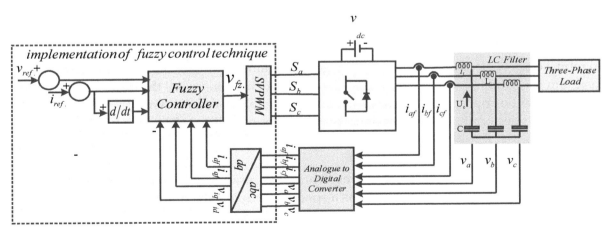

Figure 28. A Fuzzy control algorithm topology on a VSI.

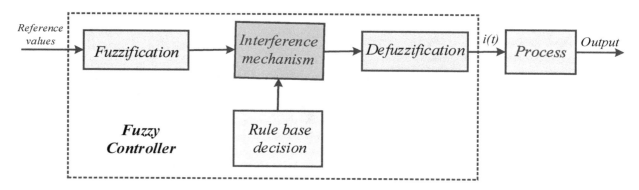

Figure 29. Block diagram of Fuzzy control algorithm.

11.7.2. Artificial Neural Network Control

The Artificial neural network (ANN) controllers are the fundamental form of the controllers based on the human-thinking mode. It consists of a number of artificial neurons to behave as a biological human brain. The reference tracking error signals are given through a suitable gain or a scaling factor (S) as input to the ANN for generating the switching signals into the power converters. This approach is used for achieving the constant switching operation in power converters. ANN can be used in both online as well offline modes while operating it on system control. It has high tolerance level to faults because of its ability to estimate the function mapping. Its topology is shown in Figure 30.

Fuzzy and ANN can be combined to achieve an optimal control performance in a power converter [113–115]. ANN does not need a converter model for its operation, however, the operational behavior of a power converter should be precisely known to the designer/operator while designing the ANN control system.

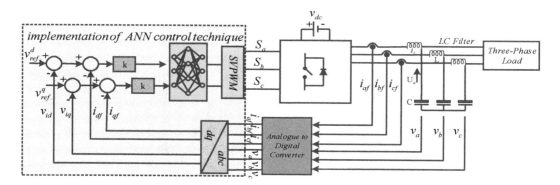

Figure 30. Schematic of Artificial Neural Network control for VSIs.

11.8. Robust Controllers

In robust control theory, a control system vigorous against uncertainties and disturbances is offered. The basic aim is to attain the stability in case of inadequate modeling. All the descriptions, criteria and limitations should be appropriately defined in order to get robust control. This controller guarantees the stability and high performance of closed-loop system even in multivariable systems [117].

11.8.1. H-Infinity Controllers

The expression H∞ control originates from the term mathematical space on which the optimization takes place: H∞ is considered as a space of matrix-valued functions that are investigative and confined in the open right-half of the complex plane. In this type of control system, first of all, the control problem is formulated and then mathematical optimization is implemented i.e., selection of the best element according to criterion from the set of obtainable alternatives. H-infinity control techniques are generally pertinent for the multivariable systems. The impact of a perturbation can be reduced by using H-infinity control techniques in a closed loop system subject to the problem formulation. The impact can be measured either in terms of performance or stabilization of the system. However, modeling of the system should be well-defined for implementation of these control techniques. Moreover, H-infinity control techniques have another discrepancy of high computational complications. In case of non-linear systems limitations, the control system cannot handle them well and response time also increases [118]. However, these controllers are implemented and well defined in [111,112,119].

11.8.2. μ-Synthesis Controllers

Mu-synthesis is based on the multivariable feedback control technique, which is used to handle the structured as well as unstructured disturbances in the system. Where μ mentions the singular value that is reciprocal of the multivariable stability margin. The basic purpose is to mechanize the synthesis of multivariable feedback controllers that are insensitive to uncertainties of the plant and be able to attain the anticipated performance objectives. This method is well described in [120,121].

11.9. Adaptive Controllers

An adaptive controller is designed to have the ability of self-tuning, i.e., to regulate itself spontaneously according to variations in the system parameters. It does not require initial conditions, system parameters or limitations for its implementation due to its ability to modify the control law according to system requirements. Recursive least squares and Gradient descent are two most commonly known technique for parameters estimation in adaptive controllers. The structure of a typical adaptive controller is shown in Figure 31. In the literature, some credible research articles and state-of-the-art techniques for adaptive controllers are found in [14,37,53,55,113,122–125]. These controllers are applicable for both dynamic as well as static processes. However, the complicated computational process leads to exertion in its implementation.

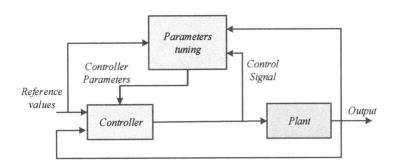

Figure 31. Block diagram of Adaptive control algorithm.

11.10. Predictive Controllers

Predictive controllers are commenced as a propitious control technique for electronics inverters. The system model is considered critically and then imminent behavior of the control variables is predicted conferring to the specified criterion. It is an uncomplicated technique and can handle multivariable systems efficiently. Moreover, it can handle the system with several limitations or non-linearities. It is generally preferred due to its prompt static as well as dynamic response and ability to handle stable errors. However, its computational analysis is complex as compared to classical controllers. It is further categorized into Deadbeat control and Model Predictive control. It can refer to literature [105,125–127] for predictive controllers. A comparison of predictive control techniques on basis of their pros. and cons. is described in Table 2.

11.10.1. Deadbeat Control

Deadbeat control technique is the most authentic, competent and attractive technique in terms of low THD value, frequency as well as rapid transient response. Differential equations are derived and discretized in this type of control system for controlling the dynamic behavior of the system. The control signal is predicted for the new sampling period for attaining the reference value. Its effective dynamic performance and high bandwidth simplify the current control for this type of controller. Error compensation is a specialty of a deadbeat controller. However, its major discrepancy is its sensitivity for network parameters and accurate mathematical filter modeling [13,54,56,128–135]. Its topology is shown in Figure 32, where a disturbance observer, a state estimator and a digital deadbeat controller are used to control the voltage and current of a VSI. The coefficient $\hat{d}$ represents the output of disturbance observer comprises of current and voltage. However, $\hat{v}_d$ and $\hat{v}_q$ represents the controlled voltage across d-axis and q-axis respectively.

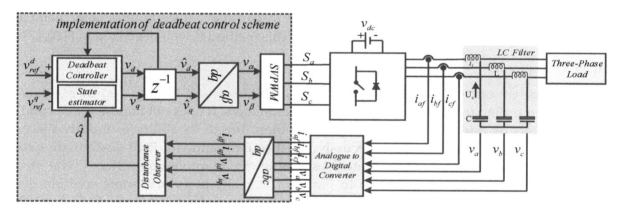

Figure 32. Deadbeat control topology for VSIs.

11.10.2. Model Predictive Control

As the name depicts, a model of the system is used to predict the behavior of the system in model predictive control (MPC) technique. A cost function criterion is defined in this type of control system, which can be minimized for optimal control actions. The controller adapts the optimal switching states according to the cost function criterion. Forecast error can be lessened for current tracking implementing. Moreover, system limitations and non-linearities, as well as multiple inputs and output systems, are handled well by MPC. Control actions of the present state are considered in order to predict the control actions of the system in the next state. Like deadbeat control, it is also found sensitive to system parameter variations [136–147]. The topology for implementation of MPC on VSI is shown in Figure 33, whereas, its control schematic is depicted in Figure 34.

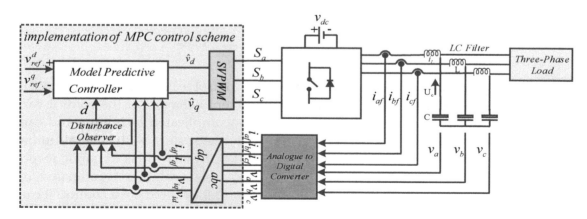

Figure 33. Model Predictive Control topology for VSIs.

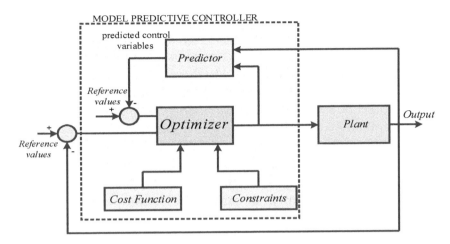

Figure 34. Block Diagram of Model Predictive Control algorithm.

11.11. Iterative Learning Scheme

Iterative Learning Scheme (ILS) is a complicated but authentic technique for attaining zero tracking error. In this scheme, each control command is executed and the system is examined and then adjusted accordingly before each repetition. Highly accurate modeling of the system is essential for the implementation of ILS; therefore, its designing technique is relatively more complicated than other schemes. ILS is capable of removing the tracking error caused by the periodic disturbances. The next cycle is predicted by considering the learning gain, system adjustment in z-transform, tracking error at each repetition, control function of the designed controller and the error function between two consecutive iterations.

12. Comparative Analysis and Future Research Goals

VSIs are specially designed for converting DC to three-phase AC, therefore, control strategies must be according to the three-leg three-phase power inverters. However, for MLIs, the control strategies must be inherited from three-leg three-phase power inverters. The control policies of VSIs in stand-alone mode can be categorized into numerous categories depending upon similar and dissimilar considerations. Considering the PWM, VSIs can be classified into two categories i.e., carrier-based modulation and carrier-less modulation. The carrier-based modulation schemes such as Selective Harmonic Elimination (SHE), 3D SVPWM, Sinusoidal Pulse Width Modulation (SPWM) and Minimum-Loss Discontinuous PWM (MLDPWM) based PWM techniques have attained significant consideration due to their constant switching frequency.

SPWM offers constant switching frequency and flexible control schemes; nevertheless, one major disadvantage of this technique is the compact efficacy of the DC voltage [148]. The 3D-SVPWM delivers an adequate DC bus utilization and a standardized load harmonic curvature as compared to the SPWM technique. However, it is complex in nature to be implemented on the digital devices. Correspondingly, the SHE-PWM suggests a flexible controller by considering the switching angle. However, the real-time enactment of this carrier-based modulation is quite difficult. The capability of the MLDPWM under nonlinear and unbalanced conditions is found relatively admissible; however, its real-time execution is found very much circuitous. However, the carrier-less modulation approaches such as flux vector and hysteresis provide a rapid-dynamic response [149]. But, they suffer from variable switching frequency [91]. Additionally, they require composite switching tables for their implementation.

The conventional PI controllers encounter problems to eliminate the steady-state error. In order to solve this problem, a PR controller is commonly used in the stationary reference frame for regulating the output voltages of the VSIs due to its sophisticated explication in eliminating the steady-state error, while controlling sinusoidal signals. Additionally, it is competent in eliminating selective harmonic uncertainties.

It is also taken into consideration that the PR controller perceives the resonant frequency to offer gains at specific frequencies. Thus, the resonant frequency should be synchronized with the frequency of the microgrid. Hence, it can be said that it is very sensitive with respect to the variations in system frequency. The PI controller is also extensively used in the *dqo* frame and performs robustly with pure DC signals. Though, in order to allocate the control variables from the *abc* to the *dqo* frame, the phase angle of the microgrid should be known. Likewise, using cross-coupling and voltage feedforward terms are the secondary problems in implementing this method.

In the stand-alone operating mode, VSIs primarily controls the power transfer, voltage and frequency of the system. Nevertheless, power quality can be enhanced by offering a suitable control technique in the inverter-based type DGs. As in VSIs, the auxiliary services for improvement in power quality embedded in the control assembly. In case of VSIs, compensation of unbalanced voltage, a lower value for total harmonic distortion, harmonic power-sharing schemes, power sharing between active/reactive powers, imbalance power, active/reactive power control and augmentation in power quality are critically considered and embedded in control schemes. VSIs are also applied on several applications in microgrid systems, extensively, for improving the power quality. This power control strategy is presented in [38]. However, a comprehensive review of various control strategies for microgrids is described in [150]. Moreover, using modular multilevel inverters can improve the modularity and scalability to meet reference voltage levels, efficiency in high power applications, reduction in harmonics in high voltage applications and size of passive filters as well as no requirement for dc-link capacitors [151].

In Table 3, different types of filters are suggested by various researchers based on the control systems. However, it is significant to use L-filters for low power applications having a simple design, nevertheless, L-filters are not found suitable in resonance state as well as for high power applications. So, LCL-filter is highly preferred in aforementioned system characteristics. The designing of this filter is comparatively complicated due to a few constrictions related to the system stability. The accuracy in designing and modeling of the system leads to better performance against resonance and harmonics. Nevertheless, the choice should be made according to the customer's demand. The prime parameters should be chosen on the basis of system condition and intended tasks to be performed by the system. Afterward, the designing of power system and control system parameters should be finalized.

This corresponding study incorporates the advantages and disadvantages of each controller in terms of stability, rapid response time, harmonic elimination, the nonlinearity of the system, unbalanced compensation and robustness against parameter variation. Various suitable control schemes for different types of VSIs are documented in this paper. However, their implementations for power generation and power quality improvement are still not perfect simultaneously. Moreover, each controller has its own benefits and obstructions. Therefore, it is not easy to decide that which control scheme is better than the others. These are significant subjects for the future research. On the basis of the analysis of former publications, appended research is suggested to be carried out in the aforementioned area.

Regardless of the several investigations in this field, none of the proposed control techniques can be selected as an immaculate solution to meet al.l the requirements of power quality, i.e., harmonic/reactive/imbalance power-sharing and voltage unbalanced/harmonic/swell/sag and Interruption compensation at the same time. Therefore, further research should be focused on the novel power-electronics topologies to fulfill all aforementioned necessities simultaneously.

Three-phase three-wire VSIs are now a well-developed and mature research topic with respect to their hierarchical control. But on the other hand, control hierarchies are not as well established for ML-VSIs, as for three-phase three-wire VSIs. It may be beneficial to consider ML-VSI system, as well as the primary, secondary and tertiary stages, whenever a control scheme is to be designed.

Substantially, a lot of work is to be done for exploiting the new control approaches for ML-VSIs. In order to achieve enhanced performance, it is compulsory to use some innovative techniques such as robust, MPC and LQR control techniques.

It is also observed through a number of studies that coupling among the phases is neglected whenever controlling an ML-VSI by means of a conventional PI controller, which results in a reduction of the system's robustness. Hence, it can also be beneficial to implement decoupled phase voltage control to realize the referred response in time domain. A comparison of the credible research articles found in literature with respect to their control techniques, modulation schemes, control parameters, loop characteristics, employed filters and applications is described in Table 3.

13. Conclusions

On the basis of research, conventional multilevel inverter topologies given in the previous sections, general and asymmetrically constituted ML-VSIs have been also reviewed in this paper. Several new hybrid topologies can be designed through the combinations of three main MLI topologies. Besides the combination of topologies, the trade-offs in MLI structures can be dealt by using H-MLIs that is formed using different DC source levels in inverter cells. PWM strategies that generate switching frequency at fundamental frequency are also introduced for H-MLIs for the switching devices of the higher voltage modules to operate at high frequencies only during some inverting instants. Due to numerous applications of conventional MLIs and flexibility to design the hybrid MLI topologies, this paper cannot cover all utilizations with MLIs but the authors intend to provide a useful basis to define the most proper control schemes and applications. In addition to these, the fundamental design and control principles of MLIs have been introduced as a result of a detailed literature survey. This paper has been destined to provide a reference to readers and the results given in this paper can also be extended with experimental studies.

Table 2. Description of predictive controllers on the basis of their pros. & cons.

	Deadbeat Control
	• -Modulator required
	• -Fixed switching frequency
	• -Low Computations
	• -Limitations not undertaken
	Trajectory Based control
	• -Modulator not required
	• -Variable frequency
	• -No cascaded structure
Predictive Control	**Hysteresis Based predictive control**
	• -Modulator not required
	• -Variable frequency
	• -Uncomplicated structure
	Model Predictive Control
	• -Modulator required in case of continuous control set (CCS) and not required in case of finite control set (FCS).
	• -Likewise, fixed switching frequency (CCS) and variable switching frequency exists in (FCS).
	• -Online optimization and simple designing is included in case of FCS.
	• -Constraints are considered in both cases

Table 3. Digital control system characteristics in numerous credible scientific proposals.

Application	Controller	Filter	Ref. Frame	Feedback	Modulation	Ctrl. Parameter	Ref.
General	adaptive	LCL	Single Phase	Multi-loop	SPWM	V, I	[51]
General	Classic, PR	LCL	Single Phase	Multi-loop	PWM	I	[152]
General	Adp., Rpt.	L	Single Phase	Single-loop	SPWM	I	[105]
UPS	DB	LC	Single Phase	Multi-loop	PWM	V, I	[153]
General	Rpt.	LC	Single-Phase	Single-loop	PWM	V	[101]
General	Rpt C	LCL	Single Phase	Single-loop	PWM	I	[104]
DG	Classic	LCL	abc, αβ	Single-loop	PWM	V, P	[47]
DG	Classic	LC	abc, αβ	Multi-loop	SVPWM	V, I	[48]
DG	Classic	LC	abc, αβ	Multi-loop	SPWM	V, I	[49]
DG	Classic	L	abc, αβ	Single-loop	VLUT	V	[50]
General	DB	L	abc, αβ	Single-loop	PWM	I	[52]
APF	Adp., Rpt.	LC	abc, dq	Multi-loop	SVPWM	I	[53]
General	DB	LCL	abc, dq	Multi-loop	PWM	I	[54]
DG	Adp., MPC	LCL	abc, αβ	Multi-loop	SVPWM	I	[55]
General	LQG	LCL	abc, dq	Single-loop	PWM	I	[66]
PV	PR, LQG	L	abc, αβ	Single-loop	SVPWM	I	[64]
General	Adp.	L	abc, αβ	Multi-loop	PWM	I	[107]
UPS	Pred.	LC	abc, dq	Single-loop	SVPWM	V	[109]
PV, APF	Pred., Fuzzy	L	abc, αβ	Multi-loop	PWM	P	[108]
PV, APF	SMC, Pred.	L	abc, αβ	Multi-loop	PWM	P	[111]
General	DB	L	abc, dq	Single-loop	SVPWM	I	[113]
General	Adp., DB	L	abc, dq	Single-loop	PWM	I	[112]
DG	DB	L	abc, dq	Single-loop	SVPWM	I	[116]
UPS	DB, Rpt.	LC	abc, αβ	Single-loop	PWM	V	[115]
General	MPC	LCL	abc, abc	Single-loop	PWM	V, I	[122]
General	MPC	L	abc, αβ	Single-loop	PWM	I	[125]
General	MPC	L	abc, dq	Single-loop	SVPWM	I	[128]
PV	MPC	L	abc, dq	Single-loop	SVPWM	I	[130]
PV	Classic, Rpt.	L	abc, dq	Single-loop	SVPWM	I	[97]

Acknowledgments: The National Natural Science Foundation of China supported this research work under Grant No. 61374155. Moreover, the Specialized Research Fund for the Doctoral Program of Higher Education PR China under Grant No. 20130073110030 is highly acknowledged.

Author Contributions: Sohaib proposed the idea for writing the manuscript. Wang suggested the literature and supervised in writing the manuscript. Mazhar helped Sohaib in writing and formatting. Sarwar helped in modifying the figures and shared the summary of various credible articles to be included in this manuscript.

References

1. Blaabjerg, F.; Teodorescu, R.; Liserre, M.; Timbus, A.V. Overview of control and grid synchronization for distributed power generation systems. *IEEE Trans. Ind. Electron.* **2006**, *53*, 1398–1409. [CrossRef]

2. European Commission Directorate-General for Energy. *DG ENER Work in the Paper—The Future Role and Challenges of Energy Storage*; European Commission Directorate-General for Energy: Brussel, Belgium, 2013.

3. Loh, P.C.; Newman, M.J.; Zmood, D.N.; Holmes, D.G. A comparative analysis of multiloop voltage regulation strategies for single and three-phase UPS systems. *IEEE Trans. Power Electron.* **2003**, *18*, 1176–1185.

4. Kassakian, J.G.; Schlecht, M.F.; Verghese, G.C. *Principles of Power Electronics*; Addison-Wesley: Reading, PA, USA, 1991; Volume 1991.

5. Mohan, N.; Undeland, T.M. *Power Electronics: Converters, Applications, and Design*; John Wiley & Sons: New Delhi, India, 2007.

6. Abdel-Rahim, N.M.; Quaicoe, J.E. Analysis and design of a multiple feedback loop control strategy for single-phase voltage-source UPS inverters. *IEEE Trans. Power Electron.* **1996**, *11*, 532–541. [CrossRef]

7. Lee, T.-S.; Chiang, S.-J.; Chang, J.-M. H/sub/spl infin//loop-shaping controller designs for the single-phase UPS inverters. *IEEE Trans. Power Electron.* **2001**, *16*, 473–481.

8. Willmann, G.; Coutinho, D.F.; Pereira, L.F.A.; Libano, F.B. Multiple-loop H-infinity control design for uninterruptible power supplies. *IEEE Trans. Ind. Electron.* **2007**, *54*, 1591–1602. [CrossRef]

9. Kawabata, T.; Miyashita, T.; Yamamoto, Y. Dead beat control of three phase PWM inverter. *IEEE Trans. Power Electron.* **1990**, *5*, 21–28. [CrossRef]

10. Ito, Y.; Kawauchi, S. Microprocessor based robust digital control for UPS with three-phase PWM inverter. *IEEE Trans. Power Electron.* **1995**, *10*, 196–204. [CrossRef]
11. Cho, J.-S.; Lee, S.-Y.; Mok, H.-S.; Choe, G.-H. Modified deadbeat digital controller for UPS with 3-phase PWM inverter. In Proceedings of the Thirty-Fourth IAS Annual Meeting, Conference Record of the 1999 IEEE Industry Applications Conference, Phoenix, AZ, USA, 3–7 October 1999; Volume 4.
12. Lee, T.S.; Tzeng, K.S.; Chong, M.S. Robust controller design for a single-phase UPS inverter using μ-synthesis. *IEE Proc. Electr. Power Appl.* **2004**, *151*, 334–340. [CrossRef]
13. Tahir, S.; Wang, J.; Kaloi, S.G.; Hussain, M. Robust digital deadbeat control design technique for 3 phase VSI with disturbance observer. *IEICE Electron. Express* **2017**, *14*, 20170351. [CrossRef]
14. Jung, J.-W.; Vu, N.T.-T.; Dang, D.Q.; Do, T.D.; Choi, Y.S.; Choi, H.H. A three-phase inverter for a standalone distributed generation system: Adaptive voltage control design and stability analysis. *IEEE Trans. Energy Convers.* **2014**, *29*, 46–56. [CrossRef]
15. Bogosyan, S. Recent advances in renewable energy employment. *IEEE Ind. Electron. Mag.* **2009**, *3*, 54–55. [CrossRef]
16. Liserre, M.; Sauter, T.; Hung, J.H. Future energy systems: Integrating renewable energy sources into the smart power grid through industrial electronics. *IEEE Ind. Electron. Mag.* **2010**, *4*, 18–37. [CrossRef]
17. Kim, M.-Y.; Song, Y.-U.; Kim, K.-H. The advanced voltage regulation method for ULTC in distribution systems with DG. *J. Electr. Eng. Technol.* **2013**, *8*, 737–743. [CrossRef]
18. Mokhtarpour, A.; Shayanfar, H.; Bathaee, S.M.T.; Banaei, M.R. Control of a single phase unified power quality conditioner-distributed generation-based input output feedback linearization. *J. Electr. Eng. Technol.* **2013**, *8*, 1352–1364. [CrossRef]
19. He, J.; Li, Y.W. An enhanced microgrid load demand sharing strategy. *IEEE Trans. Power Electron.* **2012**, *27*, 3984–3995. [CrossRef]
20. Marwali, M.N.; Jung, J.-W.; Keyhani, A. Stability analysis of load sharing control for distributed generation systems. *IEEE Trans. Energy Convers.* **2007**, *22*, 737–745. [CrossRef]
21. Zhang, Y.; Yu, M.; Liu, F.; Kang, Y. Instantaneous current-sharing control strategy for parallel operation of UPS modules using virtual impedance. *IEEE Trans. Power Electron.* **2013**, *28*, 432–440. [CrossRef]
22. Vechiu, I.; Curea, O.; Camblong, H. Transient operation of a four-leg inverter for autonomous applications with unbalanced load. *IEEE Trans. Power Electron.* **2010**, *25*, 399–407. [CrossRef]
23. Kasal, G.K.; Singh, B. Voltage and frequency controllers for an asynchronous generator-based isolated wind energy conversion system. *IEEE Trans. Energy Convers.* **2011**, *26*, 402–416. [CrossRef]
24. Nian, H.; Zeng, R. Improved control strategy for stand-alone distributed generation system under unbalanced and non-linear loads. *IET Renew. Power Gener.* **2011**, *5*, 323–331. [CrossRef]
25. Karimi, H.; Nikkhajoei, H.; Iravani, R. Control of an electronically-coupled distributed resource unit subsequent to an islanding event. *IEEE Trans. Power Deliv.* **2008**, *23*, 493–501. [CrossRef]
26. Karimi, H.; Yazdani, A.; Iravani, R. Robust control of an autonomous four-wire electronically-coupled distributed generation unit. *IEEE Trans. Power Deliv.* **2011**, *26*, 455–466. [CrossRef]
27. Escobar, G.; Valdez, A.A.; Leyva-Ramos, J.; Mattavelli, P. Repetitive-based controller for a UPS inverter to compensate unbalance and harmonic distortion. *IEEE Trans. Ind. Electron.* **2007**, *54*, 504–510. [CrossRef]
28. Yazdani, A. Control of an islanded distributed energy resource unit with load compensating feed-forward. In Proceedings of the 2008 IEEE Power and Energy Society General Meeting-Conversion and Delivery of Electrical Energy in the 21st Century, Pittsburgh, PA, USA, 20–24 July 2008.
29. Dasgupta, S.; Sahoo, S.K.; Panda, S.K. Single-phase inverter control techniques for interfacing renewable energy sources with microgrid—Part I: Parallel-connected inverter topology with active and reactive power flow control along with grid current shaping. *IEEE Trans. Power Electron.* **2011**, *26*, 717–731. [CrossRef]
30. Dai, M.; Marwali, M.N.; Jung, J.-W.; Keyhani, A. A three-phase four-wire inverter control technique for a single distributed generation unit in island mode. *IEEE Trans. Power Electron.* **2008**, *23*, 322–331. [CrossRef]
31. Delghavi, M.B.; Yazdani, A.N. Islanded-mode control of electronically coupled distributed-resource units under unbalanced and nonlinear load conditions. *IEEE Trans. Power Deliv.* **2011**, *26*, 661–673. [CrossRef]
32. Delghavi, M.B.; Yazdani, A.N. An adaptive feedforward compensation for stability enhancement in droop-controlled inverter-based microgrids. *IEEE Trans. Power Deliv.* **2011**, *26*, 1764–1773. [CrossRef]
33. Prodanovic, M.; Timothy, C.; Green, T.C. Control and filter design of three-phase inverters for high power quality grid connection. *IEEE Trans. Power Electron.* **2003**, *18*, 373–380. [CrossRef]

34. Mattavelli, P.; Escobar, G.; Stankovic, A.M. Dissipativity-based adaptive and robust control of UPS. *IEEE Trans. Ind. Electron.* **2001**, *48*, 334–343. [CrossRef]

35. Valderrama, G.E.; Stankovic, A.M.; Mattavelli, P. Dissipativity-based adaptive and robust control of UPS in unbalanced operation. *IEEE Trans. Power Electron.* **2003**, *18*, 1056–1062. [CrossRef]

36. Escobar, G.; Mattavelli, P.; Stankovic, A.M.; Valdez, A.A.; Leyva-Ramos, J. An adaptive control for UPS to compensate unbalance and harmonic distortion using a combined capacitor/load current sensing. *IEEE Trans. Ind. Electron.* **2007**, *54*, 839–847. [CrossRef]

37. Do, T.D.; Leu, V.Q.; Choi, Y.-S.; Choi, H.H.; Jung, J.-W. An adaptive voltage control strategy of three-phase inverter for stand-alone distributed generation systems. *IEEE Trans. Ind. Electron.* **2013**, *60*, 5660–5672. [CrossRef]

38. Rasheduzzaman, M.; Mueller, J.; Kimball, J.W. Small-signal modeling of a three-phase isolated inverter with both voltage and frequency droop control. In Proceedings of the 2014 Twenty-Ninth Annual IEEE Applied Power Electronics Conference and Exposition (APEC), Fort Worth, TX, USA, 16–20 March 2014.

39. Rodriguez, J.; Lai, J.-S.; Peng, F.Z. Multilevel inverters: A survey of topologies, controls, and applications. *IEEE Trans. Ind. Electron.* **2002**, *49*, 724–738. [CrossRef]

40. Nabae, A.; Takahashi, I.; Akagi, H. A new neutral-point-clamped PWM inverter. *IEEE Trans. Ind. Appl.* **1981**, *5*, 518–523. [CrossRef]

41. Franquelo, L.G.; Rodriguez, J.; Leon, J.I.; Kouro, S.; Portillo, R.; Prats, M.A.M. The age of multilevel converters arrives. *IEEE Ind. Electron. Mag.* **2008**, *2*, 28–39. [CrossRef]

42. Lai, J.-S.; Peng, F.Z. Multilevel converters—A new breed of power converters. *IEEE Trans. Ind. Appl.* **1996**, *32*, 509–517.

43. Patel, H.S.; Hoft, R.G. Generalized techniques of harmonic elimination and voltage control in thyristor inverters: Part I—Harmonic elimination. *IEEE Trans. Ind. Appl.* **1973**, *3*, 310–317. [CrossRef]

44. Khomfoi, S.; Tolbert, L.M. Multilevel power converters. In *Power Electronics Handbook*; Elsevier/Academic Press: Burlington, VT, USA, 2007; pp. 451–482.

45. Zhang, R.; Prasad, V.H.; Boroyevich, D.; Lee, F.C. Three-dimensional space vector modulation for four-leg voltage-source converters. *IEEE Trans. Power Electron.* **2002**, *17*, 314–326. [CrossRef]

46. Lohia, P.; Mishra, M.K.; Karthikeyan, K.; Vasudevan, K. A minimally switched control algorithm forthree-phase four-leg VSI topology tocompensate unbalanced and nonlinear load. *IEEE Trans. Power Electron.* **2008**, *23*, 1935–1944. [CrossRef]

47. Zhong, Q.-C.; Liang, J.; Weiss, G.; Feng, C.M.; Green, T.C. H∞ Control of the Neutral Point in Four-Wire Three-Phase DC–AC Converters. *IEEE Trans. Ind. Electron.* **2006**, *53*, 1594–1602. [CrossRef]

48. Liang, J.; Green, T.C.; Feng, C.; Weiss, C. Increasing voltage utilization in split-link, four-wire inverters. *IEEE Trans. Power Electron.* **2009**, *24*, 1562–1569. [CrossRef]

49. Miret, J.; Camacho, A.; Castilla, M.; de Vicuña, L.G.; Matas, J. Control scheme with voltage support capability for distributed generation inverters under voltage sags. *IEEE Trans. Power Electron.* **2013**, *28*, 5252–5262. [CrossRef]

50. Liu, Z.; Liu, J.; Zhao, Y. A unified control strategy for three-phase inverter in distributed generation. *IEEE Trans. Power Electron.* **2014**, *29*, 1176–1191. [CrossRef]

51. Li, Y.; Jiang, S.; Cintron-Rivera, J.G.; Peng, F.Z. Modeling and control of quasi-Z-source inverter for distributed generation applications. *IEEE Trans. Ind. Electron.* **2013**, *60*, 1532–1541. [CrossRef]

52. Ebadi, M.; Joorabian, M.; Moghani, J.S. Voltage look-up table method to control multilevel cascaded transformerless inverters with unequal DC rail voltages. *IET Power Electron.* **2014**, *7*, 2300–2309. [CrossRef]

53. Eren, S.; Pahlevani, M.; Bakhshai, A.; Jain, P. An adaptive droop DC-bus voltage controller for a grid-connected voltage source inverter with LCL filter. *IEEE Trans. Power Electron.* **2015**, *30*, 547–560. [CrossRef]

54. Abu-Rub, H.; Guzinski, J.; Krzeminski, Z.; Toliyat, H.A. Predictive current control of voltage-source inverters. *IEEE Trans. Ind. Electron.* **2004**, *51*, 585–593. [CrossRef]

55. Espi, J.M.; Castello, J.; Garcia-Gil, R.; Garcera, G.; Figueres, E. An adaptive robust predictive current control for three-phase grid-connected inverters. *IEEE Trans. Ind. Electron.* **2011**, *58*, 3537–3546. [CrossRef]

56. Moreno, J.C.; Espi Huerta, J.M.; Gil, R.G.; Gonzalez, S.A. A robust predictive current control for three-phase grid-connected inverters. *IEEE Trans. Ind. Electron.* **2009**, *56*, 1993–2004. [CrossRef]

57. Ahmed, K.H.; Massoud, A.M.; Finney, S.J.; Williams, B.W. A modified stationary reference frame-based predictive current control with zero steady-state error for LCL coupled inverter-based distributed generation systems. *IEEE Trans. Ind. Electron.* **2011**, *58*, 1359–1370. [CrossRef]

58. Miveh, M.R.; Rahmat, M.F.; Ghadimi, A.A.; Mustafa, M.W. Control techniques for three-phase four-leg voltage source inverters in autonomous microgrids: A review. *Renew. Sustain. Energy Rev.* **2016**, *54*, 1592–1610. [CrossRef]

59. Chen, H. Research on the control strategy of VSC based HVDC system supplying passive network. In Proceedings of the 2009 PES'09 IEEE Power & Energy Society General Meeting, Calgary, AB, Canada, 26–30 July 2009.

60. Qian, C.; Tang, G.; Hu, M. Steady-state model and controller design of a VSC-HVDC converter based on dq0-axis. *Autom. Electr. Power Syst.* **2004**, *16*, 015.

61. Hoffmann, N.; Fuchs, F.W.; Dannehl, J. Models and effects of different updating and sampling concepts to the control of grid-connected PWM converters—A study based on discrete time domain analysis. In Proceedings of the 2011 14th European Conference on Power Electronics and Applications (EPE 2011), Birmingham, UK, 30 August–1 September 2011.

62. Büyük, M.; Tan, A.; Tümay, M.; Bayındır, K.Ç. Topologies, generalized designs, passive and active damping methods of switching ripple filters for voltage source inverter: A comprehensive review. *Renew. Sustain. Energy Rev.* **2016**, *62*, 46–69. [CrossRef]

63. Peña-Alzola, R.; Liserre, M.; Blaabjerg, F.; Ordonez, M.; Yang, Y. LCL-filter design for robust active damping in grid-connected converters. *IEEE Trans. Ind. Inform.* **2014**, *10*, 2192–2203. [CrossRef]

64. Zhang, C.; Dragicevic, T.; Vasquez, J.C.; Guerrero, J.M. Resonance damping techniques for grid-connected voltage source converters with LCL filters—A review. In Proceedings of the 2014 IEEE International Energy Conference (ENERGYCON), Cavtat, Croatia, 13–16 May 2014; pp. 169–176.

65. Jia, Y.; Zhao, J.; Fu, X. Direct grid current control of LCL-filtered grid-connected inverter mitigating grid voltage disturbance. *IEEE Trans. Power Electron.* **2014**, *29*, 1532–1541.

66. Teodorescu, R.; Liserre, M.; Rodriguez, P. *Grid Converters for Photovoltaic and Wind Power Systems*; John Wiley & Sons: Chichester, UK, 2011; Volume 29.

67. Jalili, K.; Bernet, S. Design of LCL filters of active-front-end two-level voltage-source converters. *IEEE Trans. Ind. Electron.* **2009**, *56*, 1674–1689. [CrossRef]

68. Kazmierkowski, M.P.; Malesani, L. Current control techniques for three-phase voltage-source PWM converters: A survey. *IEEE Trans. Ind. Electron.* **1998**, *45*, 691–703. [CrossRef]

69. Kumar, K.V.; Michael, P.A.; John, J.P.; Kumar, S.S. Simulation and comparison of SPWM and SVPWM control for three phase inverter. *ARPN J. Eng. Appl. Sci.* **2010**, *5*, 61–74.

70. Lim, J.S.; Park, C.; Han, J.; Lee, Y.I. Robust tracking control of a three-phase DC–AC inverter for UPS applications. *IEEE Trans. Ind. Electron.* **2014**, *61*, 4142–4151. [CrossRef]

71. Huerta, J.M.E.; Castello, J.; Fischer, J.R.; García-Gil, R. A synchronous reference frame robust predictive current control for three-phase grid-connected inverters. *IEEE Trans. Ind. Electron.* **2010**, *57*, 954–962. [CrossRef]

72. Samui, A.; Samantaray, S.R. New active islanding detection scheme for constant power and constant current controlled inverter-based distributed generation. *IET Gener. Transm. Distrib.* **2013**, *7*, 779–789. [CrossRef]

73. Yuan, X.; Merk, W.; Stemmler, H.; Allmeling, J. Stationary-frame generalized integrators for current control of active power filters with zero steady-state error for current harmonics of concern under unbalanced and distorted operating conditions. *IEEE Trans. Ind. Appl.* **2002**, *38*, 523–532. [CrossRef]

74. Miret, J.; Castilla, M.; Matas, J.; Guerrero, J.M.; Vasquez, J.C. Selective harmonic-compensation control for single-phase active power filter with high harmonic rejection. *IEEE Trans. Ind. Electron.* **2009**, *56*, 3117–3127. [CrossRef]

75. Beza, M.; Bongiorno, M. Improved discrete current controller for grid-connected voltage source converters in distorted grids. In Proceedings of the 2012 IEEE Energy Conversion Congress and Exposition (ECCE), Raleigh, NC, USA, 15–20 September 2012.

76. Kandil, M.S.; El-Saadawi, M.M.; Hassan, A.E.; Abo-Al-Ez, K.M. A proposed reactive power controller for DG grid connected systems. In Proceedings of the 2010 IEEE International Energy Conference and Exhibition, Manama, Bahrain, 18–22 December 2010; pp. 446–451.

77. Radwan, A.A.A.; Abdel-Rady, I.M.Y. Power Synchronization Control for Grid-Connected Current-Source Inverter-Based Photovoltaic Systems. *IEEE Trans. Energy Convers.* **2016**, *31*, 1023–1036. [CrossRef]

78. Chilipi, R.; Sayari, N.A.; Hosani, K.A.; Beig, A.R. Control scheme for grid-tied distributed generation inverter under unbalanced and distorted utility conditions with power quality ancillary services. *IET Renew. Power Gener.* **2016**, *10*, 140–149. [CrossRef]

79. Busada, C.; Jorge, S.G.; Leon, A.E.; Solsona, J. Phase-locked loop-less current controller for grid-connected photovoltaic systems. *IET Renew. Power Gener.* **2012**, *6*, 400–407. [CrossRef]

80. Athans, M. The role and use of the stochastic linear-quadratic-Gaussian problem in control system design. *IEEE Trans. Autom. Control* **1971**, *16*, 529–552. [CrossRef]

81. Huerta, F.; Pizarro, D.; Cobreces, S.; Rodriguez, F.J.; Giron, C.; Rodriguez, A. LQG servo controller for the current control of *LCL* grid-connected voltage-source converters. *IEEE Trans. Ind. Electron.* **2012**, *59*, 4272–4284. [CrossRef]

82. Hossain, M.A.; Azim, M.I.; Mahmud, M.A.; Pota, H.R. Primary voltage control of a single-phase inverter using linear quadratic regulator with integrator. In Proceedings of the 2015 Australasian Universities Power Engineering Conference (AUPEC), Wollongong, Australia, 27–30 September 2015; pp. 1–6.

83. Ahmed, K.H.; Massoud, A.M.; Finney, S.J.; Williams, B.W. Optimum selection of state feedback variables PWM inverters control. In Proceedings of the IET Conference on Power Electronics, Machines and Drives, York, UK, 2–4 April 2008; pp. 125–129.

84. Xue, M.; Zhang, Y.; Kang, Y.; Yi, Y.; Li, S.; Liu, F. Full feedforward of grid voltage for discrete state feedback controlled grid-connected inverter with LCL filter. *IEEE Trans. Power Electron.* **2012**, *27*, 4234–4247. [CrossRef]

85. Lalili, D.; Mellit, A.; Lourci, N.; Medjahed, B.; Boubakir, C. State feedback control of a three-level grid-connected photovoltaic inverter. In Proceedings of the 2012 9th International Multi-Conference on Systems, Signals and Devices (SSD), Chemnitz, Germany, 20–23 March 2012; pp. 1–6.

86. Jaen, C.; Pou, J.; Pindado, R.; Sala, V.; Zaragoza, J. A linear-quadratic regulator with integral action applied to PWM DC-DC converters. In Proceedings of the IECON 2006-32nd Annual Conference on IEEE Industrial Electronics, Paris, France, 6–10 November 2006; pp. 2280–2285.

87. Bose, B.K. *Power Electronics and Motor Drives: Advances and Trends*; Academic Press: Oxford, UK, 2010.

88. Kaźmierkowski, M.P.; Krishnan, R.; Blaabjerg, F. (Eds.) *Control in Power Electronics: Selected Problems*; Academic Press: New York, NY, USA, 2002.

89. Shukla, A.; Ghosh, A.; Joshi, A. Hysteresis modulation of multilevel inverters. *IEEE Trans. Power Electron.* **2011**, *26*, 1396–1409. [CrossRef]

90. Prabhakar, N.; Mishra, M.K. Dynamic hysteresis current control to minimize switching for three-phase four-leg VSI topology to compensate nonlinear load. *IEEE Trans. Power Electron.* **2010**, *25*, 1935–1942. [CrossRef]

91. Zhang, X.; Wang, J.; Li, C. Three-phase four-leg inverter based on voltage hysteresis control. In Proceedings of the 2010 International Conference on Electrical and Control Engineering (ICECE), Wuhan, China, 25–27 June 2010.

92. Verdelho, P.; Marques, G.D. Four-wire current-regulated PWM voltage converter. *IEEE Trans. Ind. Electron.* **1998**, *45*, 761–770. [CrossRef]

93. Ali, S.M.; Kazmierkowski, M.P. Current regulation of four-leg PWM/VSI. In Proceedings of the 1998 IECON'98 24th Annual Conference of the IEEE Industrial Electronics Society, Aachen, Germany, 31 August–4 September 1998; Volume 3.

94. Rodriguez, J.; Cortes, P. *Predictive Control of Power Converters and Electrical Drives*; John Wiley & Sons: Chichester, UK, 2012; Volume 40.

95. Wu, B.; Narimani, M. *High-Power Converters and AC Drives*; John Wiley & Sons: Hoboken, NJ, USA, 2017.

96. Martins, C.A.; Roboam, X.; Meynard, T.A.; Carvalho, A.S. Switching frequency imposition and ripple reduction in DTC drives by using a multilevel converter. *IEEE Trans. Power Electron.* **2002**, *17*, 286–297. [CrossRef]

97. Hu, J.; Zhu, Z.Q. Investigation on switching patterns of direct power control strategies for grid-connected DC–AC converters based on power variation rates. *IEEE Trans. Power Electron.* **2011**, *26*, 3582–3598. [CrossRef]

98. Bouafia, A.; Gaubert, J.-P.; Krim, F. Predictive direct power control of three-phase pulsewidth modulation (PWM) rectifier using space-vector modulation (SVM). *IEEE Trans. Power Electron.* **2010**, *25*, 228–236. [CrossRef]

99. Kazmierkowski, M.P.; Jasinski, M.; Wrona, G. DSP-based control of grid-connected power converters operating under grid distortions. *IEEE Trans. Ind. Inform.* **2011**, *7*, 204–211. [CrossRef]

100. Hung, J.Y.; Gao, W.; Hung, J.C. Variable structure control: A survey. *IEEE Trans. Ind. Electron.* **1993**, *40*, 2–22. [CrossRef]

101. Tsang, K.M.; Chan, W.L. Adaptive control of power factor correction converter using nonlinear system identification. *IEE Proc. Electr. Power Appl.* **2005**, *152*, 627–633. [CrossRef]

102. Massing, J.R.; Stefanello, M.; Grundling, H.A.; Pinheiro, H. Adaptive current control for grid-connected converters with LCL filter. *IEEE Trans. Ind. Electron.* **2012**, *59*, 4681–4693. [CrossRef]

103. Herran, M.A.; Fischer, J.R.; Gonzalez, S.A.; Judewicz, M.G.; Carrica, D.O. Adaptive dead-time compensation for grid-connected PWM inverters of single-stage PV systems. *IEEE Trans. Power Electron.* **2013**, *28*, 2816–2825. [CrossRef]

104. Mohamed, Y.A.-R.I. Mitigation of converter-grid resonance, grid-induced distortion, and parametric instabilities in converter-based distributed generation. *IEEE Trans. Power Electron.* **2011**, *26*, 983–996. [CrossRef]

105. Athari, H.; Niroomand, M.; Ataei, M. Review and Classification of Control Systems in Grid-tied Inverters. *Renew. Sustain. Energy Rev.* **2017**, *72*, 1167–1176. [CrossRef]

106. Niroomand, M.; Karshenas, H.R. Hybrid learning control strategy for three-phase uninterruptible power supply. *IET Power Electron.* **2011**, *4*, 799–807. [CrossRef]

107. Mahmud, M.A.; Hossain, M.J.; Pota, H.R.; Roy, N.K. Robust nonlinear controller design for three-phase grid-connected photovoltaic systems under structured uncertainties. *IEEE Trans. Power Deliv.* **2014**, *29*, 1221–1230. [CrossRef]

108. Baloch, M.H.; Wang, J.; Kaloi, G.S. Dynamic Modeling and Control of Wind Turbine Scheme Based on Cage Generator for Power System Stability Studies. *Int. J. Renew. Energy Res.* **2016**, *6*, 599–606.

109. Baloch, M.H.; Wang, J.; Kaloi, G.S. Stability and nonlinear controller analysis of wind energy conversion system with random wind speed. *Int. J. Electr. Power Energy Syst.* **2016**, *79*, 75–83. [CrossRef]

110. Baloch, M.H.; Wang, J.; Kaloi, G.S. A Review of the State of the Art Control Techniques for Wind Energy Conversion System. *Int. J. Renew. Energy Res.* **2016**, *6*, 1276–1295.

111. Hornik, T.; Zhong, Q.-C. A Current-Control Strategy for Voltage-Source Inverters in Microgrids Based on H^{∞} and Repetitive Control. *IEEE Trans. Power Electron.* **2011**, *26*, 943–952. [CrossRef]

112. Hornik, T.; Zhong, Q.-C. H∞ repetitive current controller for grid-connected inverters. In Proceedings of the 2009 IECON'09 35th Annual Conference of IEEE Industrial Electronics, Porto, Portugal, 3–5 November 2009.

113. Guo, Q.; Wang, J.; Ma, H. Frequency adaptive repetitive controller for grid-connected inverter with an all-pass infinite impulse response (IIR) filter. In Proceedings of the 2014 IEEE 23rd International Symposium on Industrial Electronics (ISIE), Istanbul, Turkey, 1–4 June 2014.

114. Bose, B.K. *Modern Power Electronics and AC Drives*; Bose, B.K., Ed.; Prentice Hall PTR: Upper Saddle River, NJ, USA, 2002.

115. Cirstea, M.; Dinu, A.; McCormick, M.; Khor, J.G. *Neural and Fuzzy Logic Control of Drives and Power Systems*; Elsevier: Oxford, UK, 2002.

116. Vas, P. *Artificial-Intelligence-Based Electrical Machines and Drives: Application of Fuzzy, Neural, Fuzzy-Neural, and Genetic-Algorithm-Based Techniques*; Oxford University Press: New York, NY, USA, 1999; Volume 45.

117. Damen, A.; Weiland, S. *Robust Control*; Measurement and Control Group Department of Electrical Engineering, Eindhoven University of Technology: Eindhoven, the Netherlands, 2002.

118. Zames, G. Feedback and optimal sensitivity: Model reference transformations, multiplicative seminorms, and approximate inverses. *IEEE Trans. Autom. Control* **1981**, *26*, 301–320. [CrossRef]

119. Yang, S.; Lei, Q.; Peng, F.Z.; Qian, Z. A robust control scheme for grid-connected voltage-source inverters. *IEEE Trans. Ind. Electron.* **2011**, *58*, 202–212. [CrossRef]

120. Chhabra, M.; Barnes, F. Robust current controller design using mu-synthesis for grid-connected three phase inverter. In Proceedings of the 2014 IEEE 40th Photovoltaic Specialist Conference (PVSC), Denver, CO, USA, 8–13 June 2014.

121. Chen, S.; Malik, O.P. Power system stabilizer design using/SPL MU/synthesis. *IEEE Trans. Energy Convers.* **1995**, *10*, 175–181. [CrossRef]

122. Mascioli, M.; Pahlevani, M.; Jain, P.K. Frequency-adaptive current controller for grid-connected renewable energy systems. In Proceedings of the 2014 IEEE 36th International Telecommunications Energy Conference (INTELEC), Vancouver, BC, Canada, 28 September–2 October 2014.

123. Jorge, S.G.; Busada, C.A.; Solsona, J.A. Frequency-adaptive current controller for three-phase grid-connected converters. *IEEE Trans. Ind. Electron.* **2013**, *60*, 4169–4177. [CrossRef]

124. Timbus, A.V.; Ciobotaru, M.; Teodorescu, R.; Blaabjerg, F. Adaptive resonant controller for grid-connected converters in distributed power generation systems. In Proceedings of the 2006 APEC'06. Twenty-First Annual IEEE Applied Power Electronics Conference and Exposition, Dallas, TX, USA, 19–23 March 2006; p. 6.

125. Zeng, Q.; Chang, L. Improved current controller based on SVPWM for three-phase grid-connected voltage source inverters. In Proceedings of the 2005 PESC'05 36th IEEE Power Electronics Specialists Conference, Recife, Brazil, 16 June 2005.

126. Ouchen, S.; Betka, A.; Abdeddaim, S.; Menadi, A. Fuzzy-predictive direct power control implementation of a grid connected photovoltaic system, associated with an active power filter. *Energy Convers. Manag.* **2016**, *122*, 515–525. [CrossRef]

127. Ouchen, S.; Abdeddaim, S.; Betka, A.; Menadi, A. Experimental validation of sliding mode-predictive direct power control of a grid connected photovoltaic system, feeding a nonlinear load. *Sol. Energy* **2016**, *137*, 328–336. [CrossRef]

128. Mohamed, Y.A.-R.I.; El-Saadany, E.F. An improved deadbeat current control scheme with a novel adaptive self-tuning load model for a three-phase PWM voltage-source inverter. *IEEE Trans. Ind. Electron.* **2007**, *54*, 747–759. [CrossRef]

129. Bode, G.H.; Loh, P.C.; Newman, M.J.; Holmes, D.G. An improved robust predictive current regulation algorithm. *IEEE Trans. Ind. Appl.* **2005**, *41*, 1720–1733. [CrossRef]

130. Zeng, Q.; Chang, L. An advanced SVPWM-based predictive current controller for three-phase inverters in distributed generation systems. *IEEE Trans. Ind. Electron.* **2008**, *55*, 1235–1246. [CrossRef]

131. Mattavelli, P. An improved deadbeat control for UPS using disturbance observers. *IEEE Trans. Ind. Electron.* **2005**, *52*, 206–212. [CrossRef]

132. Kim, J.; Hong, J.; Kim, H. Improved Direct Deadbeat Voltage Control with an Actively Damped Inductor-Capacitor Plant Model in an Islanded AC Microgrid. *Energies* **2016**, *9*, 978. [CrossRef]

133. Zhang, X.; Zhang, W.; Chen, J.; Xu, D. Deadbeat control strategy of circulating currents in parallel connection system of three-phase PWM converter. *IEEE Trans. Energy Convers.* **2014**, *29*, 406–417.

134. Timbus, A.; Liserre, M.; Teodorescu, R.; Rodriguez, P.; Blaabjerg, F. Evaluation of current controllers for distributed power generation systems. *IEEE Trans. Power Electron.* **2009**, *24*, 654–664. [CrossRef]

135. Song, W.; Ma, J.; Zhou, L.; Feng, X. Deadbeat predictive power control of single-phase three-level neutral-point-clamped converters using space-vector modulation for electric railway traction. *IEEE Trans. Power Electron.* **2016**, *31*, 721–732. [CrossRef]

136. Hu, J.; Zhu, J.; Dorrell, D.G. Model predictive control of grid-connected inverters for PV systems with flexible power regulation and switching frequency reduction. *IEEE Trans. Ind. Appl.* **2015**, *51*, 587–594. [CrossRef]

137. Cortés, P.; Kazmierkowski, M.P.; Kennel, R.M.; Quevedo, D.E.; Rodríguez, J. Predictive control in power electronics and drives. *IEEE Trans. Ind. Electron.* **2008**, *55*, 4312–4324. [CrossRef]

138. Mariéthoz, S.; Morari, M. Explicit model-predictive control of a PWM inverter with an LCL filter. *IEEE Trans. Ind. Electron.* **2009**, *56*, 389–399. [CrossRef]

139. Sosa, J.M.; Martinez-Rodriguez, P.R.; Vazquez, G.; Serrano, J.P.; Escobar, G.; Valdez-Fernandez, A.A. Model based controller for an LCL coupling filter for transformerless grid connected inverters in PV applications. In Proceedings of the IECON 2013 39th Annual Conference of the IEEE Industrial Electronics Society, Vienna, Austria, 10–13 November 2013; pp. 1723–1728.

140. Tan, K.T.; So, P.L.; Chu, Y.C.; Chen, M.Z.Q. Coordinated control and energy management of distributed generation inverters in a microgrid. *IEEE Trans. Power Deliv.* **2013**, *28*, 704–713. [CrossRef]

141. Rodriguez, J.; Pontt, J.; Silva, C.A.; Correa, P.; Lezana, P.; Cortés, P.; Ammann, U. Predictive current control of a voltage source inverter. *IEEE Trans. Ind. Electron.* **2007**, *54*, 495–503. [CrossRef]

142. Tan, K.T.; Peng, X.Y.; So, P.L.; Chu, Y.C.; Chen, M.Z.Q. Centralized control for parallel operation of distributed generation inverters in microgrids. *IEEE Trans. Smart Grid* **2012**, *3*, 1977–1987. [CrossRef]

143. Ayad, A.F.; Kennel, R.M. Model predictive controller for grid-connected photovoltaic based on quasi-Z-source inverter. In Proceedings of the 2013 IEEE International Symposium on Sensorless Control for Electrical Drives

and Predictive Control of Electrical Drives and Power Electronics (SLED/PRECEDE), München, Germany, 17–19 October 2013.

144. Trabelsi, M.; Ghazi, K.A.; Al-Emadi, N.; Ben-Brahim, L. An original controller design for a grid connected PV system. In Proceedings of the IECON 2012 38th Annual Conference on IEEE Industrial Electronics Society, Montreal, QC, Canada, 25–28 Octorber 2012; pp. 924–929.

145. Lee, K.-J.; Park, B.-C.; Kim, R.-Y.; Hyun, D.-S. Robust predictive current controller based on a disturbance estimator in a three-phase grid-connected inverter. *IEEE Trans. Power Electron.* **2012**, *27*, 276–283. [CrossRef]

146. Krishna, R.; Kottayil, S.K.; Leijon, M. Predictive current controller for a grid connected three level inverter with reactive power control. In Proceedings of the 2010 IEEE 12th Workshop on Control and Modeling for Power Electronics (COMPEL), Boulder, CO, USA, 28–30 June 2010.

147. Sathiyanarayanan, T.; Mishra, S. Synchronous reference frame theory based model predictive control for grid connected photovoltaic systems. *IFAC-PapersOnLine* **2016**, *49*, 766–771. [CrossRef]

148. Zeng, Z.; Yang, H.; Zhao, R.; Cheng, C. Topologies and control strategies of multi-functional grid-connected inverters for power quality enhancement: A comprehensive review. *Renew. Sustain. Energy Rev.* **2013**, *24*, 223–270. [CrossRef]

149. Patel, D.C.; Sawant, R.R.; Chandorkar, M.C. Three-dimensional flux vector modulation of four-leg sine-wave output inverters. *IEEE Trans. Ind. Electron.* **2010**, *57*, 1261–1269. [CrossRef]

150. Andishgar, M.H.; Gholipour, E.; Hooshmand, R.-A. An overview of control approaches of inverter-based microgrids in islanding mode of operation. *Renew. Sustain. Energy Rev.* **2017**, *80*, 1043–1060. [CrossRef]

151. Debnath, S.; Qin, J.; Bahrani, B.; Saeedifard, M.; Barbosa, P. Operation, control, and applications of the modular multilevel converter: A review. *IEEE Trans. Power Electron.* **2015**, *30*, 37–53. [CrossRef]

152. Mahmud, M.A.; Pota, H.R.; Hossain, M.J. Nonlinear controller design for single-phase grid-connected photovoltaic systems using partial feedback linearization. In Proceedings of the 2012 2nd Australian Control Conference (AUCC), Sydney, Australia, 15–16 November 2012.

153. Vas, P.; Stronach, A.F.; Neuroth, M. DSP-controlled intelligent high-performance ac drives present and future. In Proceedings of the IEE Colloquium on Vector Control and Direct Torque Control of Induction Motors, London, UK, 27 October 1995.

12

Analyzing the Profile Effects of the Various Magnet Shapes in Axial Flux PM Motors by Means of 3D-FEA

Emrah Cetin * and Ferhat Daldaban

Engineering Faculty, Electrical and Electronics Engineering, Erciyes University, 38039 Kayseri, Turkey; daldaban@erciyes.edu.tr
* Correspondence: emrahcetin@erciyes.edu.tr

Abstract: Axial flux machines have positive sides on the power and torque density profile. However, the price of this profile is paid by the torque ripples and irregular magnetic flux density production. To gather higher efficiency, torque ripples should close to the zero and the stator side iron should be unsaturated. Torque ripples mainly occur due to the interaction between the rotor poles and the stator teeth. In this study, different rotor poles are investigated in contrast to stator magnetic flux density and the torque ripple effects. Since the components of the axial flux machines vary by the radius, analysis of the magnetic resources is more complicated. Thus, 3D-FEA (finite element analysis) is used to simulate the effects. The infrastructure of the characteristics which are obtained from the 3D-FEA analysis is built by the magnetic equivalent circuit (MAGEC) analysis to understand the relationships of the parameters. The principal goal of this research is a smoother distribution of the magnetic flux density and lower torque ripples. As the result, the implementations on the rotor poles have interesting influences on the torque ripple and flux density profiles. The MAGEC and 3D-FEA results validate each other. The torque ripple is reduced and the magnetic flux density is softened on AFPM irons. In conclusion, the proposed rotors have good impacts on the motor performance.

Keywords: axial flux machines; magnetic equivalent circuit; torque ripple; back EMF; permanent-magnet machines

1. Introduction

Axial flux permanent magnet machines (AFPM) are one of the futuristic candidates for the higher performance aspiration. AFPM machines have high power/torque density, light mass/volume. It is applicable for many systems as researched in the literature. Mignot et al. designed an AFPM motor with magnetic equivalent circuit [1]. Kierstead et al. studied an in-wheel AFPM motor with non-overlapping windings [2]. Fei et al. researched an AFPM in-wheel motor with two air gap. They compared an approximation method with the 3D-FEA according to the calculation of the back EMF and cogging torque values [3]. Caricchi et al. suggested AFPM motors for direct-drive in-wheel applications with slotted windings. They considered the mitigation of the undesired effects, such as cogging torque and power loss due to flux pulsation in the core teeth, winding conductors, and rotor magnets [4]. Additionally, AFPM machines are investigated for many applications. Seo et al. studied robotic applications by using an analytic model and numeric analysis [5]. Parviainen et al. designed a generator in a small-scale wind-power applications [6]. Di Gerlando et al. focused on wind power generation after defining a general analysis of the model and design features of the AFPM machine [7]. De et al. proposed an ironless AFPM motor with low inductance for the aerospace industry in their paper [8]. One of the common points of these applications is sensitivity with the torque performance. Thus, torque ripples need to be as low as possible. Torque ripples mainly occur due to two main constituents, which are ripple torque and the cogging torque. The cogging torque is cultured by the mutual effects of the reluctance variation in stator and rotor magnetic flux. The ripple torque is mainly constituted by the coaction of the stator current magnetomotive force and rotor magnetic flux distribution in the surface permanent magnet (SPM) machines [9]. Both the cogging and ripple torques are related to rotor magnetic flux distribution, which is manipulated by shapes of PM in the SPM machines. In addition, the back EMF is one of the most crucial characteristics of the AFPM machine profile which is affected by the winding configurations as described by Saavedra et al. in [10].

Magnetic flux is the one basic characteristic of electric machines. There exist so many techniques to analyze the magnetic flux. One of the most prevailing methods is the MAGEC. Analyzing the magnetic flux by using MAGEC is one of the easiest analytical tools in comparison of the other methods given in the literature. Since magnetic flux paths simply turn into the circuit components and the problem is solved by the circuit analysis easily [11–14]. If the solution of the MAGEC is done, air gap magnetic flux density, permanent magnet axial length, the total permeance of the machine, back EMF value, winding resistance, self-inductance, and torque and output power can be composed according to the study of Mignot et al. [1]. The stator winding resistance, eddy current resistance, end winding resistance, power factor, phase voltages, output power, and steady-state torque are counted by the MAGEC perusal in the research of Wang et al. [15]. Parviainen et al. [11], Tiegna et al. [12], Bellara et al. [13], and Lubin et al. [14], mentioned different analytical calculation techniques for analyzing the characteristics of the AFPM machine in the literature.

In the literature, various topologies were investigated to reduce torque ripple including the shaping of rotor magnet pole and stator slots. Aydin et al. studied the shaping of rotor magnet pole which was realized by skewing or displacing magnet poles [16]. Sung et al. proposed the shaping of stator slots by recasting slot or teeth numbers in [17]. Saavedra et al. researched the effects of the magnet shaping under demagnetization fault conditions by means of 3D-FEA. The research aimed to determine a more efficient magnet geometry [18]. Kahourzade et al. summarized the torque ripple reducing methods by means of the classification of the AFPM machines [19].

This paper suggests analyzing the three-phase single air gap AFPM machine as given in Figure 1 by means of the 3D-FEA analysis. Five different rotor designs are investigated to achieve the goals. The first goal is to use MAGEC to observe the magnetic flux path and to describe the magnetic events in an analytic way. Another goal is to investigate the torque ripple, back EMF, and air gap magnetic flux distribution results of the proposed rotors in 3D-FEA. Two of the five designs are the novel magnet shapes, which are mainly developed for mitigating the torque ripples. The proposal is stepping and shifting the rotor magnets. Due to this action affects the magnetic flux path, back EMF, and torque waveforms are changed. All of the designs are simulated in 3D-FEA to compare the novel topologies and exciting results are obtained.

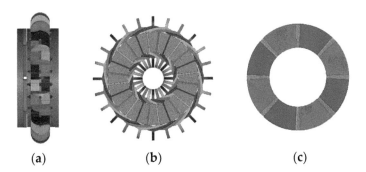

(a)   (b)   (c)

Figure 1. The single gap AFPM machine structure (**a**) total stack; (**b**) the stator with distributed windings; and (**c**) the rotor.

2. MAGEC Design and Analysis

MAGEC design is composed of the magnetic flux path at the machine given in Figure 2. Each definition of the flux sources and the permeance are situated by considering this path.

The describing of the elements of MAGEC eases the analytical solutions by obtaining the parameters easily.

An interesting specialty of the AFPM machine is that most of the parameters vary by the radius. The produced torque is defined by the radius, too. Due to this, the MAGEC is designed by considering the single air gap AFPM motor in this section. Conceptual 2D and 3D representations are seen in Figure 2.

The rotor and the stator back irons are produced from ferromagnetic steel, and these steels are designed from layers, like round strips, which are laminated in the circumference direction. Permanent magnets are placed on the surface of the rotor back iron. There are gaps between each pole to minimize the permanent magnet flux leakage. As seen from Figure 2, flux paths are the same for each permanent magnet pole. Since the flux divides by two for each pole, just one closed loop is modeled in the MAGEC Equations.

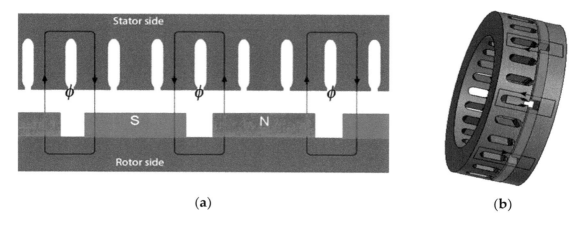

(a)   (b)

Figure 2. Magnetic flux path of the AFPM machine in (**a**) the 2D view and (**b**) the 3D view.

The developed MAGEC is shown in Figure 3, which is designed by minding the flux path illustrated in Figure 2a,b. The two permanent magnet halves, rotor and stator back iron, air gap, and the gap between poles are included in the MAGEC.

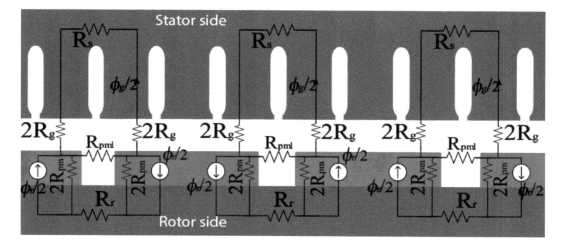

Figure 3. MAGEC design of the studied single air gap, slotted AFPM machine.

Three stator teeth per one pole are determined for the studied single air gap AFPM motor, as demonstrated in Figure 2. Thus, the stator has 24 teeth and the rotor has eight magnets. Other parameters are listed in Table 1.

Table 1. The parameters of the single air gap AFPM motor.

Parameter	Value	Unit
Inner radius (D_i)	40	mm
Outer radius (D_o)	75	mm
Slot/Pole	24/8	
Magnet height	5	mm
Magnet fill factor	0.8722	
Air gap	1	mm
Winding	Distributed, overlapping	
Turns	40	
Stator width	50	mm
Rotor back iron width	10	mm
Rated Speed	2200	rpm
Rated current	175	A

Since air gap permeability μ_0 is much lower than iron permeability, the air gap reluctance is much higher than the rotor and stator back iron reluctances. Due to this, the rotor and stator reluctances can be neglected to have an easier solution. Thus, the MAGEC can be simplified, as in Figure 4. In the end, the permeance values are taken into account instead of the reluctances.

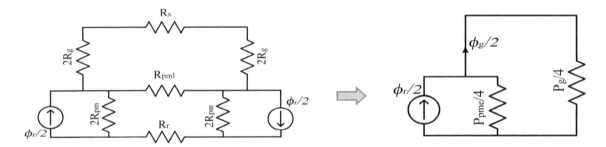

Figure 4. The simplified MAGEC.

The simplified MAGEC is located to the 2D design of the AFPM machine as given in Figure 5. The relationship between the air gap flux and the rotor flux is pointed out in the Equation (1).

$$\phi_g = \frac{1}{1 + P_{pme}/P_g}\phi_r \tag{1}$$

where P_{pme} is the effective permanent magnet permeance, and P_g is the air gap permeance.

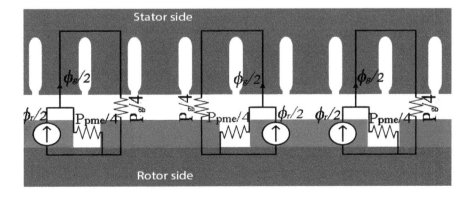

Figure 5. Simplified MAGEC design in the 2D view of the studied AFPM machine.

Permanent magnet and steel data are given in Table 2. Additionally, permeance of the magnet is defined in Equation (2):

$$P_{pm} = \frac{\mu_r \mu_0 A_{pm}}{L_{PM}}, \tag{2}$$

Table 2. Data of the permanent magnet and the steel.

Permanent Magnet: NdFe–N35	
B_r (T)	1.17
μ_r	1.099
H_{cb} (kA/m)	868
H_{cj} (kA/m)	955
Steel: M250-35A	
B_{ref} (T)	1.5
μ_r	660
Loss (W/kg)	2.5
f (Hz)	50

Here, L_{PM} is the permanent magnet's height and the A_{PM} is the surface area of the permanent magnet. The height of the permanent magnet can be determined by Equation (3) [20]:

$$L_{PM} = \frac{\mu_r B_g}{B_r - \left(\frac{K_f}{K_d} B_g\right)} (g K_c), \tag{3}$$

The permanent magnet surface area is calculated in contrast to the inner and outer radii, as given by Equation (4):

$$A_{pm} = \alpha_{pm} \frac{\pi}{N_{pm}} \left(D_o^2 - D_i^2\right), \tag{4}$$

Here, N_{pm} is the number of the pole of the AFPM machine. The pole area A_p is necessary to find the magnet fill factor α_{pm}:

$$A_p = \frac{\pi}{N_{pm}} \left(D_o^2 - D_i^2\right) \tag{5}$$

$$\alpha_{pm} = \frac{A_{pm}}{A_p} \tag{6}$$

A magnetic flux leakage occurs between the adjacent magnets on the rotor. The path of this flux leakage draws an arc between two magnets. When this path is accounted, the obtained leakage permeance is calculated by Equation (7). P_{pml} is the permeance of the gap between the two adjacent permanent magnets. If the simplification of the MAGEC is taken into account by applying $P_{pme} = P_{pm} + 4P_{pml}$, Equation (8) can be derived to simplify the equation by a coefficient (K_{pml}) which is given in Equation (9). The effective permanent magnet permeance (P_{pme}) is defined by the multiplication of the permanent magnet permeance P_{pm} and K_{pml} in Equation (8):

$$P_{pml} = \frac{\mu_0 (D_o - D_i)}{\pi} \ln\left(1 + \pi \frac{g}{d_f}\right) \tag{7}$$

$$P_{pme} = K_{pml} P_{pm} \tag{8}$$

$$K_{pml} = 1 + 4 \frac{L_{PM} N_{pm}}{\pi^2 \mu_r \alpha_{pm} (D_o + D_i)} \ln\left(1 + \pi \frac{g}{d_f}\right) \tag{9}$$

Equations (7)–(9) allows simplifying the MAGEC, as seen in Figure 4. In addition, air gap permeance can be calculated correctly by defining the effective air gap $g_e = K_c g$, and the air gap area [21]. Thus, the interaction between the air gap flux and the rotor flux becomes as specified in Equation (11):

$$P_g = \frac{\mu_0 A_g}{g_e} \tag{10}$$

$$\phi_g[Wb] = \frac{1}{1 + 2\frac{\mu_r \alpha_{pm} K_{pml} K_c g}{(1+\alpha_{pm}) L_{PM}}} \phi_r[Wb] \tag{11}$$

One of the main subjects to create the MAGEC is the defining the air gap magnetic flux density. In the light of the Equation (11), the magnetic flux density B_g can be calculated as stated in the Equation (12) where $K_{k\varphi} = A_{pm}/A_g$ and $C_p = L_{PM}/gK_{k\varphi}$:

$$B_g[T] = \frac{K_{k\phi}}{1 + \frac{\mu_r K_{pml} K_c}{C_p}} B_r[T] \tag{12}$$

As given in Equation (13), permanent magnet flux produces the air gap flux density and results in the voltage induction, called back EMF, in the stator windings. This can be seen from the MAGEC depicted in Figure 4.

$$e_{ind} = w N_{pm} N_w B_g (D_o - D_i) \tag{13}$$

The force Equation (15) is composed of the electric-magnetic loads from Equation (14) and the total area of the magnets from Equation (4). If these Equations are applied from the inner to the outer radius, the electromagnetic torque equation becomes that shown by Equation (16):

$$Q_{load} = B_g J_{in} \tag{14}$$

$$F_{emr_i} = \pi B_g J_{in} (D_o{}^2 - D_i{}^2) \tag{15}$$

$$T_{em} = \pi B_g J_{in} \int_{D_i}^{D_o} D_i \mathbf{r} d\mathbf{r} = \pi B_g J_{in} (D_o)^3 \lambda (1 - \lambda^2) \tag{16}$$

Here, J_{in}, is the current density at the inner radius D_i, and λ is the rate of the radiuses which is counted by D_i/D_o.

3. Analyzed Rotor Pole Designs

The single air gap, slotted AFPM motor is taken into account as the reference design structure, which is demonstrated in Figure 1. The studied motor parameters are given in Table 1. Five different rotor pole designs are investigated in this research. Design I is a conventional rotor pole design of an AFPM machine model, taken as the reference model for this study, which can be seen in Figure 6a. It has sharp edges. This type of magnet can be easily found on the market. Design II is an improved rotor pole model for an AFPM machines, as seen in Figure 6b. It has sinusoidal edges. This design is studied to reduce the cogging torque in the literature [22]. Design III is one of the proposed rotor pole models for this research, and is shown in Figure 6c. This design is the novel proposal for AFPM motors. It is studied to reduce the torque ripples. Design IV is another proposed rotor model for this research, as seen in Figure 6d. This design is a novel proposal for axial flux machines. Design V is developed for validating the FEA model. Aydin and Gulec proposed that cogging torque has minimum values when skewing angle is 18.75°, such as that used in this study for design V, given in Figure 6e. The FEA simulation of designs I and V prove the validation of the FEA model in comparison with [23].

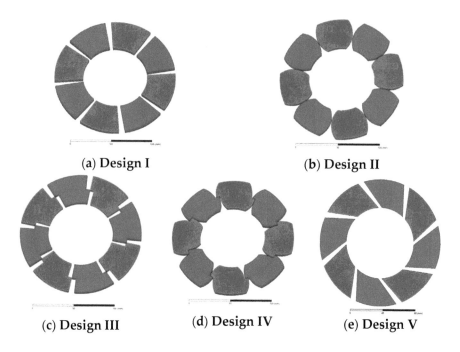

(a) Design I   (b) Design II

(c) Design III   (d) Design IV   (e) Design V

Figure 6. The researched designs of the magnets.

All of the permanent magnet poles are magnetized in the z-axis and the total value of the inner and outer diameters are the same for each pole designs. The MAGECs of each design do not change in majority due to the constant magnet fill factor α_{pm}, which is 0.8722 for each design. Additionally, all permanent magnet pole designs have symmetry in the radial direction.

4. 3D-FEA Analysis

The back EMF, torque, and flux density distribution waveforms are obtained from the three-dimensional finite element analysis. Both transient and static analyses are performed. 3D-FEA simulations are performed for the $\frac{1}{4}$ of the AFPM motor designs, as given in Figure 7, in order to shorten the simulation time. A runtime process of 10 milliseconds is chosen, thus, the motor turns more than one time during the simulation. M250-35A steel and NdFe magnet specifications are given in Table 2. The values are defined in 3D-FEA. Additionally, the cylindrical coordinate system is used to define the axial flux steel orientation.

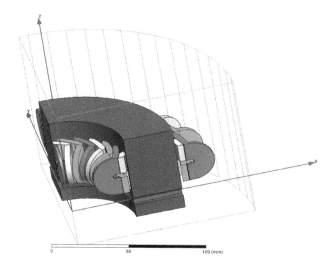

Figure 7. $\frac{1}{4}$ part of the simulated AFPM motor designs.

Before starting the comparison of five designs, the optimum shifting angles of the suggested rotor poles must be specified. The shifting angle means that inner rotor step magnets are displaced by an angle from the outer rotor magnets, as seen in Figure 6c,d. One of the aims of this shifting method is the mitigation of the torque ripples. There are some analytical methods to define the best shift angle in the literature. One of them is the cogging torque period method that is described in [11], but this method does not give the best results for the AFPM machines. In this research, parametric analysis with 3-D FEA is used to find the optimum shifting angle.

The shifting angle is defined as a variable and differs from 0° to 14° by 1° steps. The third magnet design is used to perform this analysis. The average torque and the torque ripple values are taken into account for each result in order to mitigate the total torque ripples. Figure 8 demonstrates the results of the parametric analysis. Table 3 shows each peak-to-peak torque ripple and the average torque value for each shifting angle.

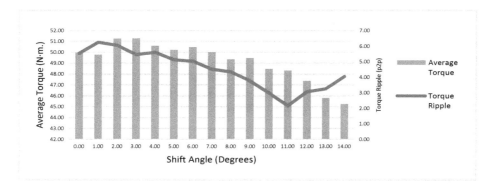

Figure 8. Defining the shifting angle by 3D-FEA parametric analysis.

The simulations gave interesting results from the total parametric analysis. If the torque ripple is the most important anchor of the application, the best result is the 11° shifting angle which gives 2.16 N·m. of peak-to-peak cogging torque. However, if the average torque value is the most valued parameter, the 3° shifting angle has the highest average torque of 51.27 N·m., which is 1.3 N·m. higher than the 0° shifting angle. Torque ripple drops from a shifting angle of 1° to 11°, but after 11° it starts to rise again.

Table 3. The results of the 3D-FEA parametric analysis of the shifting angle.

Shift Angle (Degree)	Average Torque (N·m.)	Torque Ripple (p2p) (N·m.)
0	49.95	5.52
1	49.77	6.23
2	51.24	6.04
3	51.27	5.44
4	50.58	5.59
5	50.21	5.11
6	50.46	5.02
7	50.00	4.51
8	49.35	4.34
9	49.47	3.77
10	48.46	2.98
11	**48.32**	**2.16**
12	47.37	3.06
13	45.81	3.25
14	45.25	4.04

After defining the shifting angle of the third and fourth designs, the magnetic simulations are completed for each design in both static and dynamic conditions. The stator was split into four identical parts and one of them was investigated due to the symmetrical geometry to reduce the simulation time.

5. Comparison of the Results of the Proposed Designs

In the 3D-FEA analysis, PMs have an 11° shifting angle in designs III and IV due to the seeking of the lowest torque ripple. Torque and back EMF waveforms are taken from the dynamic simulations. Figure 9 illustrates the electromagnetic torque results of the five designs that are shown in Figure 6. As seen from the torque results, the lowest torque ripple is in the third design, with 62.4% mitigation, despite a 4.3% reduction on the average torque compared to the design I. The table of the comparison is demonstrated in Table 4. Although it has step and shift on the magnets, design IV has some of the worst data in the view of the torque ripple in this study. This is because of the magnet edges. Since some arrays are sinusoidal, some arrays are sharp. Thus, magnetic flux distribution is unsteady. Additionally, design V has a 43.2% reduction in torque ripple with a 0.8722 magnet fill factor (pole-arc ratio), as validated by Aydin and Gulec, although with some different characteristics of the simulated motors, like magnet thickness, air gap, and the dimensions in [23]. As seen from the simulations, the skewing process has a lower reduction effect than the stepping and shifting process effect on the torque ripple, as demonstrated in Table 4.

Back EMF depends directly on the speed, number of turns, pole numbers, inner and outer radii, and the magnetic flux density, as given in the Equation (13). All of these parameters are stationary without the magnetic flux density in this study. Magnetic flux density depends on the permanent magnet magnetic flux. The average values of the parameters are given in Table 5. Analytical results are obtained by calculating the MAGEC Equations (12), (13), and (16) developed earlier. The constant values are given in Table 6. Figure 10 illustrates the back electromotive force waveforms of each design. The smoothness of these waveforms is crucial to have more constant torque. That means lower torque ripple. Hence, design III has smoother back EMF and lower torque ripple waveforms.

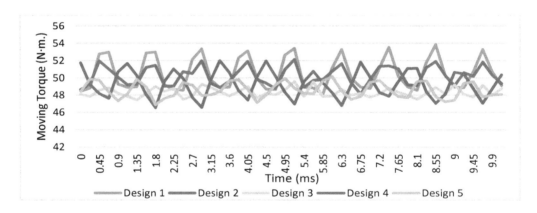

Figure 9. Electromagnetic torque results.

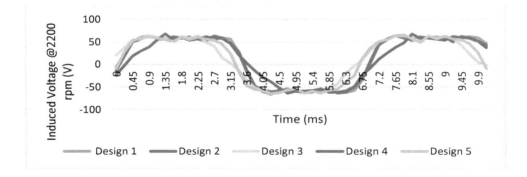

Figure 10. Back EMF results.

The flux density distributions are given in Figure 11 for five designs which demonstrate the radial components of the flux density between the magnets and the stator steel. Since the simulations are conducted for $\frac{1}{4}$ of the motor, the waveforms are produced in 90 degrees. The simulation has interesting results that have caused by the proposed rotor pole designs. The rotor permanent magnet flux does not drop under 0.2 T in designs III and IV, unlike designs I, II, and V. The geometry of the proposition provides these conclusions. Figure 12 shows the magnetic flux densities on the surfaces of the AFPM motor. The first and fifth designs have too high a magnetic flux leakage between the magnets, as seen in Figure 12. The high magnetic flux causes the saturation of the iron. Saturation is an undesired situation which may cause heat and unsteady inductance. Stepping and shifting of magnets allow resistance to the leakage flux. Designs I and V have strong magnetic saturation between the adjacent magnets due to the magnet shapes. Constant width and straight edges of the adjacent magnets ease the magnetic flux leakage. Due to the sinusoidal shape, design II has a lower magnetic saturation than designs I and V. Figure 13 shows the prototype of the AFPM machine.

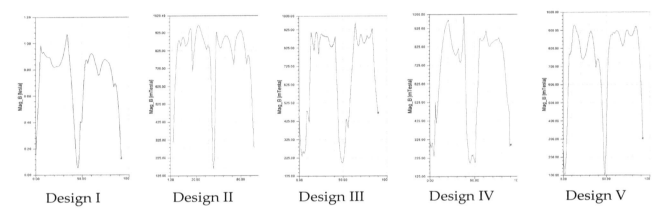

Figure 11. Flux density distributions for $\frac{1}{4}$ machine (for 90 mechanical degrees).

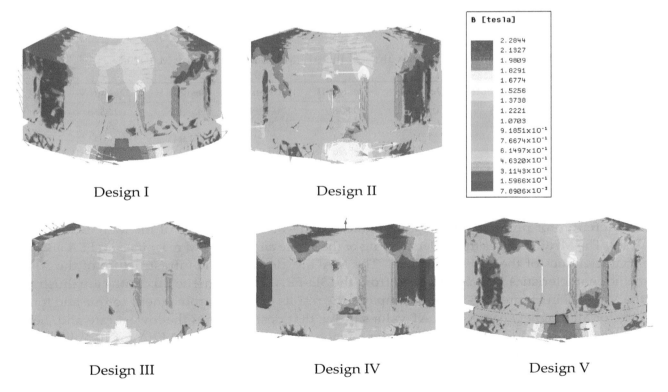

Figure 12. Magnetic flux densities on the surfaces.

Figure 13. The prototype machine.

Table 4. The total comparison of the 3D-FEA results.

	Design I	Design II	Design III	Design IV	Design V
Average Torque (AT)	50.521	50.25	48.32	49.28	48.59
Torque Ripple (TR)	5.744	3.935	**2.159**	5.456	3.264
Rate (TR/AT)	0.114	0.078	**0.045**	0.111	0.067
AT Reduction	ref.	−%0.1	−**%4.3**	−%2.4	−%3.8
TR Reduction	ref.	−%31.5	−**%62.4**	−%5.1	−%43.2

Table 5. The average values of parameters by means of the MAGEC and 3D-FEA.

Average Values of	Simulation	Design I	Design II	Design III	Design IV	Design V
Magnetic Flux	MAGEC	0.7	0.69	0.62	0.59	0.61
Density, B_g (T)	3D-FEA	0.72	0.7	0.63	0.60	0.63
Back EMF (V)	MAGEC	17.24	16.58	14.97	14.24	14.23
	3D-FEA	17.643	16.878	15.1502	14.358	14.866
Torque (N·m.)	MAGEC	49.117	49.532	47.553	48.458	47.047
	3D-FEA	50.521	50.25	48.32	49.28	48.59

Table 6. Some of constant values of the AFPM motor counted by MAGEC.

Values of the Constants	Design I	Design II	Design III	Design IV	Design V
$K_{k\varphi}$	0.8386	0.8386	0.8386	0.8386	0.8386
K_{pml}	2.09	2.2	3.04	3.46	3.17
K_c	1.04	1.04	1.04	1.04	1.04
C_p	5.96	5.96	5.96	5.96	5.96
λ	0.533	0.533	0.533	0.533	0.533

6. Conclusions

Different rotor pole designs are investigated in this study by means of the MAGEC and FEA analyses. The MAGEC gives the understanding of the single air gap AFPM machine and FEA analyzes the characteristics of the AFPM machine. The MAGEC describes the infrastructure of the AFPM machine characteristics that are obtained from the 3D-FEA. The magnetic flux paths are illustrated by the MAGEC in Figures 2–5. Table 4 compares the results of the electromagnetic torque and torque ripples for all magnet shapes. TR/AT values prove that design III has the lowest rate and, hence, an average torque reduction of 4.3%. Transient analysis is performed by 3D-FEA for 10 milliseconds. Each design is discussed in contrast to the simulation results. Additionally, a parametric analysis is fulfilled to determine the best solution for the shifting angle. The electromagnetic torque and the back EMF waveforms are demonstrated in Figures 9 and 10, which are obtained from the transient analysis. Table 5 is demonstrated to prove the methods. The MAGEC and 3D-FEA results are compared in

Table 5 in terms of air gap magnetic flux density, back EMF, and torque characteristics. The 3D-FEA and MAGEC results validate each other.

Furthermore, the magnetic flux density distribution waveforms are given in Figure 11 and the surface magnetic flux density profiles are given in Figure 12. The air gap magnetic flux density is not collapsed under 0.2 T by the permanent magnets in designs III and IV, unlike design numbers I, II, and V. However, Figure 12 gives information for the saturation of the irons. Design I and V have strong flux leakage between the adjacent magnets, but the saturation points are mostly in the rotor iron, hence, the results are not affected much at the 3D-FEA simulation time as given in the Table 5. Additionally, the MAGEC results do not contain saturation effects. Thus, the heat effects are neglected in the 3D-FEA results in Table 5. If the permanent magnets are damaged by the heat caused by the saturation, all characteristics in Table 5 could be changed dramatically. Resultantly, the third design has the best results in contrast to the precision on the stability of moving torque. Additionally, the results show that stepping and shifting method has better results compared with the skewing method in the view of torque ripple mitigation. The magnets will be produced privately for designs III and IV. Hence, the costs may be higher for the prototype, but the magnet costs of each design will be the same for mass production since the magnet weights being the same. Moreover, a prototype machine can be seen in Figure 13 which is manufactured in the light of this paper for further studies.

Acknowledgments: This study researched by the supports of the TUBITAK (The Scientific and Technical Research Council of Turkey), and WEMPEC. (Wisconsin Electric Machines and Power Electronics Concorsium).

Author Contributions: E.C. and F.D. conceived and designed the MAGEC and 3D-FEA; E.C. performed the simulations; E.C. and F.D. analyzed the data; E.C. wrote the paper.

Nomenclature

R_s	The reluctance of the stator back iron,
R_r	The reluctance of the rotor back iron,
R_g	The reluctance of the air gap between the stator and the rotor,
R_{pm}	The reluctance of the permanent magnet,
R_{pml}	The reluctance of the air gap between the two permanent magnets,
φ_r	Magnetic flux flows from the rotor pole,
φ_g	Flux flow passed from the air gap into the stator
P_{pm}	The permeance of the permanent magnet
P_{pml}	The permeance of the gap between adjacent magnets
K_c	Carter's coefficient
K_f	The correction factor of the air gap magnetic flux density in radial direction
K_d	Flux leakage coefficient
K_{pml}	Leakage coefficient between the magnets
$K_{k\varphi}$	Flux density coefficient
d_f	The distance between adjacent magnets
N_{pm}	Number of the magnets
A_p	Area of a pole
C_p	Permeance factor
N_w	Number of turns
J_{in}	Current density
N_s	Slot number

References

1. Mignot, R.B.; Dubas, F.; Espanet, C.; Chamagne, D. Design of Axial Flux PM Motor for Electric Vehicle via a Magnetic Equivalent Circuit. In Proceedings of the First International Conference on REVET-2012 Renewable Energies and Vehicular Technology, Hammamet, Tunisia, 26–28 March 2012; pp. 212–217.
2. Kierstead, H.; Wang, R.; Kamper, M. Design optimization of a single sided axial flux permanent magnet in-wheel motor with non-overlap concentrated winding. In Proceedings of the 18th Southern African Universities Power Engineering Conference, Stellenbosch, South Africa, 28–29 January 2009; pp. 36–40.

3. Fei, W.; Luk, P.; Jinupun, K. A new axial flux permanent magnet segmented-armature-torus machine for in-wheel direct drive applications. In Proceedings of the Power Electronics Specialists Conference, Rhodes, Greece, 15–19 June 2008; pp. 2197–2202.

4. Caricchi, F.; Capponi, F.G.; Crescimbini, F.; Solero, L. Experimental study on reducing cogging torque and no-load power loss in axial-flux permanent-magnet machines with slotted winding. *IEEE Trans. Ind. Appl.* **2004**, *40*, 1066–1075. [CrossRef]

5. Seo, J.M.; Rhyu, S.; Kim, J.; Choi, J.; Jung, I. Design of Axial Flux Permanent Magnet Brushless DC Motor for Robot Joint Module. In Proceedings of the IEEE International Power Electronics Conference, Sapporo, Japan, 21–24 June 2010; pp. 1336–1340.

6. Parviainen, A.; Pyrhönen, J.; Kontkanen, P. Axial flux permanent magnet generator with concentrated winding for small wind power applications. In Proceedings of the IEEE International Conference on Electric Machines and Drives, San Antonio, TX, USA, 15 May 2005; pp. 1187–1191.

7. Di Gerlando, A.; Foglia, G.; Iacchetti, M.F.; Perini, R. Axial flux pm machines with concentrated armature windings: Design analysis and test validation of wind energy generators. *IEEE Trans. Ind. Electron.* **2011**, *58*, 3795–3805. [CrossRef]

8. De, S.; Rajne, M.; Poosapati, S.; Patel, C.; Gopakumar, K. Low inductance axial flux BLDC motor drive for more electric aircraft. *IET Power Electron.* **2012**, *5*, 124–133. [CrossRef]

9. Jahns, T.M.; Soong, W.L. Pulsating torque minimization techniques for permanent magnet ac motor drives—A review. *IEEE Trans. Ind. Electron.* **1996**, *43*, 321–330. [CrossRef]

10. Saavedra, H.; Urresty, J.-C.; Riba, J.-R.; Romeral, L. Detection of interturn faults in PMSMs with different winding configurations. *Energy Convers. Manag.* **2014**, *79*, 534–542. [CrossRef]

11. Parviainen, A.; Niemela, M.; Pyrhonen, J. Modeling of axial flux permanent-magnet machines. *IEEE Trans. Ind. Appl.* **2004**, *40*, 1333–1340. [CrossRef]

12. Tiegna, H.; Bellara, A.; Amara, Y.; Barakat, G. Analytical modeling of the open-circuit magnetic field in axial flux permanent-magnet machines with semi-closed slots. *IEEE Trans. Magn.* **2012**, *48*, 1212–1226. [CrossRef]

13. Bellara, A.; Amara, Y.; Barakat, G.; Dakyo, B. Two-dimensional exact analytical solution of armature reaction field in slotted surface mounted pm radial flux synchronous machines. *IEEE Trans. Magn.* **2009**, *45*, 4534–4538. [CrossRef]

14. Lubin, T.; Mezani, S.; Rezzoug, A. 2-d exact analytical model for surface-mounted permanent-magnet motors with semi-closed slots. *IEEE Trans. Magn.* **2011**, *47*, 479–492. [CrossRef]

15. Wang, R.J.; Kamper, M.J.; Westhuizen, K.V.; Gieras, J.F. Optimal Design of a Coreless stator Axial Flux Permanent Magnet Generator. *IEEE Trans. Magn.* **2005**, *41*, 55–64. [CrossRef]

16. Aydin, M.; Zhu, Z.Q.; Lipo, T.A.; Howe, D. Minimization of Cogging Torque in Axial-Flux Permanent-Magnet Machines: Design Concepts. *IEEE Trans. Magn.* **2007**, *43*, 3614–3622. [CrossRef]

17. Sung, S.J.; Park, S.J.; Jang, G.H. Cogging torque of brushless DC motors due to the interaction between the uneven magnetization of a permanent magnet and teeth curvature. *IEEE Trans. Magn.* **2011**, *47*, 1923–1928. [CrossRef]

18. Saavedra, H.; Riba, J.-R.; Romeral, L. Magnet shape influence on the performance of AFPMM with demagnetization. In Proceedings of the 39th Annual Conference of the IEEE Industrial Electronics Society, Vienna, Austria, 10–13 November 2013; pp. 973–977.

19. Kahourzade, S.; Mahmoudi, A.; Ping, H.W.; Uddin, M.N. A Comprehensive Review of Axial-Flux Permanent-Magnet Machines. *Can. J. Electr. Comput. Eng.* **2014**, *37*, 19–33. [CrossRef]

20. Mahmoudi, A.; Kahourzade, S.; Abd Rahim, N.; Hew, W.P. Design, Analysis, and Prototyping of an Axial-Flux Permanent Magnet Motor Based on Genetic Algorithm and Finite-Element Analysis. *IEEE Trans. Magn.* **2013**, *49*, 1479–1492. [CrossRef]

21. Qishan, G.; Hongzhan, G. Effect of Slotting in PM Electric Machines. *Electr. Mach. Power Syst.* **1985**, *10*, 273–284. [CrossRef]

22. Shokri, M.; Rostami, N.; Behjat, V.; Pyrhönen, J.; Rostami, M. Comparison of performance characteristics of axial-flux permanent-magnet synchronous machine with different magnet shapes. *IEEE Trans. Magn.* **2015**, *51*. [CrossRef]

23. Aydin, M.; Gulec, M. Reduction of Cogging Torque in Double-Rotor Axial-Flux Permanent-Magnet Disk Motors: A Review of Cost-Effective Magnet-Skewing Techniques With Experimental Verification. *IEEE Trans. Ind. Electron.* **2014**, *61*, 5025–5034. [CrossRef]

Permissions

The contributors of this book come from diverse backgrounds, making this book a truly international effort. This book will bring forth new frontiers with its revolutionizing research information and detailed analysis of the nascent developments around the world.

We would like to thank all the contributing authors for lending their expertise to make the book truly unique. They have played a crucial role in the development of this book. Without their invaluable contributions this book wouldn't have been possible. They have made vital efforts to compile up to date information on the varied aspects of this subject to make this book a valuable addition to the collection of many professionals and students.

This book was conceptualized with the vision of imparting up-to-date information and advanced data in this field. To ensure the same, a matchless editorial board was set up. Every individual on the board went through rigorous rounds of assessment to prove their worth. After which they invested a large part of their time researching and compiling the most relevant data for our readers.

The editorial board has been involved in producing this book since its inception. They have spent rigorous hours researching and exploring the diverse topics which have resulted in the successful publishing of this book. They have passed on their knowledge of decades through this book. To expedite this challenging task, the publisher supported the team at every step. A small team of assistant editors was also appointed to further simplify the editing procedure and attain best results for the readers.

Apart from the editorial board, the designing team has also invested a significant amount of their time in understanding the subject and creating the most relevant covers. They scrutinized every image to scout for the most suitable representation of the subject and create an appropriate cover for the book.

The publishing team has been an ardent support to the editorial, designing and production team. Their endless efforts to recruit the best for this project, has resulted in the accomplishment of this book. They are a veteran in the field of academics and their pool of knowledge is as vast as their experience in printing. Their expertise and guidance has proved useful at every step. Their uncompromising quality standards have made this book an exceptional effort. Their encouragement from time to time has been an inspiration for everyone.

The publisher and the editorial board hope that this book will prove to be a valuable piece of knowledge for researchers, students, practitioners and scholars across the globe.

List of Contributors

Jhon Bayona
Facultad de Ingeniería, Universidad ECCI, Bogotá 111311, Colombia

Nancy Gélvez and Helbert Espitia
Facultad de Ingeniería, Universidad Distrital Francisco José de Caldas, Bogotá 11021-110231588, Colombia

Qian Cheng, Chenchen Wang and Jian Wang
School of Electrical Engineering, Beijing Jiaotong University, Beijing 100044, China

Yifan Zhang, Xiaodong Li, Chuan Sun and Zhanhong He
Faculty of Information Technology, Macau University of Science and Technology, Macau, China

Alberto Sanchez and Angel de Castro
HCTLab Research Group, Universidad Autonoma de Madrid, 28049 Madrid, Spain

Elías Todorovich
Facultad de Ciencias Exactas, Universidad Nacional del Centro de la Provincia de Buenos Aires, Tandil B7001BBO, Argentina
Faculty of Engineering, FASTA University, Mar del Plata B7600, Argentina

Milan Srndovic and Gabriele Grandi
Department of Electrical, Electronic, and Information Engineering, University of Bologna, 40136 Bologna, Italy

Rastko Fišer
Department of Mechatronics, Faculty of Electrical Engineering, University of Ljubljana, 1000 Ljubljana, Slovenia

Ahmet Aksoz and Ali Saygin
Department of Electrical and Electronics Engineering, Faculty of Technology, Gazi University, 06500 Ankara, Turkey

Yipeng Song, Frede Blaabjerg and Pooya Davari
Department of Energy Technology, Aalborg University, AAU 9220 Aalborg East, Denmark

Jiazheng Lu, Qingjun Huang, Xinguo Mao, Yanjun Tan, Siguo Zhu and Yuan Zhu
State Key Laboratory of Disaster Prevention and Reduction for Power Grid Transmission and Distribution Equipment, State Grid Hunan Electric Company Limited Disaster Prevention and Reduction Center, Changsha 410129, China

Zachary Bosire Omariba
National Center for Materials Service Safety, University of Science and Technology Beijing, Beijing 100083, China
Computer Science Department, Egerton University, Egerton 20115, Kenya

Lijun Zhang and Dongbai Sun
National Center for Materials Service Safety, University of Science and Technology Beijing, Beijing 100083, China

Nagaraja Rao Sulake and Sai Babu Choppavarapu
Department of Electrical and Electronics Engineering, Jawaharlal Nehru Technological University, Kakinada 533003, Andhra Pradesh, India

Ashok Kumar Devarasetty Venkata
Department of Electrical and Electronics Engineering, Rajeev Gandhi Memorial College of Engineering and Technology, Nandyal 518501, Andhra Pradesh, India

Sohaib Tahir
School of Electronic, Information and Electrical Engineering, Shanghai Jiao Tong University, Shanghai 200000, China
Department of Electrical Engineering, COMSATS Institute of Information Technology, Sahiwal 58801, Pakistan

Jie Wang
School of Electronic, Information and Electrical Engineering, Shanghai Jiao Tong University, Shanghai 200000, China

Mazhar Hussain Baloch
School of Electronic, Information and Electrical Engineering, Shanghai Jiao Tong University, Shanghai 200000, China
Department of Electrical Engineering, Mehran University of Engineering & Technology, Khairpur Mirs 67480, Pakistan

Ghulam Sarwar Kaloi
School of Electronic, Information and Electrical Engineering, Shanghai Jiao Tong University, Shanghai 200000, China
Department of Electrical Engineering, Quaid e Awam University of Engineering & Technology, Larkana 77150, Pakistan

Emrah Cetin and Ferhat Daldaban
Engineering Faculty, Electrical and Electronics Engineering, Erciyes University, 38039 Kayseri, Turkey

Index